The Potential of Earth-Sheltered and Underground Space:
Today's Resource for Tomorrow's Space and Energy Viability

Pergamon Titles of Related Interest

Related Journals*

*Free specimen copies available upon request.

THE POTENTIAL OF

EARTH- SHELTERED

AND

UNDERGROUND SPACE:

Today's Resource for
Tomorrow's Space
and Energy Viability

Proceedings of the Underground Space
Conference and Exposition, Kansas City,
Missouri, June 8-10, 1981

Edited by
T. Lance Holthusen

Pergamon Press
New York Oxford Toronto Sydney Paris Frankfurt

Pergamon Press Offices:

U.S.A. Pergamon Press Inc., Maxwell House, Fairview Park,
Elmsford, New York 10523, U.S.A.

U.K. Pergamon Press Ltd., Headington Hill Hall,
Oxford OX3 0BW, England

CANADA Pergamon Press Canada Ltd., Suite 104, 150 Consumers Road,
Willowdale, Ontario M2J 1P9, Canada

AUSTRALIA Pergamon Press (Aust.) Pty. Ltd., P.O. Box 544,
Potts Point, NSW 2011, Australia

FRANCE Pergamon Press SARL, 24 rue des Ecoles,
75240 Paris, Cedex 05, France

FEDERAL REPUBLIC Pergamon Press GmbH, Hammerweg 6, Postfach 1305,
OF GERMANY 6242 Kronberg/Taunus, Federal Republic of Germany

Printed in the United States of America

SPONSORED BY

The American Underground-Space Association

PLANNING COMMITTEE

Donald R. Woodard, General Chairman
Charles Fairhurst, Program Chairman
Michael B. Barker, Session Developer
David R. Mosena, Session Developer
Ray Sterling, Session Developer
William (Tom) Thomas
Richard Vasatka
J. Gavin Warnock, Session Developer
Thomas C. Atchison, Executive Director, AUA
T. Lance Holthusen, Conference Director
Irwin I. Chaitin, Exposition Director

LOCAL ARRANGEMENTS COMMITTEE CO-CHAIRPEOPLE

Truman Stauffer, Associate Professor Geoscience, University of Missouri,
    Kansas City
Donald R. Woodard, Vice President, Planning and Development, The Great
    Midwest Corporation

SPECIAL ASSISTANCE FROM The Kansas City Underground Developers Association

The USCE Planning Committee and Local Arrangements Committee are particularly indebted to the Kansas City Underground Developers Association for assistance at all
levels of planning for this event.

CONFERENCE AND EXHIBITION STAFF

Thomas C. Atchison, Executive Director, American Underground-Space Association
T. Lance Holthusen, Conference Director
Irwin I. Chaitin, Exhibition Director
Susan Bette Taylor, Program Coordinator
Mary Rollwagen, Membership and Marketing Coordinator
Jeanne Severson, Exhibition and Tours Coordinator and Registrar
Carol Mulligan, Local Arrangements and Publicity Coordinator
Marian Thomas, Local Arrangements Assistant
Marjory Christensen, Assistant to the Executive Director, AUA
Lucille LaFave, Julie Gill, Dorothy McNaughton, Secretaries

TABLE OF CONTENTS

Petroleum Storage Policies And Problems

FOREWORD

This volume of Proceedings contains papers which have been prepared for the Underground Space Conference and Exposition held in Kansas City, Missouri, June 8, 9 and 10, 1981. Papers are printed directly from the author's original. Papers received after our printing deadline will be included with the post-Conference Summary.

The Introduction is taken in its entirety from the editorial introducing the Conference published in UNDERGROUND SPACE, Volume 5, pages 262-263, 1981. Organization of this volume follows the organization of the Conference Agenda.

As editor of these Proceedings and Conference Director I want to take this opportunity to give my thanks to all of the authors who have participated in this volume and as Presenters at the Conference. I would also like to recognize the invaluable time and hard work of the USCE Planning Committee listed on the preceding page. Further, both the Planning Committee and I would like to thank the additional staff of TLH Associates, Inc. also identified on the preceding page for their long hours and skills in making the content and logistics of this Conference all come together.

T. Lance Holthusen

T. Lance Holthusen
Conference Director and Editor of the Proceedings

St. Paul, June 1981

UNDERGROUND SPACE CONFERENCE AND EXPOSITION (USCE-81)
AN INTRODUCTION BY SESSION DEVELOPERS
Donald R. Woodard, General Chairman
Charles Fairhurst, Program Chairman
Session Developers:
Michael Barker, Public Policy
Raymond L. Sterling, Earth-Sheltered Buildings
David Mosena, Urban Planning
J. Gavin Warnock, Deep Underground Space Use

PUBLIC POLICY

The increasing development of the subsurface raises many questions of public
policy. In most cases, the successful implementation of an underground project requires both public consensus on the merit of that project, and long-term planning.
Some of these subsurface projects include:

> .transportation and utility systems in major urban centers;
> .storage of toxic or nuclear wastes;
> .civil defense shelters for protection against war and
> natural disasters;
> .energy-conserving housing and storage facilities; and
> .expansion--underground--in urban areas that have limited
> surface space.

These issues, and others, will be discussed in the various conference sessions.

In the next twenty years, increasingly limited energy supplies will have a profound effect on lifestyles throughout the world. In the plenary session, William
R. Gibbs, past president of the American Society of Civil Engineers, will explore
the implications of subsurface use for energy conservation -- both its domestic
and global implications. Alan Muir Wood, honorary president of the International
Tunnelling Association, will discuss how underground space use can be integrated
into long-range planning to achieve maximum social and economic benefit. R.
Randall Vosbeck, president of the American Institute of Architects, will discuss
the design of the human environment as it relates to energy and subsurface development. These speakers will set the stage for more detailed discussions that will
follow later in the conference.

Transportation has always been fundamental to the shape of cities and cultures. A
special panel of practitioners and academics will examine the ways subsurface
transportation systems can potentially shape urban form. The prospects for underground transit systems in the U.S. -- in light of cost, management, and construction problems -- will be addressed. The audience will be invited to discuss the
merits of low-density urban sprawl and its alternative: high-density cities
oriented around underground transportation systems.

Protection from war and civil disaster is another major policy issue facing the

country. The U.S. is one of the few nations that has not developed a vigorous
civil defense program. Experts in the field -- both from the U.S. and abroad --
will examine the capacity of the U.S. to survive various forms of disaster, and
the potential of underground shelters to provide protection. The audience will be
invited to debate the extent to which the government should provide protection for
its population and vital functions from foreign attack, natural disaster, or
nuclear accident.

We hope the public policy session of the conference will attract a great deal of
audience participation; recommendations from the session may be channeled to the
appropriate government agencies. AUA will closely monitor these sessions for input into its positions on policy related to the use of the subsurface.

                        EARTH-SHELTERED BUILDINGS

The earth-sheltered building session is designed to gather and update information
that has been presented in the many conferences on earth-sheltered construction
held during the past three years. In all, a plenary session, eight technical
sessions, and over twelve workshops are planned.

In the plenary session, Michael F. Kelly, president of the Urban Land Institute,
will speak on behalf of that organization. John Millhone, director of the Office
of Buildings and Community Systems in the Department of Energy (DOE) will give the
keynote address. Millhone has been responsible for the development of the Building Energy Performance Standards (BEPS). He also directs DOE's program for earth-
sheltered buildings within the Innovative Structures Program.

Two brief technical presentations have been selected for the plenary session.
David Bennett of Myers and Bennett Architects/BRW will speak on the design issues
involved in earth-sheltering. Myers and Bennett/BRW has designed several out-
standing earth-sheltered buildings, including the award-winning Williamson Hall at
the University of Minnesota, and the new visitor's center at the U.S. Air Force
Academy in Colorado Springs. George Meixel, a research associate at the University of Minnesota's Underground Space Center, will discuss the state of the art in
heat-transfer research on earth-sheltered buildings. Meixel is the principal in-
vestigator on a major research study, funded by DOE, of the passive cooling
effects of earth-sheltered construction.

The remainder of the plenary session will feature short reviews of current construction activity and innovation in the different regions of the U.S. Written
papers in the conference proceedings will provide this information in depth; the
presentations are intended to allow a rapid grasp of which design techniques are
popular and successful in each region.

On two afternoons, technical sessions will be devoted to earth-sheltered buildings.
Presentations for these sessions will be selected from submitted abstracts, and
hence will cover a wide range of research and construction activities. The evening workshops, intended to provide design guidance for professional architects and
builders and interested lay people, will have a less structured format, allowing
plenty of time for questions.

As session chairman, I have worked hard to make this a comprehensive and valuable
session to all people interested in the design of earth-sheltered buildings. I
hope this and future AUA conferences will provide a regular forum for broadcasting
advances in the design and analysis of this rapidly developing building form.

URBAN PLANNING

In this part of the conference program, keynote presentations on urban planning
will address the planning and development of major underground residential, commercial and industrial projects. Kansas City is the site of one of the country's
leading underground parks, which houses a variety of activities. Local public
officials will discuss how the projects were planned -- specifically, how building
codes and ordinances were revised to accommodate this underground development, and
how further use of the subsurface will be integrated into surface development.
Leading industry representatives from Kansas City, including Forrest Browne,
president of Great Midwest Corporation, will describe the economic and other benefits the private sector can realize through locating facilities underground. A
delegation from the People's Republic of China will discuss the extensive use of
the underground in their country.

The series of afternoon theme sessions will explore these issues in greater depth.
One session will examine earth-sheltered residential projects, including development strategies, site design, and landscaping techniques. Cornelius Wood of
Architerra, Inc. will dicuss his firm's hillside residential developments, which
are moving into the U.S. market from France.

Environmental and health problems that have occurred in some underground residential and office developments will be the topic of another session; engineering
and design techniques to mitigate these problems will be presented.

A third session will highlight cities with extensive underground commercial property and pedestrian systems, such as the Royal Bank Plaza in Toronto, and the
Underground Pedestrian and Business Center in Dallas. The economics of these
commercial projects will be emphasized together with the institutional constraints
inherent in their planning and development.

DEEP UNDERGROUND SPACE USE

Coverage of deep space use at USCE-81 will emphasize policies and practices that
can enhance or impede development of deep space projects in North America. The
plenary session on Wednesday afternoon will be introduced by Jack K. Lemley, vice
president of the International Tunnelling Association, who will outline trends in
the planning and management of major underground projects on this continent and
around the world. The use of the underground for supplying, storing, and conserving energy will then be discussed from the public interest viewpoint.

The last segment of the plenary session will be devoted to exploring the problems
of permanent storage of nuclear wastes in America. Our interest is not to debate
the validity of nuclear power, but rather to explain what needs to be done to
develop an equitable waste isolation solution, recognizing that this is a major
deep space application.

The panel of speakers include: Colin A Heath, director of the Department of
Energy's Office of Nuclear Waste Isolation, who will present DOE's National Plan
for Nuclear Waste Management; Margaret Maxey, South Carolina Energy Research
Institute, who will address the "ethics" questions involved; and a representative
from state government who will discuss the problem of siting and the progress
being made in reconciling local and national government positions.

The first of two afternoon sessions, moderated by Magnus Bergman, secretary
general of the Rockstore Conferences in Sweden, will examine institutional issues

arising from deep space use. Contractual, legal, insurance, and labor issues will be covered by experts in each of these fields. A second session will be concerned with policies and problems of storing petroleum underground. Representatives of industry and government will describe the plans, progress, and impediments that have developed in the U.S. program for a strategic petroleum reserve and will probe the reasons for the difference in approach to underground oil storage in North America and Europe. These two sessions will provide opportunities for questions and comments from all participants.

CHAPTER I
PUBLIC POLICY

SESSION DEVELOPER: Michael Barker

THE ENGINEERING PROFESSIONS' ROLE IN THE ENERGY CRISIS

William R. Gibbs

Past President, American Society of Civil Engineers

Member, ASCE National Energy Policy Committee

Ladies and gentlemen, I am pleased to have the opportunity to speak to you today on a subject of great importance to all of us. I am hopeful that this gathering of persons for the purpose of discussing the subject of underground space utilization and earth-sheltered buildings, and the concomitant concept of energy savings inherent therein may signify that there is beginning an awakening realization of the critical nature of the energy crisis which is facing the United States at this time. What makes the matter a crisis is the fact that in the United States and in the world, we are facing a shortage of energy—and there is nothing that can be done about this prospective shortage.

At some time in the not too distant future, we are going to experience a shortage or shortages of energy which probably will continue for a period or periods of not months, but literally, years. It is already too late for any constructive action to be taken which would enable us to avoid this situation. Is this forecast of the future of energy too dark; is this prediction for the future too bleak? Unfortunately, I think not. Dr. John McKetta, Professor of Chemical Engineering at the University of Texas, energy advisor to President Nixon and energy advisor during the current transition period in our federal government, recently made that same prediction at a luncheon meeting among members of Congress, congressional staff assistants, and members of the engineering profession. According to McKetta, and I agree with him, it is already too late for action to avoid energy shortages. McKetta said that 4,000 square miles of solar reflectors would be needed, based on presently known technology, for the U.S. to realize the goal for.solar energy development established by the last administration. Is that a reasonable figure? I don't know, but based on the reflector area involved in a 5 megawatt solar facility designed by my firm, I should say that is not unreasonable. Based on estimates from some quarters, the number of nuclear plants required to generate the electricity that could, or should, be generated from this source by the year 2000, McKetta stated would be 400. How many nuclear plants have gone into service, placed on-line prepared to deliver electrical power to its customers, in the last 5 years? None that I can name at the present time. If there have been any, the number is negligible compared to the demand.

In Munich last September, the World Energy Conference verified and reiterated the general finding of its Commission on Conservation which in its report in 1978 "World Energy, Looking Ahead to 2020" said that coal and nuclear energy would be the only sources in the next two decades that would be able to replace our imported oil and gas. Replacement of our imported oil and natural gas is an imperative if

3

it is desired to avoid the disastrous consequences to our economy if our supply of oil and gas from abroad was reduced or terminated. Even if the supply is not lost, there must be a point in time and cost when it is decided that the dollar drain caused by buying from outside the country is contributing to a deficit in our balance of payments which is too great to continue.

Regional black-outs and "rolling black-outs" could foreshadow a general shortage of electric power during the 1980's according to Floyd Culler, president of the Electric Power Research Institute, who was testifying in Washington before the Nuclear Regulatory Commission. Nuclear power is intrinsically less expensive than coal-fired generation despite high capital costs, but the U.S. is rapidly being left behind in development of nuclear power with vigorous nuclear power programs under-way in Western Europe, Asia, and Latin America. Continuous regulatory changes which occur during the construction period of nuclear plants are counterproductive not only adding substantially to the cost of the plants but often degrading their safety and encouraging a proliferation of additional safety standards. Culler also said that only coal and nuclear power could provide additional electrical power for the U.S., but if coal were to provide all the additional electricity, 16 out of every 100 men in the West would have to be coal miners, and the railroads would not be able to move the coal.

Gordon C. Hurlburt, president of the Westinghouse Power Systems Company, testify-ing at the same hearing, said other countries have recognized that the develop-ment of nuclear power is needed to raise the standard of living and are moving ahead rapidly. Westinghouse is building 41 nuclear reactors in the U.S. and 24 overseas, but the pace of foreign orders is rising while no new orders have been placed in the U.S. in more than 2 years. The situation could have the effect of forcing Westinghouse gradually to send its engineering expertise to other coun-tries and cause the U.S. to lose its lead in reactor technology and safety. Hurlburt warned that other nations are beginning to set their own standards in reaction to the rapidly changing safety regulations in the U.S. Reactor safety standards will become less safe with a continued world-wide proliferation of standards. Westinghouse plans to compete for reactor sales in Argentina, France, Italy, Korea, Mexico, Spain, Taiwan, and other countries in the next few years but will not build additional nuclear plants in the U.S. until the uncertainty is removed from the licensing procedure.

The construction license in most foreign countries, Hurlburt said, is based on safety standards that rarely are changed before completion of the reactor as compared to the U.S. where the standards are in a continuously changing state. As an example of the economic advantage of such an arrangement, a reactor built recently in Korea without changes in standards during construction, recouped its construction cost in 31 months by replacing costly imported oil.

Several plants whose applications have been pending for more than 7 years were cited by Herman R. Hill, executive vice-president of the General Electric Power Systems Sector. Overseas, he said at the hearing, a nuclear power plant can come on-line within 7 years of the initial licensing application. Can a new administra-tion make the necessary changes to alter this situation? That remains to be seen. At any rate, it will be a tremendous task, but if all the changes were in place today to reach the point of having a plant on-stream in 7 years, we would still face shortages of energy. If it were possible to effect all the necessary changes this year, and if it were possible to convince the manufacturers that they could rely on this situation to continue, and if it were possible that the electric utilities would decide to start placing orders for new reactors, it would be almost 1990 before any new plants could be placed on-line. It is doubtful if we have the financing available or the qualified personnel available to produce or operate the reactors in this period of time. It is supposing a great deal about the proposed changes, however, when one remembers that the Democratic party at its last major convention placed in its platform a statement that nuclear power should be phased out of the energy system of the U.S. as quickly as possible.

The annual report of Exxon on the world energy outlook stated that coal and nuclear energy would replace much of the oil used in commercial and industrial applications, including electric power generation, but the U.S. will remain dependent on oil for transportation and specialized non-energy uses. Oil imports are likely to rise further to a peak of about 8.5 million barrels per day in 1985. The American Petroleum Institute recently revised its figures of 8.5 million barrels per day in imports since U.S. imports of crude and petroleum products averaged only 6.5 million barrels per day, but the Exxon report said this trend will be reversed bringing imports back to the original figure. Net oil imports might decline to 400 million barrels per day by the year 2000, but only if synthetic fuels can be gotten into production sufficiently to make up the other 4.5 million barrels per day. Combined oil and gas use is expected to decline significantly, falling from about 75 per cent of total energy use in 1979 to about 50 per cent of total use in 2000. Coal use should rise to account for 66 per cent of the projected growth in U.S. energy supplies for domestic consumption, increasing its share from 19 per cent to 31 per cent by 2000. Total coal consumption is expected to double by 2000 to about 14 million barrels per day of oil equivalent, including 2 million barrels per day oil equivalent, consumed in synthetic fuels production. Nuclear's share of total energy supply is projected to increase from 3 per cent in 1979 to 11 per cent by 2000. Despite the delays, high capital costs, and other problems facing the industry, nuclear energy is the most economic form of base load electricity in most locations. Assuming a major national commitment to synthetic fuels, which we seem to have made, Exxon reports production of synthetic oil and gas from coal and oil shale may reach about 1 million barrels per day of oil equivalent by 1990. This will increase to 4 million barrels per day or 9 per cent of total energy supply by 2000. Solar energy, including hydropower, is expected to increase from 4 per cent to 6 per cent.

Exxon said real energy costs are likely to rise throughout the period, and the world and the industrial countries, in particular, will remain dependent on internationally traded oil for many years, thus remaining vulnerable to oil supply disruptions. Except for that last statement, it seems to me that Exxon's predictions are too optimistic. I am sure that all of you have heard of Murphy's Law which states, "If anything can go wrong, it will!" Around my firm, we have Gibbs' Corollary to Murphy's Law which says, "Murphy was too optimistic." So it is with the Exxon report.

The question has been brought up, "Can transnational cooperation solve our energy problems?" The answer frankly is no. Transnational cooperation cannot solve our energy problems. Not to the extent that we can hope to avoid shortages or even more significant problems that would arise if we attempted to solve our shortage by other means. Transnational cooperation can, I believe, serve to alleviate our problems and, therefore, deserves our utmost study and consideration.

Before further discussion, however, I should like to review with you some of the circumstances which lead to my appearance before you today. I am a civil engineer and member of a large consulting firm which is a designer of electrical energy generation plants and their ancillary facilities. However, I am not a designer of electrical facilities even to the extent of doing the design work normally performed by civil engineers of which my firm has several hundred. My specialty is water resources and water and wastewater treatment. I was concerned with EPA which, believe me, is concern enough for anyone. I know because I served EPA for 4 years as a member of its volunteer management advisory group. Although I was aware of the concerns of others about our energy situation, I was pleased to leave that subject to my brethren who were working in the field. However, when I became president of the American Society of Civil Engineers in 1977, I rather quickly became completely aware, from the vantage point of my office with its associations with the political and other leadership of the U.S., of the acute situation in which we already found ourselves with prospects of no near-term future improvement. That same year we met with one of our members who is very active in the energy field and sought his advice. He assembled a group of energy leaders who discussed and debated all aspects of the energy situation. All of the problems of the energy industry were laid before us, and the potential results of various alternative actions were considered. As a result, the American Society of Civil Engineers went on record with a resolution approved by its Board of Directors establishing energy as our most important subject of concern. The Board of Directors further dedicated a society-wide effort to urge the government of the U.S. and the industry itself to take all of the necessary actions to avoid the catastrophic results which might result if we continued to hide our heads in the sand instead of facing our problems and overcoming them.

At the same time, I acceded to the importunings of friends who urged me to enter a political race against a less than completely popular U.S. Senator from Kansas. The investigations and study I did to educate myself on the issues of the day led me to base my campaign on four issues which I believed could lead to the impairment of our country's role in the world or to its complete downfall. One of these was the recognition of, and the determination to undertake, the solution of the energy problems.

Suffice it to say that I did not win my political race. The senator announced that he would not run for reelection which threw the field wide open in the primary election and candidates came forth from every direction. The very charming lady who won the primary election and ultimately the general election and an important role in today's Senate was Nancy Landon Kassebaum. She became a good friend who is doing a tremendous job in the Senate and who has provided support for continuing efforts to move the U.S. government in the right direction. Just two months ago, we attended together the luncheon meeting in Washington at which John McKetta was the speaker.

In 1979, with our plans formulated, the American Society of Civil Engineers was attempting to push the federal government into the adoption of a definite plan for energy by contracts with President Carter and every member of Congress, but especially with those whose committee assignments were related to the energy field. We were greatly pleased and felt rewarded when similar bills were introduced in both houses that year. The Senate bill was introduced by Senator Bennett Johnston of Louisiana and its companion in the House by Congressman Tom Corcoran. Our concerns, for the establishment of energy targets, were addressed in Title III of the Energy Security Act which has become law. The administration each year in February is required to submit for congressional review a series of targets or goals which it seeks together with its plan for reaching these goals. Such goals and plans will be reviewed each February and a report made on their accomplishments.

In late January this year, DOE Secretary James B. Edwards notified the Congress that the new administration could not meet the February 1 deadline but would report by the end of February. In fact, he reported on March 11, 1981, with a complete set of targets for energy. His plans have not yet been submitted.

In the meantime, ASCE and the American Association of Engineering Societies under the active personal leadership of Ellis Armstrong and Robert Jaske, immediate past chairman of the AAES' Coordinating Committee on Energy, were developing their own targets for submittal to the federal government. These were presented on March 20 at the congressional luncheon. The two sets of figures are not identical, but they are close enough for our purposes. The principal reason for establishing them is to keep everyone aware of the critical nature of the problems and ensure that it is foremost in the minds of all concerned.

Many other activities were also started by the Carter Administration following the enactment of the Energy Security Act. Prominent among these were the establishment of the Synfuels Corporation and the review and offering of grants for numerous projects concerning the production of synthetic fuels. Also included were research and demonstration grants for many alternative fuel sources such as solar, wind, biomass, geothermal and hydraulic as well as a number of programs to advance conservation measures. Sadly neglected, however, were nuclear energy and energy from the direct use of coal. No new nuclear energy power stations were authorized, work on the breeder reactor continued to be abated, and no great efforts were made to improve or expand coal production and utilization which continued to be hampered by environmental concerns and regulations. The previous administration continued to stay with its announced policy to phase out all nuclear energy activity. Generally, it has been concluded by knowledgeable persons that the United States must make maximum utilization of both nuclear energy and coal in order to achieve independence from reliance on imported oil and gas.

Although it is still too early to make a complete analysis and evaluation of the intent of the Reagan Administration's plans for energy development, there is every indication that it will follow a more reasoned approach to the avoidance of a crisis in energy. It is expected that the licensing procedure for nuclear energy plants will be shortened and that coal production will be stimulated. There are indications also that the breeder reactor program will be revived and that increased attention will be given to new forms of energy. It would appear from discussions of the Reagan budget that an effort will be made to move energy activities at a faster rate but with more risk capital to be provided from industry rather than from the federal government.

In addition to the continuing effort of the American Society of Civil Engineers to publicize the need for energy actions and to carry on a strong program of insistence on reasonable approaches to solution of the nation's energy problem, the profession, through the Coordinating Committee on Energy of the American Association of Engineering Societies, has been active in demanding from both the federal government and the energy industry a rational and timely approach to solutions. The program recommended has the following major areas of concern.

Conservation is an important element of any national energy program. Energy conservation information and incentives need to be developed and disseminated. Systems, appliances, and engines need to be developed that will improve energy efficiency. Artificial constraints on energy prices need to be reduced and eventually removed.

Solid fuel utilization must be encouraged. Coal washing and cleaning should be provided for at the mines. Coal production and leasing of mineral rights on federally owned land should be increased. Practical procedures for land restoration and environmental protection should be developed. Research, development, and demonstration efforts for all sectors of coal utilization should be accelerated. Environmental controls on coal utilization procedures should be moderated so that the public health and safety are protected without damaging the economy.

Oil and gas development requires exploration for new sources, improvement of techniques for recovery from current sources, development of new techniques for recovery of known oil and gas deposits, and increased leasing of federal lands. An attempt should be made to develop more international gas and oil sources which will have secure supply lines to avoid reliance on present sources which might be interrupted.

Shale oil recovery projects should be developed as early as possible even if an infusion of federal money along with private risk capital is necessary.

Nuclear energy must provide a major amount of electrical energy in the next two to three decades and must be accelerated immediately in order to hold future shortages to the minimum possible. Energy produced from nuclear fission should be established as national policy. Public information activities should be expanded. Present technology should be employed for spent fuel reprocessing and for disposal of nuclear wastes. Research and development of the breeder reactor and of fusion energy possibilities should be reactivated.

Solar and other renewable energy sources should be stimulated for maximum exploitation. Economic utilization of these options must be stressed with consideration given to methods of integration into existing utility networks and establishment of sound standards for performance.

Deficiencies in the licensing process for electric power generation and transmission facilities must be corrected to avoid delays and higher costs. Electric utility planning should be in public view; licensing procedures should be completed quickly and before heavy financial commitment by utilities; state and federal regulations which overlap and cause duplication should be eliminated; and there should be timely resolution of environmental conflicts.

Long-range planning for energy should be established and followed in the industry and in the federal government. All supply options, demands, research, international as well as national implications, and long-term as well as short-term impacts must be considered.

In the foregoing program, the well known fact that underground and earth-sheltered space utilization is extremely efficient in conservation of energy can very well play an important role in a national energy conservation plan. Underground construction in the Kansas City area gives an example of energy savings which are possible. Underground sites are not affected by windstorms, blizzards, heavy rains, extremely hot or cold periods, and the radical changes in temperature which such events can cause.

Approximately 40 million square feet of space are already developed for secondary use with a constant year-round temperature of $56-60^{o}F$ which cuts heating and cooling costs dramatically. A recent survey comparing energy use of above grade facilities with underground facilities used for similar purposes found that savings in using underground facilities averaged 47-60 per cent for manufacturing operations; 60 per cent for service facilities; and 70 per cent for warehousing.

Inland Storage Company began developing underground freezer space in the 1950's at about the same time as the advent and subsequent very rapid growth of the frozen foods industry. Inland has 4.4 million square feet developed and another 18 million square feet of potentially developable space. About 90 per cent of the space is used for warehousing with the remainder used as office space. There are 24 million cubic feet of refrigerated storage and 3 million cubic feet of cooler rooms. It is estimated that Inland saves at least 15 per cent on energy costs, although continuous refrigeration is needed to lower temperatures from the ambient. The rock walls are "frozen" to $-5^{o}F$ initially so the temperature will remain relatively constant even during a power interruption.

The Downtown Industrial Park has a positive air system which provides preconditioned air to raise the natural temperature of $58^{o}F$ to $76^{o}F$ and to maintain 50 per cent relative humidity.

These are only a few examples of the many situations that could be described in which the energy demand for underground space can contribute to the solution of the nation's energy problems. Significant savings also may be obtained when utilizing earth-sheltered space, particularly when such use is combined with solar devices for heating or for hot water and with reflective insulating blinds. Public buildings of this nature are the National Art Headquarters in Reston, Virginia; the Los Alamos Laboratory Support Complex in Los Alamos, New Mexico; and a State Office Building in Sacramento, California.

With the contribution that may be made in energy conservation, promotion of the use of underground and earth-sheltered space by the American Underground Space Association will contribute materially to the alleviation of future energy shortages experienced in the United States. As the general public becomes increasingly aware of the dangers of such shortages, there should be placed on underground space utilization a premium for the conservation which naturally occurs. Preference should be given to energy service on underground facilities during shortages.

With such conservation, the AUA will be assisting the nation so that it will not be tempted to alleviate future shortages by the use of more adventuresome and perhaps far riskier methods. It is with this hope that all efforts toward the solution of the energy problem of the United States should be advanced.

UNDERGROUND SPACE : ITS CONTRIBUTION TO THE SUSTAINABLE SOCIETY

A.M. Muir Wood

Sir William Halcrow & Partners, 45 Notting Hill Gate,
London, W.11, UK

ABSTRACT

The impact of the finite nature of the world and its resources requires a radical reappraisal of economics as presently understood and practised, with practical consequences in support of the better use of the space beneath the surface.

KEYWORDS

Socio-economics, energy economics, underground planning

INTRODUCTION

Several of the forces which dominate the growing support for an increased and improved use of the subsurface are I believe exemplars of an overall change in attitude to many social and economic philosophies until recently little questioned. This is the general theme of my talk and I cannot expect the substance to be accepted without first an explanation of my personal view of the factors involved.

THE FIRST STAGE OF THE INDUSTRIAL REVOLUTION

Adam Smith (1723-90) epitomises the fundamental characteristics of what I will call the first industrial revolution, which was well into its swing by the time of the 'Wealth of Nations' in 1776. He here clearly depicted the direct economic benefits of the initial steps in a new socialised form of industry, whereby the several operations in making a particular article are subdivided and undertaken by individuals or groups who each work on a single process, yielding increased efficiency and reduced cost of the finished article. He chose the manufacture of a pin as a particularly evident example of the benefit in cost and time. The reasons that gave rise to the division, and hence greater efficiency, of labour can be discussed from many view-points. Here it is only necessary to emphasize the growing pressures on land and the consequent need to satisfy the new demands of an increasing urban population who were unable to continue to enjoy an essentially rural way of life of largely self-supporting communities.

This first stage of the industrial revolution can be seen as an essential step

11

towards the industrial society which alone would enable a growing population to exploit new resources and to meet the basic needs of a prosperous and powerful nation, using relatively simple implements and machines. Initially the diversion of resources to capital accumulation, including urban and infrastructural development, contributed to a lowering of living standards for the mass of the people. The extent of this lowering is considerably masked when assessed by statistics of Gross Domestic Product which exclude the contribution from self-sufficiency to the standard of living of a rural community. The uncosted elements of rapid urbanisation in the associated squalor, disease and destabilisation of settled social units must have entailed yet more important economic consequences, and these experienced over an extended period. The first industrial revolution was a necessary, if blunt and often cruel, means of satisfying new imperitives of rising population and diminishing ability for survival of the local 'sustainable society' of John Stuart Mill. The notion of the 'sustainable society' is related in the first place to the degree of economic self-sufficiency without drawing upon wasting resources from outside. The concept also has philosophical connotations but it is the material aspect which concerns me in this account. The new way of life brought with it new opportunities, first for the relatively few, but with the proportion gradually increasing with improved education, acquisition of skills and upwelling of new demand.

THE SECOND STAGE

The second stage of the revolution, not sudden but representing to me such a change in emphasis as to warrant the term, may be described as corresponding to the consumer society. We should not delude ourselves as to suppose that it was conceived primarily for the benefit of the consumer; but to stoke his (or just as often her) appetite for ephemeral consumer goods. Whereas Adam Smith saw expanding needs, as a result of social change, for goods made to last the consumer society was concerned to stimulate continuous growth of demand associated with more stable social conditions.

While Karl Marx saw the inherent contradictions of capitalist society - unaware of the extent to which he was conditioned by the ferments of the time in which he lived - he did not foresee the supreme contradiction of a society whose industry thrives upon stimulations of dissatisfaction with the status quo as it encounters a finite world in resource and technology terms. This metamorphosis has been and yet is occurring against an unprecedented rate of population growth with increasing competition for those same limited resources. The consumer society is irreconcilable with limitations to growth and to continued improvement.

The notions of continuous economic growth have been attacked for many years; I have the impression nevertheless that those who command economic policy of the world continue to look upon such attacks as a regrettable scourge, only to concentrate the minds yet more resolutely towards revival of economic growth on the old pattern with the economists primitive faith in extrapolation. It is fashionable to refer to the ideas of Thomas Kuhn and I cannot here forbear to remark that we have here all the elements of a paradigm of economics, self-consistent (to a degree that economics ever have been consistent), based upon a set of axioms which appear to fit selected evidence of the past and to support dismissal of evidence of the present, incapable of change to meet new conditions by accretion and assimilation of new knowledge and evidence. Change in such circumstances has to occur as a step-function rather than as the steady curve of a continuous function. In fact the term 'economic growth' and the manner of its measurement need to be redefined and, incidentally in so doing, made practically attainable.

The general recent trend in capital-intensive industry may be summarised as requiring increasing magnitudes of investment before a commercial pay-back may be achieved on investment. This is no longer a situation in which even manufacturing industry can meet the real needs of society for general improvements in living standards against declining availability and consequent increasing cost of natural resources. The countries of centralised economies with the ironing out of risk-takers, initiatives and responsibility, provide no solution and are indeed already manifestly leading to a low-level quasi-stability (and probably low adaptability to change, i.e. lacking the essential evolutionary feature of 'development potential') and hence are not promising survivors in the world of today.

THE PRESENT DAY

We are now entering the phase of the third industrial revolution applicable to both developed and developing countries, where skills and powers of invention remain important but no longer the prime determinant in a recognisably finite universe. What is principally wrong is that technology has advanced but economics have stuck fast in highly convoluted cerebral debate, comparable to the 'schoolmen' in the evolution of mediaeval philosophy, but still in Adam Smith's environment without much understanding in the contribution of economics to questions of quality and satisfaction. What we urgently need is a new flexibility in approach on economic affairs conducive to encouraging initiatives and enterprise, with solicitation for skills and all sorts of dexterity so that the notion of 'quality of life' loses its hollow ring. We need far greater subtlety, humility and wisdom than we find in any present school of economics. At the extremes monetarism corresponds to the stage in medicine of the application of leeches, with more than a touch of sympathetic magic and the same lack of concern for the irreversability of mortality. Welfare economics, on the other hand, as presently developed, erodes self-reliance and engenders chronic valetudinarianism to be treated largely be placebos. The revolution, when and where it comes, will represent an explosion from suffering humanity subjected to vivisectionary experiments, to satisfy curiosity of blinkered political economists.

The foundation of an acceptable regimen at the present day must recognise first that economic growth at the expense of consumption of irreplaceable resources is no solution. Indeed part of the distortion in present rates of foreign exchange arises because the commercial world, unlike national budgets, recognises the dominant influence of a nation's capital, inter alia, in unexploited reserves. We can no longer afford to discuss economics in the abstract; it must be related to individual people. People will not docilely conform to economic theories unless they are tolerably content. There is no other word than scandalous at the present day for the manner in which large sums by any standards can be found for destructive purposes while the seed corn, represented by education and by the maintenance of the essential social and material fabric of urban living, is dissipated, short-term expediency having taken over from statemanship.

A NEW DIRECTION FOR ECONOMICS

The main barrier to relating economics to the needs of the present day lies in its inability to comprehend questions of quality. As an example, the ancient Greeks distinguished between work (εργον) and labour (βανασος), the former entailing skill and satisfaction, the latter the tedious underpinning of a prosperous society to be undertaken by slaves. The former clearly requires encouragement; the latter requires to be reduced to a minimum. A single statistic of unemployment has no reference to such a distinction. Nor is it by any means certain that much of the mechanisation of skilled work, with its attendant diminution of interest and adaptability to change, has ever been in the ultimate interest of the consumer. What is certain is that the smallholding, the low-capital informal fall-back of the rural unemployed of the early industrial revolution, urgently requires a modern day equivalent, whose products would rapidly repay capital and much more than compensate for costs extra over subsistence. Possibly the solution will include low capital cost small-scale workshops with simple tools and equipment for a particular trade, which would for example charge for their use but permit the craftsman to take the profit on the product, without affecting eligibility for unemployment benefit. Such an arrangement would raise many problems for solution and would possibly tend towards a simple form of co-operative. The point of substance is that 'value added' would considerably exceed direct cost but the products would nevertheless be moderately priced, and wider benefit would come from the 'Keynsian multiplier'. In substance the problem of mass unemployment is soluble by such stratagems for low capital cost but only if we escape from the flat earth of the consumer society economists. Such contributions as these to the 'alternative economy' are attracting increasing support although there is a depressing absence of the 'alternative economist' in position of influence. The notion of 'from each according to his ability' and 'to each according to his need' is far more fundamental than any political dogma. Economic and social solutions need to recognise the very different needs of people considered as ethnic and other groups and as individuals. Perhaps wise government would include such a feature as Sir Peter Parker's 'Council of Industry', an upper house of government representative across those who are in positions to influence major contributions to a prosperous nation, somewhat similar to a proposition by Sir Winston Churchill many years previously.

Posterity will condemn each part of the world of the present day, whether divided East and West or North and South, for failing to recognise that each provides example in particular respects to the other; meanwhile each is desperately far from Voltaire's 'best of all possible worlds'. Major social change usually demands a severe external factor and the energy crisis, foreseen by the scientific world but not by the economists, may belatedly jolt us from the Second to the Third Industrial Revolution.

No present-day nation can afford to devise economic policies which do not take full cognisance of social consequences and of an acceptable set of energy scenarios. The former may seem an obvious criterion but it is remarkable that governments appear to become blinded to the personal consequences of their concepts, and even to forget that their ultimate justification is that of service to the individual members of the community. The function of a Council of Industry would be mainly to define the boundaries of the real options open to government, avoiding the effects of harmful dogma of all types and to open the eyes of legislators to the probable, as opposed to the wished for, consequences of their enactments.

It is becoming increasingly evident that every human activity is attuned to certain acceptable limits of scale. Even on the industrial scene the benefits of large-scale production are now seen to have been greatly over-rated where they entailed a massive labour force. It is no coincidence that the emphasis on management techniques in the abstract, and their attendant academic circuses, have built up industries which have fallen first victims to the cold winds of recession, largely because they concentrate upon the system rather than the product, the routine of technique rather than the adaptability to change.

When one attempts such a brief survey of a vast and complex subject, it is impossible to avoid accusation of over-simplification and of course this is justified. If we reflect for a moment we have to accept that our achievements in understanding nature and in science and engineering have advanced ahead, and nearly out of sight, of our demonstrated incapacity to build a well ordered society. Why is this?

THE THIRD, COMPLEX, STAGE

We have to accept that all governments, policy makers and philosophers are baffled by the sheer complexity of the problems. Perhaps the greatest challenge of all is that the problems are all inter-related. Donne's observation that 'no man is an island' has developed today to the realisation that every local part-solution has its national and international connotation. A feature of particular significance, in recognising the dominance of resource limitation, is that the prosperity of one community or nation is in the longer term enhanced by the increasingly effective solution of the common resource problems by all nations of the world. Profligate waste of resources enhances their scarcity to all, contributes to consequent increases in living costs and eventually to the prospects of conflict. So we have to find, in all respects, means of combining incentive and encouragement of initiative with a world-wide sharing of their benefits. Thinking along this line must lead to the conclusion that we have to learn to build upon the best of a planned society with a market economy. So many of the examples of such hybrid experiments to the present day have led to the worst of all possible worlds, because of short-term expediency and selfishness, and this must not deter us from following this essential direction. The failures have sprung from an inability to accept the essential disciplines needed in obtaining the equitable equation between individual and community interest. As I see it there is no other acceptable choice.

This lengthy introduction of my subject, with more hint of the underworld of the economist than the subsurface of the engineer, needs some further justification. I hope the argument has a ring of conviction but the obvious objection is to ask why it is not the 'normal science' of the day. The answer must stem I think from the narrow teaching that pervades all economics, based on the two basic ingredients of capital and labour, excluding the increasingly over-whelming claims of energy and making the other unacceptable simplification, utterly warping in its consequences, of neglecting all matters of quality and ingenuity not expressible in direct money terms. So far as the first is concerned Samuelson in his influential book "Economics", observes that the social sciences recognise no principle like the conservation of energy, a remarkable but true and indeed alarming basis indeed for the guiding principle for all governmental policy in today's finite world. So far as the neglect of quality is concerned not only does it deny that the civilised behaviour of a community depends upon the degree of inspiration they can draw from their surroundings (How many city dwellers could today echo Wordsworth's view on London as seen from Westminster Bridge in 1802 "Earth has not anything to show more fair") but confines consideration for the future to the ridiculously short-term of the discounted cost/benefit analysis. On the other hand, the increasing dichotomy between the physical and the social sciences has discouraged all but the intrepid few, of whom I am one, to stray across the frontier. We had virtually no audience until the foreseeable increase of oil prices occured in 1973 since when the annual rate of increase in constant price terms has exceeded any realistic long-term figure for the economist's rate of growth.

ENTROPICS

In thermodynamics we recognise the concept of entropy whereby the increase in entropy of a substance is represented by the quantity of heat taken up by the substance divided by its temperature. On account of the second law of thermodynamics any transfer of heat is accompanied by overall increase in entropy and overall thermodynamic efficiency may be equated to a minimum rate of gain of entropy of the total system. Economics for the sustainable society may then be conceived in terms of 'entropics' following the lead of Rifkin, recognising that unless this concept controls economy policy all the efforts for economic recovery and assistance to the developing world will be frustrated. An increase in energy costs of 1% in real cost terms has the effect of depressing the nett GNP by about $\frac{1}{4}$% at the present day and this fraction will increase with time unless the efficiency in energy use is designed to compensate energy cost increases. Energy subsidies provide no answer to the problems; they simply divert attention from the urgency of the situation.

BEARING UPON THE SUBSURFACE

Much of this conference will be concerned with the assessment of subsurface projects in terms of energy savings. At Rockstore 80 I emphasised that urban planning
based on the efficient use of energy was also consistent with designing for civilised living in the future, a question of quality largely overlooked by the domina-
tion of orthodox economics. Appraisal of projects in terms of opportunity cost
only has meaning if the 'do nothing' alternative is a real one. I suggest that
chronic under-employment coupled with a genuine responsibility for posterity allows
a new look at the need for overall planning of the infrastructure of a habitable
world of the future if we are not to lead our successors to the edge of a cliff
with only a bottomless pit ahead. Clearly a revolution in thinking is needed but
this has little to do with the left and right of politics and involves a new dimension in political thought. Fundamentally we need to plan the mechanics of living
in order to be able to enjoy freedom in our style of life. For this we need to
make better use of the third dimension of physical planning.

WOMB OR TOMB?
THE DESIGNER'S ROLE IN THE ENERGY CRISIS

R. Randall Vosbeck, FAIA, President
VVKR Inc., Principal

ABSTRACT

Since America's buildings consume approximately 40 per cent of the energy used in
this country, it is clear that architects through energy-conscious design can have
a significant effect in reducing America's dependence on non-renewable sources of
energy while at the same time providing a built environment not of less, but
better. Nowhere is the opportunity and achievement more visible than in the area
of earth sheltered design. This paper goes on to cite specific projects in which
the design skill of the architect has, by responding creatively to the particular
concerns of underground space (such as engineering, lighting and psychological
issues), achieved award-winning projects. In so doing, architects illustrate the
difference between a defensive strategy that would drag out society into a post-
petroleum era, and a strategy of design that leads Americans to a better quality of
life. From the standpoint of underground space, the difference might be described
as the distinction between a hole in the ground and an attractive, habitable space.

KEY WORDS

Energy, Earth Sheltered Design, Architecture, Design, Housing Architects

INTRODUCTION

We have a curiously ambivalent attitude toward the earth. On the one hand, it's
our mother, our source of strength, in classical mythology -- a goddess. On the
other, we're constantly trying to blast away from terra firma whether in the
resurrection motifs of religious allegories or in rocket ships and space shuttles.

Being in on the "ground" floor of a hot investment scheme is a good thing; but for
the economy to be in the "pits" is something else again. "Cellar door" is, I am
told, one of the most beautiful sounding combinations of words in the language.
But being in the cellar in the pennant race is to out of the running.

So we have to recognize that in public discussions of earth sheltered design --
which is a euphemistic way of talking about "underground space" -- there is a great

prejudice that all of us have to overcome in broadcasting an important message. And what is that message? That what architects, engineers, and contractors are doing today below the surface of our earth is neither exotic nor gimcrackery now unprecedented.

We need to impress upon the public the fact that developing habitable underground space is not necessarily a perverse preoccupation with tomb building, but rather, a recognition that the earth is a womb, a source of life, a shelter from those forces that might threaten the high quality of life most of us in this society have come to expect. We need to impress upon the public the fact that for a wide range of reasons -- available space, site considerations, economics, climate, and most recently energy -- the underground is (to use a popular cliche) "where it's at."

But how do we get that message across? How do we reinforce that positive image of the earth that sees the space below the surface not as a tomb, but as a nurse, a shelter, a home? How do we transform the necessity of human activity that already takes place underground (such as storage facilities and transportation systems as well as habitable space) into a positive experience?

This is the role of design. This is the special skill of architects who are trained to solve such problems by a creative marriage of technology with those concerns and humanistic values traditionally associated with art. This is the message of Washington's new Metro system and Atlanta's MARTA. This is the message of Terraset in Reston, Virginia, and the Underground Space Center at the University of Minnesota.

This is the message of a past chairman of the AIA's Design Committee, Bill Morgan and his innovative earth sheltered houses in Florida. And design is at the heart of what I want to say to this Convention of the American Underground-Space Association.

The job of persuading the public to a more positive appreciation of the past achievements and current potential of underground space has recently received a boost from an unexpected ally: energy. Of course, earth sheltered design did not have to wait for the energy crunch to come along to be an option on the architect's palette. But there is no question that what the media call the "energy crisis" has thrown a spotlight on what has been achieved and what is anticipated from those whose expertise is earth sheltered design.

Why is this so?

Consider the facts. Americans may have turned down their thermostats and elected to buy smaller cars. But relative to the other industrialized nations of the world, we still gorge non-renewable resources. One response to the crisis has been to campaign for a loosening of fuel efficiency and pollution standards as well as lobbying for import restrictions that will deny Americans fuel-efficient Japanese cars.

To focus for a moment on oil is to be made aware of just how critical is the situation we find ourselves in. The United States continues to import nearly one-half of its oil. And with every barrel of OPEC crude, it imports a broad range of dangerous international problems. These include a limp dollar, deficit trade balances and inflationary pressures.

In 1972, the oil import bill for the U.S. was $4.5 billion. In 1980, the tab jumped to approximately $90 billion. By the late 1980's, the red ink may spill to the tune of over $125 billion. This is private capital that is being squeezed out of our economy and sent overseas instead of being channeled right here at home to revitalize our crumbling cities and industrial plants while meeting the needs of all our people.

Let's translate those billions into a language all of us can understand. In 1980, $90 billion would have bought more than the entire state of Iowa at $2,000 an acre. In 1981, OPEC can have Wisconsin. In 1982, we could trade our entire steel industry for part, only part of the oil we will be importing; we'd have to give them Sears and A&P to pay for the rest -- except that now the Germans own A&P.

How does the cost of our energy bill concern architects? The relevance of the current energy situation to the design professions may be seen in two broad areas. The first is obvious: there is a finite amount of capital. Any extensive hemorrhaging of dollars from the economy is bound to ripple down to the construction industry and hit it hard.

However, the energy crunch is relevant to architects and the construction industry in a second and ultimately more challenging way. Although architects cannot increase the supply of energy -- at least not directly -- we can reduce demand. The key is, once again, design: design that makes a creative and pleasing use of the four primary elements of air, fire, water, and, of course, earth.

How much of a contribution design professionals can make is signaled by the fact that right now, in 1981, nearly 40 percent of the energy produced in this country is going into heating, cooling and lighting this nation's buildings. There is no question that if we take carefully-thought-out and effective steps to reduce energy consumption in new and existing buildings, we can ease much of the squeeze of the current energy crunch. We can do this, and in doing so, give our society the necessary breathing space to move confidently into a post-petroleum world.

In responding to the challenge and opportunities inherent in making energy a key consideration of the design process, architects have, however, run into some popular misconceptions. Far too many people are convinced that energy-conscious design has to be curious, unpleasant, bizarre -- and expensive.

We can make the greatest inroads in dispelling this misconception by two effective and related strategies. First, we need to point out that energy as a key determinant of form is not new; it had been central to the look of architecture until abundant, cheap fossil fuel in the twentieth century revolutionized the design process. The New England saltbox, the sunny Italian villages that cascade down to the sea, the desert pueblos, the Southern mansion with its wide, open stairways, high ceilings and airy verandas, the sod houses of America's Great Plains -- all these traditional residential forms are very much a product of energy consciousness informing the design process.

The second strategy is more obvious and forward looking. America's architects are from coast to coast and in every climate region designing energy-conscious commercial and residential space that is both cost effective and attractive. This is what design does. And it's very different from a defensive, non-design approach that merely piles on the insulation and reworks the mechanical systems without any significant rethinking of those design considerations that affect a building's energy consumption, such as site, siting, landscaping, daylighting, shading, thermal mass, and earth sheltering.

All of which is to say that the popular misconceptions that tend to cloud energy-conscious design will evaporate in the light of the accumulating evidence and achievement of America's architects. But this will happen only if the design process is truly informed by the best acquired and intuitive skills of architects as they draw from the lessons of the past and learn about emerging technologies.

The evidence of the achievement of those who specialize in earth sheltered design is to be found not only within the pages of those magazines and other publications that are specifically aimed at these professionals, but also within the general architectural press and the national media. In July, 1979, passengers aboard all Eastern flights would have read a piece by David Martindale, "Down-to-Earth Architecture". Here's his conclusion: "Practical, imaginative architecture is hardly a panacea for the ills of urban society. Yet earth sheltered buildings do suggest that we could just be standing on part of the solution." Educators reading the May 1980 issue of American School and University would have found these words about the University of Minnesota's proposed engineering building, a strikingly innovative design that makes creative use of both solar and earth sheltered technologies: "Although initial costs are expected to be par with those of conventional buildings, operating costs are expected to be far below."

Moving somewhat afield of habitable earth sheltered design, here are the lead sentences that appeared in an article in the Detroit Free Press about a recent underground project of the architects William Kessler and Associates: "Kessler's firm designed this city's prettiest tunnel -- the 450-foot-long futuristic tube that links five hospital facilities at the Detroit Medical Center. The tunnel uses skylights with mirrors to reduce the claustrophobic sense created by such spaces." The reporter goes on to describe those design strategies that transformed a tunnel into an attractive, yet functional space, or, as the headline reads, the "Prettiest Tube in town."

Early in 1978, the AIA's Journal published two major reports on underground architecture, in the April and November issues. "The concept is not a new one," wrote the reporter. "In fact", she goes on, "it is as old as the hills in which underground buildings have nestled for centuries in Spain, Turkey, China and elsewhere. In America, Frank Lloyd Wright in 1950 built the Cabaret-Theater at Taliesen West -- into, rather than upon, the desert. And in the mid-1960s, Philip Johnson, FAIA, went underground with the art gallery next to his glass house in New Canaan, Conn., and the less celebrated Geier house near Cincinnati.

"What is new in its intensity", she continued, "is architectural concern for energy conservation and for the landscape." This writer's perception was sound because by 1980, the number of earth sheltered houses had ballooned (or should I say mushroomed) to something on the scale of 3,000 to 5,000 units, while other underground projects from libraries to transportation systems enjoyed a similar growth pattern.

But, as I have been arguing, the mere fact of rising concern for the environment and the troubling energy question would not have persuaded Americans in increasing numbers to turn to earth sheltered structures. What turned a mere hole in the ground into a pleasant habitable space was the critical impact of design in addressing the engineering, site, energy and psychological concerns unique to underground architecture. The achievement of America's architects in meeting these concerns has been recognized in a string of award-winning projects culminating in 1979 in an AIA Honor Award to Herbert S. Newman Assoc., for Yale University's Center for American Arts, which the jury evaluating this project described as a sensitive design solution that demonstrates the efficacy of providing much needed expansion in dense, urban college campuses by using underground adjacent space that preserves valuable landscaped courtyards.

But clearly much more needs to be done to accelerate the public acceptance of earth sheltered design as an attractive option for a work or home environment. The awards help, together with their attendant publicity. The growing body of literature on the subject including how-to books also broadcasts the message. But it is the rising concern for the environment and the shadow cast by the energy question that provide the designer who specializes in underground space the opportunity to reeducate the public to the advantages of earth sheltered design. And here research into the psychology of underground space is called for so that the designer has this vital information during the design process.

Already other industrialized nations are far ahead of us in reminding their citizens that the earth is even more a source of growth than it is a tomb. Sweden alone has spent some $2 billion on underground installations, about half for civilian uses. Just across our northern border, in Montreal, is the extensive undercover complex at Place Ville Marie and Place Bonaventure, a system of attractive underground shops, offices, restaurants, cinemas and pedestrian walkways fully integrated with above-ground facilities and a modern underground transportation system. Altogether, very energy-conscious responses to an increasingly energy-starved world.

In all of the projects I have cited above, both in this country and abroad, design has led rather than dragged our people into the new realities of our evolving civilization. That's the historic role of design. And that's why the AIA's theme this year -- "A Line on Design and Energy" -- leads off with this key word. Design is what the profession of architect is all about. Today the nation's attention is focused, as it should be, on energy. Earth sheltered design is one creative response. No doubt other issues will emerge tomorrow. But as long as there are architects, future generations will have a potent tool in continuing the evolution of humanity. And that tool will be the unique expertise of this profession -- in a word, design.

PLANNING RAPID TRANSIT SYSTEMS

Walter L. Dougherty, P.E.
Urban Mass Transit Authority
Washington, DC

ABSTRACT

This paper briefly indicates the type of planning questions that should be
addressed in assessing the urban impact of a proposed rapid transit system.
Construction costs in the urbanized areas for both Highways and rapid transit
systems are discussed and the author concludes that these costs reflect the
difficulty of construction in the urban setting. Two studies evaluating energy
usage in the construction of transportation systems are assessed with the author
concluding that additional studies are still required. The final section stresses
the importance of the management process in the delivery of a transportation
system.

KEY WORDS

Planning, trade-off, system patronage, construction costs, construction methods,
energy studies, management process.

The underground space for the New York City Transit System occupies more than
37 million square feet. That's more than twice the commercial underground space
developed here in Kansas City where the Underground Space Conference is being
held. With Rapid Transit Systems making such a significant impact on the use of
underground space, I believe it is important for us to discuss the major questions
involved in planning these systems.

In this brief paper, I intend to look at some of the planning questions that
should be asked when rapid transit systems are planned in order to assess the
systems impact on the development of an urban area. I will also cover the
construction costs associated with rapid transit systems particularly in relation
to urban highways. In addition, I will also look at whether or not rapid transit
systems actually consume more energy in construction than other transportation
systems. Finally, I will briefly cover how the development of a rapid transit
system can be managed to minimize its cost and increase its effectiveness.

Planning

Planning as used throughout this paper, refers to that stage in the development
of a mass transit system that leads to a firmly fixed configuration. It is more

23

than the selection of a preferred alternative which includes the selection of a corridor and the mode of transportation such as Bus, Light or Heavy Rail. Planning as here used includes those design studies performed during Preliminary Engineering which lead to configuration layouts.

The future development of the urban area is influenced by the mass transit mode selected, its location and layout. In this paper we are primarily concerned here with the impact of underground rapid transit systems on urban development but we should realize that even the bus mode will have a significant irreversible impact.

For example, the selection of the all bus option involves placing large numbers of buses in the city environment and possible deminution of the air quality. To be truly effective, the bus mode transit system probably requires restricting autos from full usage of main thoroughfares and may require the dedication of exclusive bus lanes on the arterial feeder highways. This option may include in some cases constructing busway tunnels in the main Central Business District (CBD) to be truly workable. These changes can require heavy capital costs which will have a significant impact on the urban scene and are essentially irreversible decisions.

The effect of even such a simple solution as bus mode for the urban mass transit system will influence the development of the city, depending on the layout of the City street grid. Of course urban growth depends on more than the street grid layout. Ease of mobility from other points in the corridor and the total peak flow past the point are strong determinents in density development. In the case of a regular grid with numerous through streets, the city may well grow in a uniform development. Where few streets run through the grid, buses will restrict traffic flow and may create denser corridors. Similarly, the choice of either light or heavy rail rapid transit systems will influence the development of the urban areas. The selection of light or heavy rail systems is generally dependent on system patronage.

System patronage is the driving force behind the transportation mode selected and the type of operation planned. Peak hour patronage and the peak on peak demand coupled with the selected train headway determine the number of trains required for the rapid transit system. For example, assume that we design the system for a peak on peak period of 15 minutes. If we select a long headway, say one train every 8 minutes, we will need longer trains to serve that peak than if we used a 6 minute headway. The closer headway allows the peak period to be served by two trains instead of one.

In the 8 minute headway example, each station on the system must now be designed for the longer train. Let us assume that an 8 car train is necessary for the longer headway but only a 6 car train in the shorter. In this example, all stations would have to be built 1/3 longer to accommodate the longer headway trains. Construction and long term maintenance costs will be higher and the system will run under capacity with a built in potential for enormous growth along the corridor.

We can look at the problem another way. Lets assume that the planners wish to restrict growth to a certain density along the rapid transit corridor. One alternative is to decrease headways and build shorter stations. When the density rises to a certain predetermined level, parallel lines could then be constructed. However, it could backfire and the density might increase along the original corridor requiring an expensive reconstruction of the existing line to accommodate longer trains.

These are but a few examples of the trade-off studies that could be carried out in the planning stages for any urban area that plans to utilize mass transit.

A logical question is "How do we know when an urban area requires a rapid transit system?" - That's a difficult question to answer. Its perhaps easier to determine when we don't need a rapid transit system than to determine when we do. The first planning question should be "What are we trying to do?" For example, are we trying to revitalize the area, assuming of course that sufficient demand can be generated to justify the system?

In the case of Buffalo, New York or the Southwest Corridor in Boston, revitalization was one of the major parameters in the selection of a rapid transit system. It is hoped that both systems will revitalize the areas they are running through. When the Queens Boulevard line in New York City was constructed in the 1930's, the area was basically vacant land. Today, it is referred to as the "Queens Canyon." Although it wasn't built to revitalize the area, the rapid transit system did stimulate enormous growth along the corridor. Is this what might happen in Buffalo or Boston? Maybe?

Does the development of a rapid transit system improve land values along the corridor? There is some evidence to support this conclusion. Among other studies, one performed for UMTA entitled "The Effect of the Washington Metro on Urban Property Values" by Steven R. Lerman et al, [1] indicated that pre metro property values increased along the route of the metro system in Washington as the planned system became a reality. A preliminary study performed by the staff of the House Subcommittee on the City, Committee on Banking, Finance and Urban Affairs and submitted on January 21, 1981 concluded that property values have continued to increase along the operating Washington Metro routes because of the rapid transit systems. Of course the most dramatic change in property values occurred in New York City along Third Avenue after the unsightly 3rd Avenue "EL" was removed.

The conclusion is obvious to me. Yes property values do increase along the rapid transit routes provided the rapid transit system doesn't adversely affect the physical environment of the street surface.

To conclude this short planning segment:

1. Planning mass transit systems requires much hard thinking about the impact it will have on the growth of the urban area.

2. Its probably easier to determine when a rapid transit system isn't needed that when it is.

3. Some recent experiences with rapid transit systems indicate that property values increase along the rapid transit corridor.

The next major question is "How much does a rapid transit system cost?"

Cost

Regardless of what type of transportation system is chosen for an urban area, the costs are major. Any construction in the Central Business District (CBD) will be expensive, whether it be highways or a rapid transit system.

[1]The Effect of the Washington Metro on Urban Property Values; Lerman, Lerner-Lam and Young; Massachusetts Institute of Technology, Center for Transportation Studies

Interstate highway systems built in the inner cities can equal and exceed the cost of rapid transit systems. Surface systems require extensive rights-of-way, at least 130 ft. wide for a standard 6 lane highway. While maximum highway loading is about 2,400 passengers per hour per lane or 7,200 max. for 3 lanes, each track of a rapid transit system can handle at least 54,000 persons per hour. (30 trains, of 8 cars with 225 passengers/car).

The property requirements for a modern highway system are large and the effect of these rights-of-way on adjacent land values in the CBD cannot be encouraging.

Tunneling affects less property but the cost of modern highway tunnels are actually higher than rapid transit tunnels because of their size and the need to continuously ventilate the structures.

The most recent experiences in highway construction in urban areas indicate costs that range from $30 million to $322 million per mile. These figures are for the "Gap" portions of the Interstate Highway system and represent the per mile costs to connect the system in urbanized areas. In Little Rock, Arkansas with a population of about 140,000 persons, the 2.25 mile segment is budgeted at $31 million in 1980 dollars. The 4.7 mile Westway segment in New York City, the most expensive in the U.S. is budgeted at $322 million per mile in 1980 dollars. These costs are indicative of the difficulty of construction in urban areas.

A more reasonable figure for comparison might be the Century Freeway in Los Angeles. It is a proposed 17.3 mile aerial and surface structure running through the urbanized area, but not extensively through the CBD. The budgeted costs in updated 1980 dollars is $55 million per mile. This cost represents extensive property taking and the difficulty of construction. As the difficulty in construction increase, so does the cost. The Seattle Interstate "Gap," the second most expensive in the U.S. is 6.2 miles long and is budgeted in 1980 dollars at $145 million per mile. This section is basically built adjacent to and through water courses and contains tunnel and bridge sections which will be costly to construct. There are numerous other portions of the Interstate system between the $30 million and $145 million cost range and their costs reflect the difficulties involved in their construction. The conclusion is obvious - construction in the denser urbanized areas is expensive.

Rapid transit systems under construction indicate a range of costs depending on the difficulty of construction. Excluding extensions to existing systems, new systems range from $37 million/mile in Miami to $88 million/mile in Baltimore. The Miami system is the only one which is exclusively aerial while the other systems include, tunnel, at grade and in some cases elevated structures. The average cost per mile excluding the Miami system is about $80 million per mile. In comparison to the cost indicated for the Interstate Highways, these costs reflect the midpoint of construction rather than a base year. Much of the rapid transit construction is under contract with the actual midpoint for the collective construction sometime after the 1980 base. Therefore in terms of the comparable 1980 Highway base, the 80 million per mile rapid transit cost should be scaled downward.

The difficulty of construction sets the bottom line on underground rapid transit costs. However, many construction methods are available to perform this work. They range from the traditional cut and cover methods to tunneling methods.

For the cut and cover method, the use of permanent support systems such as slurry wall or secant pile structures eliminate the need for temporary support systems. This permits the construction of only one box structure instead of the traditional box within a box structure. The slurry wall method of construction is now used : in the U.S. while the secant pile construction in North America has only been used in Edmonton, Canada. To date, these construction techniques and their layout and design possibilities have hardly been explored in the U.S.

Similarly U.S. rapid transit tunneling techniques are just beginning to include some of the more innovative technologies developed in other countries. A tunnel mole using the earth pressure balance shield is now undergoing its first application in San Francisco by a Japanese firm and the slurry pressure shield has yet to be used.

Innovative layouts and designs will follow from the adoption of new techniques and I believe that we can look forward to reducing construction costs because of these techniques.

Another area where rapid transit costs can significantly be reduced is the area euphemistically referred to as "Goldplating." Goldplating refers to the practice of adding frills beyond what is essentially needed to functionally operate and maintain the rapid transit system. For example, the choice of station layout, architectural style and material will greatly influence the station costs. Station costs represent about 40% of a systems construction cost. These station layouts are the product of many factors such as community pressure and historic need. Whatever the reasons driving the station layouts, effective management of the project can be a significant tool to control project costs.

To sum up this section:

1. Inner city highway costs and rapid transit system costs reflect the difficulty of construction in urban areas.

2. Innovative construction methods, layouts and designs can reduce construction costs.

3. Effective management of planned rapid transit systems can reduce costs due to construction and community pressure.

Energy Usage in the Construction of Rapid Transit Systems

This topic is really a peripheral discussion on the cost of rapid transit systems. I will briefly cover two different studies on this subject.

In a study entitled "Comparative Energy Costs of Urban Transportation Systems" by Margaret Fels,[2] the author examines the quantity of energy required to manufacture the vehicles, construct the guideways and operate several transportation systems.

The study concludes that rapid transit systems consume the largest amount of energy as a transportation system.

The energy used in the production of basic materials such as concrete, steel, aluminum, etc. is used as a base to establish the energy expended in construction. The subject materials are tabulated as percentage components in the construction of the project and the energy usage is then computed.

[2]Comparative Energy Costs of Urban Transportation Systems, Margaret Fulton Fels, Transportation Research Vol 9, No. 5 October 1975, Pergamon Press, Great Britain

Highways are taken as a sample pavement from the New Jersey DOT, a 12 foot wide mile long segment of the Interstate system. The BART system is the base for the rapid transit data. As noted earlier in this paper, costs for rapid transit systems and the gap interstate highway links are comparable units because of the difficulty of construction in the inner cities. However, I doubt that the sample New Jersey highway segment was based on the gap construction costs, since it would have been mentioned. The study didn't take into account land values etc. nor did the study analytically allow for the difficulty of construction. For example, the different costs and lost energy associated with idle equipment and manpower coping with inner city traffic and congestion vs. that encountered in the suburban or rural setting. Since the energy expended on highway construction was compared to a rapid transit system, the basis should have been the same.

The highway lifetime is estimated at 50 years with 2000 peak hours per year. Both numbers are inaccurate. Based on a work schedule of 250 work days a year, this number gives an 8 hour daily peak, which is a large number of peak hours per day. The product of these two numbers 50X2000 gives 100,000 peak hours of life for the highway system. Both the highway and rapid transit systems are amortorized over the same base period. However, rapid transit system life times are 100 years and better.

Rapid transit loading per car is taken as 1/3 of seated compacity or 25 passengers/car. However, actual rapid transit loading is 180-260 passengers per car. The study uses peak hour loading on BART of 25,000 passengers using cars with a seated capacity of 72 passengers per car. This low per car capacity penalized the BART system in the study since the analysis required 347 cars in the peak hour to handle the loading.

In conclusion: The Fels study suffers from many shortcomings in data and analysis. The study's conclusions are questionable, based on the method of analysis and are not supported by the facts.

The second study "Indirect Energy Consumption for Transportation Projects" by John Smylie[3] follows an analysis which is similar to the Fels study. However, the supporting data and assumptions underpinning the authors energy usages for each type of construction is not fully included in the main report. Nevertheless, the author tells us that the highway data combines both rural and urban highways. As indicated earlier, if comparisons are to be valid between highways and rapid transit systems, they should both use the same base, the "Gap" highway connections.

The conclusion of the study is that the construction energy used in constructing a rapid transit system, i.e. the indirect energy, is approximately equal to 10 years of direct energy consumption during operation. So What! If operating energy was inefficiently used, the construction energy might be equal to two years of operating energy. Would this make the construction more efficient. Lets assume the opposite that new operating efficiencies increases the period to 20 years instead of 10. Does this mean that the energy used in construction is less efficient? Simply put, the conclusion tells us nothing about the energy efficiency or inefficiencies involved in constructing a rapid transit system.

In concluding this section, I believe that these studies do not offer us any guidance, but only confuse our thinking about the efficiencies of different systems. Additional, realistic studies still have to be made before we can

[3]Indirect Energy Consumption for Transportation Projects, John Smylie, DeLeuw Cather and Company, Los Angeles Office, October 21, 1975.

definitely determine whether highway or rapid transit systems are the most
efficient forms of transportation in our urban areas from the energy standpoint.

Management of a Proposed Rapid Transit System.

No single activity is more important to the rational development of the future
rapid transit system than the Management Process. The process should produce a
management plan, clear lines of responsibility and a project control mechanism.
The management plan is a road map which guides the project from inception to
completion. Only with a clear well developed management plan can a cost
effective system emerge. I personally believe that the management system should
be in place at least by the beginning of the Preliminary Engineering process and
even earlier if possible.

A strong management presence is required during the planning stages while the
system configuration is taking shape. It is here that the trade off analysis
must take place. The early planning questions raised in the first section of
this paper should be thought through and rationally answered. Similarly, the
later trade offs studies and evaluations of alternatives, alignments, layouts and
construction techniques can only occur if the management process is well defined.

The project staff should be sufficiently developed to guarantee at a minimum,
strong oversight capability. This does not mean that the staff has to perform
all the studies themselves, but it does mean that the staff have a high level of
expertise in order to carry out that oversight responsibility. Regardless of the
technical competency and quantity of consultants performing the work, the project
staff will still have to evaluate the product of their work and direct them to
additional tasks.

It is in these areas that the management of the project is most important. The
management structures should have clear lines of responsibility, with built in
project control mechanisms to indicate when the critical component tasks are
slipping.

One method used on several new rapid transit projects was to fully detail the
work to be performed during Preliminary Engineering (PE). This PE work
statement detailed the specific tasks to be performed in a two year, three phased
process. A Work Breakdown Structure (WBS) can then be developed from the work
statement which assigns the tasks to specific consultant and assigns responsibility to specific staff. Priortizing the WBS enables the project manager to
control the work flow and monitor responsibility.

This particular approach is currently being used on the Los Angeles Wilshire
Boulevard Corridor Line. At this early date, it is not possible to accurately
assess how successful the process will be, only time and results will tell. If
it is successful, the approach will probably be followed by any other urban area
which may develop a rapid transit system.

In concluding this short management section, it is my belief that successful
management of a developing mass transit system can reduce cost and produce a cost
effective system. Only through a strong management process can any planned mass
transportation system be comprehensively examined to determine the most suitable
system to meet the needs of the community.

Conclusion:

In this brief paper, I have attempted to trace some of the steps that a
developing rapid transit system might follow. Starting with a rational planning

process, there are numerous valid questions which should be answered concerning the impact of the system on the development of the urbanized area. It was noted that the impacts exist regardless of the transportation system selected and these impacts should be addressed.

A brief analysis of construction cost indicates that all transportation construction in the urbanized area is costly, regardless of the amount of people moved on the particular mode of transportation. Additionally, new technologies have the potential to reduce construction costs. An analysis of existing energy studies indicated that the studies of energy usage in the construction of transportation systems do not answer the questions and are generally misleading.

The paper concluded with a short discussion on the crucial need to develop the management system early in the project. Effective management of the proposed system has the potential to reduce cost and produce a cost effective system that meets the needs of the community.

INCORPORATING CIVIL DEFENSE SHELTER SPACE IN NEW UNDERGROUND CONSTRUCTION[1]

C. V. Chester

Solar and Special Studies Section
Energy Division
Oak Ridge National Laboratory
Oak Ridge, Tennessee 37830

ABSTRACT

At the present time, the population of the U.S. is approximately ten times more vulnerable to nuclear weapons than the Soviet population. This vulnerability can be reduced rapidly by urban evacuation in a crisis. However, the need to keep the essential economy running in a crisis, as well as coping with attacks on short warning, makes the construction of shelter space where people live very desirable. This can be done most economically by slightly modifying underground construction intended for peacetime use.

The designer must consider all elements of the emergency environment when designing the space. Provisions must be made for emergency egress, light and ventilation (without electric power), blast closures, water, sanitation, and food. The option of upgrading the space in a crisis should be considered. An example is given.

Improvement of the survival capability of the West should reduce the temptation for a confrontation, and with it the risk of a consequent miscalculation leading to catastrophe.

KEY WORDS

Underground construction; earth sheltered structures; blast shelter; fallout shelter; crisis upgrading; ventilation.

INTRODUCTION

Underground construction has obvious utility as protection against nuclear weapons effects in a civil defense emergency. This paper presents a brief review of factors that should be considered by the underground space designer when considering use of the space as emergency shelter against weapons effects. It is directed

---

[1]Research sponsored by the U.S. Department of Energy under contract W-7405-eng-26 with the Union Carbide Corporation.

to members of the underground space community who have had little experience with the nuclear weapons environment.

We will attempt to provide a rationale for building shelter, an indication of the location of different types of shelters, factors that need to be considered in the design of shelter space, and we will present an illustrative example of special features in an earth-sheltered residence that make it better adapted to the civil defense environment.

RATIONALE--WHY BOTHER?

At the present time, the population of the United States is considered by knowledgeable people to be of the order of ten times more vulnerable to a nuclear war than the population of the Soviet Union. Under many pausible scenarios in which the Soviet Union has time to occupy its extensive shelter system and implement its well planned urban evacuation, their population losses in a full exchange could be a few percent -- less than they lost in World War II. In the same scenarios, casualties in the United States could range from 50 to 80 percent (CIA, 1978; Sullivan, 1978).

This disparity in civil vulnerability cannot avoid providing a temptation for the leadership of the Soviet Union to take greater risks in a confrontation to extract diplomatic and geopolitical concessions from the United States leadership. This same disparity will bear heavily on the decisions made by the U.S. leadership in a severe crisis.

The vulnerability of the U.S. population can be reduced very rapidly by a combination of evacuating our own cities and the use of what is called expedient shelter (i.e., shelter against weapons effects constructed in one to two days with materials and tools at hand). Plans for an urban evacuation in the United States, called Crisis Relocation Planning, would deal with the most worrisome aspect of the Soviet Civil Defense plans -- their use of an urban evacuation in a diplomatic confrontation. However, an evacuation in the United States would take a day or two for most of the urban population and up to a week for the very large cities such as New York and Los Angeles. This planning, which is in progress, would have no capability of protecting the population against attacks on very short warning. Perhaps more seriously, it would require that the entire U.S. economy be shut down during the crisis while the population was evacuated -- at least until shelter could be constructed in the emergency for essential workers. A program to construct shelter for high-priority populations begun in the near future could, over several years of steady effort, gradually reduce the shortcomings of the evacuation plan. Incorporating provisions for emergency shelter in new underground construction can be done much less expensively than constructing underground structures whose sole purpose is that of shelter. Typically the additional cost of the structural modifications will run only a few percent of the cost of single-purpose shelters, which in the experience of Switzerland cost from $400 to $700 per space.

WHERE TO BUILD IT?

Fallout

Fallout shelter, which is obtained essentially for free in underground construction, is needed everywhere in the United States. Figure 1 shows an idealized fallout pattern that would be obtained from one hypothetical, 5000-megaton attack

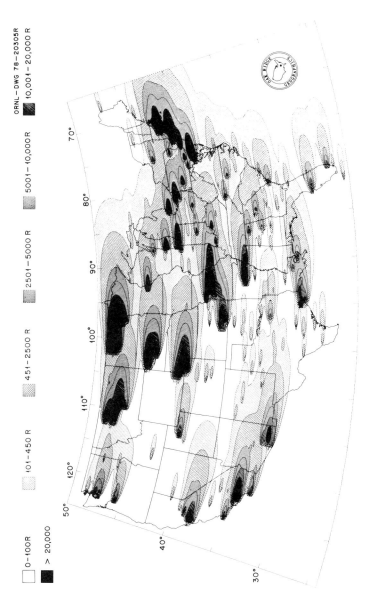

Fig. 1.   Fallout patterns from a hypothetical, 5000-MT attack, 20-mph west wind.

with a constant 20-mph west wind over the whole country (Haaland, 1976). Shelters are needed in the shaded areas which cover most of the area of the country. The location of these areas will change with the variation of wind direction and speed. There is no area of the country which can be considered safe from fallout.

Blast

Figure 2 is a map showing, in shading, areas of the country believed likely to experience at least 1/7 of an atmosphere (2 psi) of blast overpressure in a large attack on U.S. military and industrial targets. Blast must be considered a potential threat anywhere within 10 miles of:

    missile bases,
    7000-ft runways,
    submarine bases,
    conventional military bases,
    strategic and industrial assets (including 1000 MW(e) plants or
      40,000-barrel-per-day oil refineries), and
    population centers over 50,000.

Design Hardness

Moderate levels of hardness (ability to withstand 1 to 3 atmospheres of overpressure), which can be obtained quite inexpensively in underground construction, can dramatically reduce the risk area from megaton weapons. Figure 3 shows the 2-psi circles from one hypothetical attack in the New England area (Haaland, 1976). Compare this with Fig. 4 which shows the 1 atmosphere circles from the same attack. The population at risk has been reduced by at least a factor of two. From Fig. 5, it can be seen that increasing target hardness from a seventh of an atmosphere to one atmosphere reduces the effective area of the weapon by a factor of ten. Because of the overlapping of weapons effects in most targeting patterns, the net reduction of the area at risk is lower, but still very significant.

VITAL CONSIDERATIONS

Threat

How much of which weapon effects must be designed for will depend on the distance, direction, and nature of the nearest potential target. Fallout protection must be included in any case; if there are targets within ten miles, blast must be considered.

In all areas subjected to blast and in all areas within 20 miles of targets likely to be attacked by high-yield weapons, fire must be considered. While this will not affect underground structures, it may cause buildings above the underground space to burn and collapse. Particular attention must be given to ventilation air intakes and access passages in this situation. If blast is a possibility, the presence of buildings with large amounts of combustible material upwind (i.e., toward the target area) must be considered. The possibility of smoldering debris blown from a nearby structure against or over air intakes must be considered. This is usually handled by having the air intake protrude some distance above the ground. Provisions must be made for a quick temporary sortie from the shelter to deal with the problem if it arises.

ORNL-PHOTO 1866-79

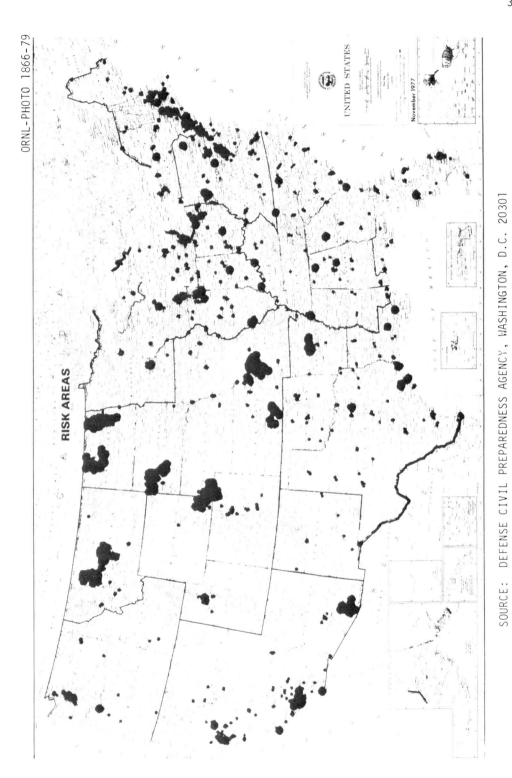

SOURCE: DEFENSE CIVIL PREPAREDNESS AGENCY, WASHINGTON, D.C. 20301

Fig. 2. Two-psi risk areas in the U.S.

36

ORNL-DWG 76-7134

NORTHEAST
2PSI

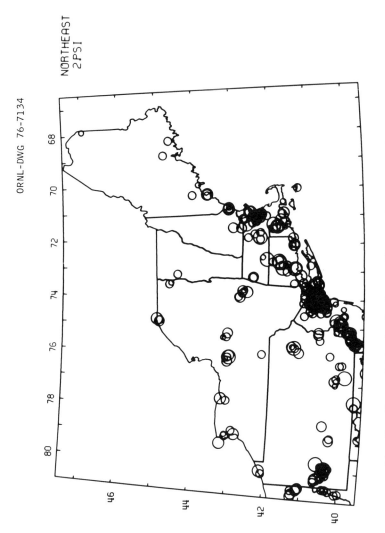

Fig. 3. Two-psi blast circles on the northeastern U.S.

37

ORNL-DWG 76-7132

NORTHEAST
15 PSI

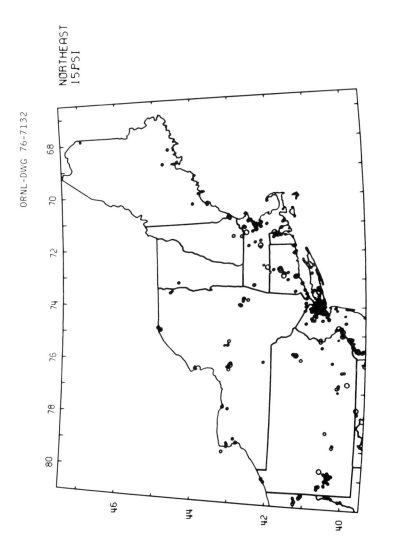

Fig. 4. Fifteen-psi blast areas in the northeastern U.S.

38

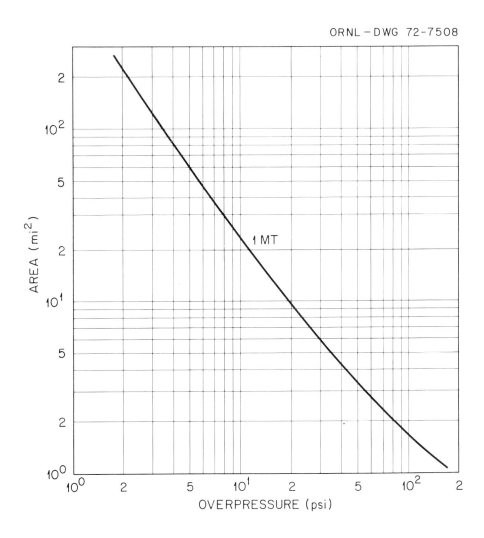

Fig. 5.   Area covered vs overpressure for a one-megaton weapon.

Loss of electric power is a very important factor in the design of underground shelter space. Most spaces for peacetime use depend on electricity for ventilation. It must be assumed that commercial electric power will not be available after a nuclear attack. The concurrent loss of illumination also must be dealt with.

Vital Requirements

Emergency Egress. As a matter of prudence and psychological acceptability, any underground space expected to be used for shelter of more than a few people should have at least two methods of egress. Only one of these need be a normal door. The other can be a hatch and ladder or can be blocked with a frangible partition or even sand-filled until needed. In many designs one ventilation port is constructed large enough to permit exit via crawling.

Ventilation With Power Out. Underground spaces normally ventilated by mechanical systems must be provided with some means of emergency ventilation if they are to be used as shelter against weapon effects. If crowded to the recommended civil defense limits (i.e., approximately one square meter of floor per person) heat prostuation can become a life-threatening factor in many parts of this country in the summertime. To meet the most severe conditions, ventilation air of up to a cubic meter per minute per occupant can be required. This can be best done if an air inlet and an air exhaust are available on opposite sides of the space. One of these can be the normal door to the space and the other can be the emergency exit or a specially designed ventilation port.

Some means for moving the air must be provided. This can be a commercially supplied, manual blower. If outside temperatures are not too high, proper arrangement of the inlet and exhaust port will provide enough ventilation for low population densities by natural thermal convection. Air can be driven through a sufficiently low pressure drop ventilation channel by proper manipulation of a stiff piece of cardboard. A device which can be fabricated in a few hours (Fig. 6) is the Kearny Air Pump (Kearny, 1979). One of these built into a door can move 20 cubic meters of air per minute (6000 cfm) when operated by a man working only moderately hard.

Air filtration is required only in areas subject to blast and then only dust filters are needed. The requirements for these are minimized if there is enough shelter volume for the occupants so that the ventilation system can be shut down for a few hours until the air clears.

Light. Habitability, safety, and efficiency of shelter space is enormously increased if there is some provision for a minimum amount of light when the power goes off. This can be provided by emergency generators in elaborate shelters, or by emergency lighting powered by large storage batteries, flashlights, candles, or improvised oil lamps. While people can survive without light, they have great difficulty in coping with unexpected conditions in total darkness.

Water. Without water in a shelter, occupants will be forced out to seek water in four days or less. At least one quart and preferably one gallon per occupant per day should be accessible to the occupants of the shelter. This can be stored in a specially designed tank or a tank with some other peacetime use, e.g., a hot water heater. In an emergency, water can be stored in garbage cans or pillow cases which have been lined with a plastic garbage bag of the appropriate (slightly larger) size. For residences, the most economical strategy would be to locate the hot water heater in the planned shelter space with provision for valving it off and venting and tapping the water in it.

ORNL-DWG 66-12320A

UNUSED PARTS OF DOORWAY COVERED

PULL CORD
(SLACK)

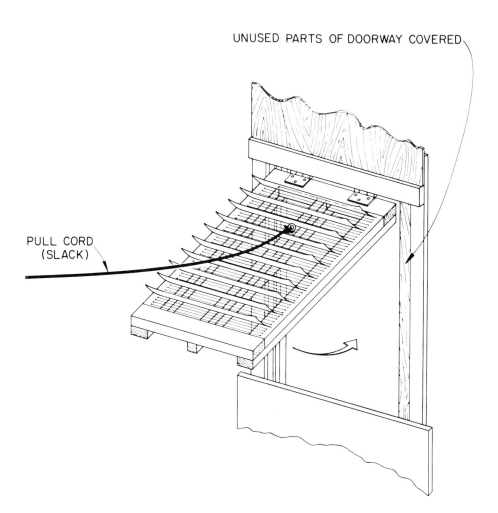

Fig. 6. Kearny Air Pump installed in a doorway.

Blast Closures. If a shelter is in a blast area, it would be much more efficient in protecting the lives and safety of the occupants if the access areas have been designed to exclude high-pressure blast waves. This is particularly true as the overpressures exceed one atmosphere and the ratio of door area to shelter volume is at the larger end of the spectrum. While a human being can survive two atmospheres of blast over-pressure, many people would be killed or injured at one atmosphere or below by being thrown against something by the blast wind or by being hit by debris. Eardrums are damaged or destroyed at approximately 1/3 of an atmosphere overpressure. The design of valves and doors is relatively straight-forward and commercial models are available from foreign countries, notably Switzerland and Finland. The thing that is sometimes overlooked is that the loads developed by doors and valves in operation must be supported by the structure.

Sanitation. Some provision for removal or storage of human waste must be made. Failure to deal with this problem in larger shelter populations can run the risk of a variety of severe health problems. While principally an aesthetic and psychological problem with small groups, failure to deal with it is a mark of very poor planning.

Food. Food is not normally considered a life-threatening factor for the period of shelter confinement. Most Americans can survive two to four weeks without food. In fact, since the majority are overweight, a week or two on half rations would be beneficial. Most Americans keep about a two-week supply of food of miscellaneous types in their house.

However, a nuclear war would disrupt the food distribution system in this country on a scale that is completely without precedent in our history. While there is ample supply of food in the country to feed the population (Haaland, 1976), reestablishment of the transportation system and an emergency distribution mecha-nism might take several weeks in some areas. There is little point in bringing people through an attack only to have them starve to death a few weeks afterwards. Prudence would suggest the storage of several weeks of supply of some very simple food such as wheat, if there are no food stocks in the area.

Peacetime Use. In order to be economical, the shelter space should have an economic peacetime function. This will require some type of convenient access, usually a horizontal entryway. If the peacetime use entails the presence of hazardous materials in the space, a provision must be made for their removal at the early stage of a crisis, or the valving off of any lines carrying them. If this can't be done, then some other space should be considered for dual use as a shelter.

Information. Shelter occupants who have some information about the hazards of nuclear war and the methods of dealing with them are going to have a better chance of survival than those who don't. A book or two with this information in it would be a small-cost addition to the shelter with potentially a large payoff. The best information we know about is Nuclear War Survival Skills by Cresson Kearny (1979), available from the American Security Council Education Foundation, Boston, Virginia 22713. While this book is intended for people in a crisis with no shelter, it has a great deal of useful information for people with shelter.

Crisis Upgrading

There are a variety of actions which can be taken during a crisis to improve the blast resistance and/or fallout protection of almost any structure. Underground structures are particularly adaptable to this strategy which can permit deferral of cost until its need is readily apparent. The structure or its access can be modified to improve greatly its shelter suitability with measures which would make the shelter much less suitable for its peacetime function.

Shoring Spans. The unit bearing capacity of a one-way span is inversely proportional to the square of the distance between supports. Cutting the unsupported span by a half or a third with the addition of columns and lintels can increase its load-bearing capacity by four to nine if the supports and the deck are designed for the loads. In wooded areas, the materials for doing this can be improvised. Specially prepared columns and lintels can be fabricated and stored. Materials which are adaptable to this function can serve some other function in peacetime use -- for example, cribbing for retaining embankments.

Adding Earth Cover. A reinforced horizontal surface may have additional earth cover added to improve its shielding factor. Earth may be piled up against exposed walls.

Closing Openings. Windows, doors, and corridors can be designed for retrofit with fabricated blast closures (i.e., blast doors and blast valves). Improvised closures to increase the shielding factor can be constructed from sand bags, earth filled boxes or other containers, concrete blocks, or shelter supplies.

AN EXAMPLE

The application of the principles above is demonstrated on a 2000-sq-ft, passively solar heated, earth-sheltered house of the general design which is being built by many contractors today. The floor plan is shown in Fig. 7. It is designed (by H. B. Shapira of ORNL) to be acceptable to most building codes. The layout is intended to minimize the perception of being underground and this eliminate one of the most common concerns about earth-sheltered construction.

Covered with two feet of earth on the roof and constructed of post-tensioned concrete, the structure as it stands is quite tornado- and fire-resistant and offers very good fallout protection. It is designed so that one or two rooms can be fitted with blast closures. Alternatively, the entire structure can be hardened against blast in a crisis if enough labor is available.

The building is designed with the minimum exposed surface consistent with passive solar heating. Only the south wall is exposed. North, east, and west walls are completely buried.

The corridor walls down the center are load-bearing walls to keep the roof span short. Partition walls are load bearing and tied to the roof, floor, and connecting walls to provide shear strength as well.

The overhang shielding the south windows against the summer sun is designed for light construction or to be a trellis. This way it provides no heat path into the building such as a solid overhang might, and it is designed to transmit no stresses to the building if hit by a blast wave from the south. It would simply blow away.

The steel in the structure can be arranged for expedient shoring. This involves putting the prestressing tendons in the midplane of the roof slab and putting negative steel at the appropriate points in the roof slab and the floor slab.

The front wall can be designed for blast by appropriate design of the steel in the columns between the windows. The windows are designed for blast closures or additional fallout protection -- this can be sand bags or specially designed concrete blocks.

ORNL-DWG 81-7475

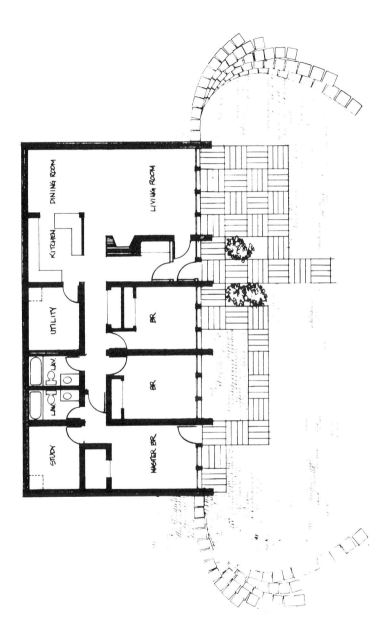

Fig. 7. Earth-sheltered structure adapted to crisis upgrading plan.

Skylights are indicated for the corners of at least two rear rooms. These are located to minimize the effects on the strength of the structure. The diameter selected, 75 cm (30 in.), was predicated on emergency exiting. Knockout panels can be incorporated in the partition walls toward the rear to allow for emergency exit and through ventilation of the rooms in the rear. The location of the skylights and the doors was selected to permit through ventilation. One rear room is designed to accept blast closures -- a blast door and a blast valve to go in the skylight.

The walkway or patio is made of concrete blocks which are sized to stack in front of the windows. By incorporating the appropriate size rebar in these blocks they can be designed to resist blast overpressure. Alternatively, a patio in front of the building can be designed with its pavement in strips which can be tilted up against the building prior to piling earth against it. The retaining walls are intended to be constructed of stacked cribbing and/or planters which can be rearranged in a crisis to support earth piled against the front of the building (Figs. 8 and 9). Ideally, grading of the berms on either side of the building would be done in a way to provide a convenient supply of soil for banking against the front of the building in a crisis.

The water heater and food storage should be located in the rear room that is intended to be used as a blast shelter.

The air intake would be low on the front door and would exhaust through the skylights. With this arrangement at low population densities or cool temperatures, natural circulation should provide enough ventilation. The low air flow rates should permit removal of any airborne particles by settling in the front of the house.

POSTSCRIPT

The foregoing illustrates the considerations that must be made in adapting underground space for nuclear shelter. This is not an exhaustive text.

It is our personal opinion that a nuclear war between the United States and the Soviet Union is extremely unlikely. Even with their shortcomings, our strategic nuclear forces could do terrible damage to the Soviet Union. While it likely the Soviets would emerge from the war vastly stronger than the United States, it is unlikely that they would willingly accept this level of damage. This is particularly so because they believe in, and their strategy is pointed to, obtaining the fruits of victory without having to fight the war.

The universal perception of the terrible state of unpreparedness of the United States and the Soviets' belief in their own ability to survive a nuclear war cannot help but affect their calculation of what is an acceptable risk in international adventures. The perception of third parties of this disparity or vulnerability must affect their perception of overall military strength. It will affect their calculation of who should have the biggest influence on their future policy. These perceptions dovetail very comfortably with the Soviet strategy of slow-but-steady, spreading influence and hegemony by a variety of means.

If we can correct our strategic vulnerability and institute a program of steady improvement in our survival capability, it will help enormously to correct the growing perception of Western weakness. This should greatly reduce the temptation to the Soviet Union to provoke a confrontation. The risk of a consequent miscalculation and catastrophe is proportionally reduced.

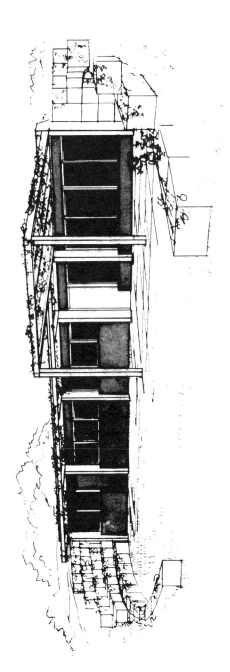

Fig. 8.   Earth-sheltered structure perspective.

46

ORNL-DWG 81-7476

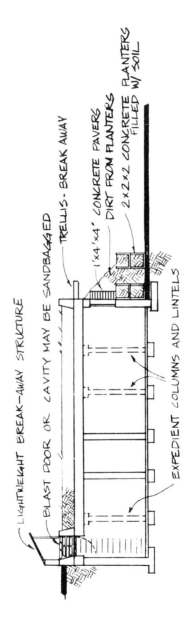

Fig. 9.   Crisis-hardened earth-sheltered structure section.

The paradoxical nature of civil defense preparedness is that if you have it you are unlikely to need it; if you don't have it, then you are more likely to need it. Incorporating shelter space in new underground construction built for other purposes is one way of improving civil defense preparedness and survival capability while making the most efficient use of resources.

REFERENCES

Director of Central Intelligence (1978). Soviet Civil Defense, NI78-10003.
Haaland, C. M., C. V. Chester and E. P. Wigner (1976). Survival of the Relocated Population of the U.S. After a Nuclear Attack, ORNL-5041. Oak Ridge National Laboratory, Oak Ridge, Tennessee.
Kearny, C. H. (1979). Nuclear War Survival Skills, ORNL-5037. Oak Ridge National Laboratory, Oak Ridge, Tennessee.
Sullivan, R. J. (1978). Candidate U.S. Civil Defense Programs, SPC-R-342. System Planning Corporation.

PEOPLE PROTECTION - SWITZERLAND'S APPROACH

H.H. Oppliger
Luwa Ltd., 8047 Zurich
Switzerland

ABSTRACT

This paper briefly describes the Swiss civil defence system, its historical development, legislation, concept, civil defence constructions and their dual purpose use
and finally the present state of Swiss civil defence.

Possible war scenarios are indicated, weapon effects and means and extent of protection described. Basic requirements on shelter construction are listed and standard shelter construction parts as approved in Switzerland are briefly described.

KEYWORDS

Civil defence, war scenarios, weapon effects, shelters, degree of protection,
shelter construction parts.

CONCEPT AND STRUCTURE OF SWISS CIVIL DEFENCE

Development of Civil Defence / Legislation

During World War II, Switzerland disposed of a civilian air-raid protection organization which based on a system of makeshift shelters, put up in the cellars of existing houses. After the end of the war, this organization was abandoned, and the
reinforcement of the cellars partly removed.

Amazingly, only five years later, i.e. 1950, and probably under the impression of
the Korean War, civil defence measures were taken anew. Founded on a very general
constitutional stipulation, a federal law on protection of the population was enacted. There was no word of a protective organization, but essentially the requirement, to incorporate compulsorily air-raid shelters in practically every new building, and in all communities counting more than 1'000 inhabitants.

The financial regulation based on the fact, that in Switzerland practically all
houses dispose of a cellar, which in most cases is underground. For the construction
of a shelter it was legally stipulated, that each constructor of a new building must

reserve at least one part of the cellar for the planned shelter (Fig. 1). The extra
cost caused by putting up such a shelter were borne up to 30 % by the confederation,
the canton and the community. The remainder had to be paid by the private house-
owner.

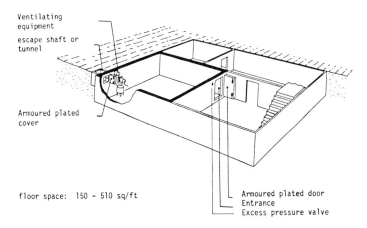

Ventilating
equipment

escape shaft or
tunnel

Armoured plated
cover

floor space: 150 - 510 sq/ft

Armoured plated door
Entrance
Excess pressure valve

Fig. 1 Private shelter

A consequence of this early resumption of shelter construction was the accustoming
of the citizen to this not always agreeable obligation. Nowadays, hardly anybody
would doubt shelter construction in Switzerland. It is taken for granted same as
pollution control or any other infrastructural task in a community. No community
liable to civil defence duty may permit the construction of a new building, if the
shelter project has not been approved before!

In 1959 civil defence was explicitly anchored in the Swiss constitution. On this
basis, two new and much more comprehensive laws were prepared and put into force
in 1962 and 1963. They are still valid and, in spite of many changes having taken
place since, form the foundation for the structural build-up of civil defence in
Switzerland.

The first law, the federal law on civil defence, stipulates compulsory civil defence service for men between the age of 20 and 60 who are not or no more drafted
for military service. This law also stipulates, that the community is the main
bearer of civil defence. The community constitutes an independent local protective
organization, equips it with material and constructs the protective buildings required for this organization.

The second law, the federal law on constructional measures in civil defence, essentially is a new edition of the law of 1950. However, it contains remarkable improve
ments and supplements. The compulsory constructions of a home shelter in all new
buildings in communities having 1'000 or more inhabitants, was retained. 1978, when
the law was revised, the obligation to build shelters was extended to all communities. Moreover, by extending this construction programme, the communities were
obliged to build public shelters in all cases, where for some reason, private
shelters could not be built.

Both laws brought Switzerland an ambitious and financially largescale programme,
whereby the chief importance is clearly with the constructional measures.

The 1971 Concept of Civil Defence / Structure of Civil Defence in the Community

An increasingly better and more detailed understanding of the different weapon
effects convinced the authorities concerned, that adequate consideration of modern
mass destruction weapons, especially nuclear ones, is indispensable in civil defence.
From the study of war scenarios and weapon effect patterns relevant to civil defence,
analysis of the optimal possibilities to minimize losses, and in this connection
the establishment of the efficiency of constructional and organizational measures,
resulted finally in a clear picture on structure and tasks of civil defence. This
picture reflects in the 1971 concept of civil defence which can be summarized as
follows:

Target
- Civil defence as part of the national defence must, in a credible manner, make an
  essential contribution to prevention of war.
- Civil defence must provide conditions for the survival of the greatest possible
  part of the population even in the event of serious armed conflicts.

Principles
- Independance of war scenario
  . Provide each inhabitant of Switzerland with a place in a shelter in the vicini-
    ty of his residence.
  . Shelter constructions to be simple and robust.
  . Large scale evacuation is not envisaged.
  . Gradual and preventive occupation of shelter when political and military ten-
    sion reaches a critical level.
  . Ensure a stay in shelters for days or weeks independant from the outside world.

- Economy
  . Deliberate renunciation of absolute protection.
  . All measures, organisational and constructional, to be well balanced.
  . Optimum use of all protective possibilities, dual use of shelter constructions.
  . Prevention is more efficient and less costly than cure.

- Consideration of psychological and physiological characteristics of man
  . Provide equal chances of survival.
  . Maintain naturally grown communities - in particular families - also in shelters.
  . Provide management and guidance for shelter inhabitants.
  . Rely on adaptability of man.

The concept described above, its target, its parameters and principles, has been
approved by the federal governement and agreed to by the parliament. Consequently,
it is the basis for the structural build-up of a modern civil defence in Switzerland.

The bearer of civil defence is the community which, with the assistance of the canton
and the confederation, enforces all measures. At the head of the civil defence organization of each community (or municipality) is the local management with its
staff. It is in direct contact with the population in the shelters, in order to
guide and inform this population continuously and to bring into action the civil
defence formations or units at its disposal.

Civil Defence Constructions

Precondition for successful civil defence measures are protective constructions. Organizational measures are to bring about optimal operation and use of this valuable protective construction system.

Backbone of this system are the private home shelters being built since many years. Next to these there are public shelters built by the communities, usualy in conjunction with public or private underground car parks (Fig. 2) and, however rarely though, in road tunnels (Fig. 3). Public shelters are to cover shelter space deficits in areas with buildings put up prior to the legislation described above (for instance old towns, which at the same time present an extreme fire hazard).

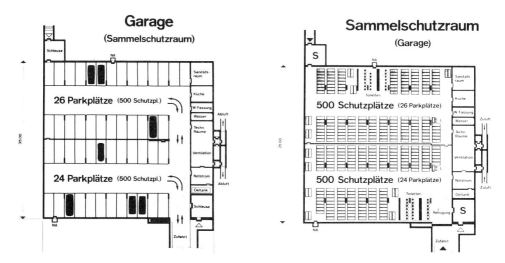

Fig. 2 Public shelter in underground car park

Protective constructions for the organizations and the civil defence medical service are mainly command posts (in the USA called "emergency operating centers") for the local management and its staff, first aid posts and medical stations for the civil defence medical service.

The technical requirements which must be met by civil defence constructions are standardized. Shelters must provide a certain protection against the effects of nuclear weapons, near hits of conventional weapons and the effects of C-weapons. The extent of such protection, called degree of protection, is measured by the blast effect of a nuclear explosion. But it also includes all other weapon effects within this pressure range.

The minimum degree of protection for private and public shelters is 14 psi (incident pressure wave which equals approx. 40 psi reflected pressure). Degree of protection for structures of the civil organization, such as command posts etc. is 42 psi (incident pressure wave which equals approx. 160 psi reflected pressure).

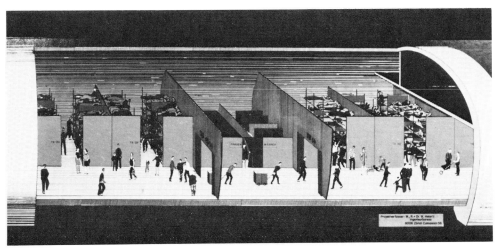

Fig. 3   Model of partitioning and furnishing of public shelter in the Sonnenberg
road tunnel, Lucerne (photo: Heierli, consulting engineers, Zurich)

Planning and Supervision of Civil Defence Measures

The realization of this nationwide constructional programme is carried out uniformly in all communities, basing on the General Civil Defence Planning (GCDP).
This planning is implemented by the local management according to standard in-
structions and verified every 2 to 3 years.

GCDP includes the following planning-work:
- Establishment and verification of the tactical structure of the entire communal
  or municipal area.
- Elimination of the danger zones (especially fire hazard, flooding and risk of
  debris accumulation) where no shelters are allowed and also no rescue operations
  are scheduled (danger plan).
- Establishment of the actual inventory of the number and distribution of population
  and shelters available. Allocation of the population to the shelters.
- Inventory of the existing protective facilities of the organization and of the
  medical service, and of the disposable strength of people assigned to CD service.

Necessary Finance and Present State of Swiss Civil Defence

The cost estimate for the 20 year period 1970 to 1990, when the 1971 concept is ex-
pected to be fully operational is as follows:
- Constructions $ 3.4 billions (SwFrs. 5.57 billions)
- Equipment, instruction
  maintenance, administration $ 0.7 billions (SwFrs. 1.20 billions)
- Research and development $ 0.02 billions (SwFrs. 0.03 billions)
            Approximate total $ 4.3 billions (SwFrs. 6.80 billions)

This corresponds to an approximate yearly expenditure of approx. US $ 33 per capi-
tal over a 20 year period.

Present state of Swiss civil defence for a total population of 6.35 millions.

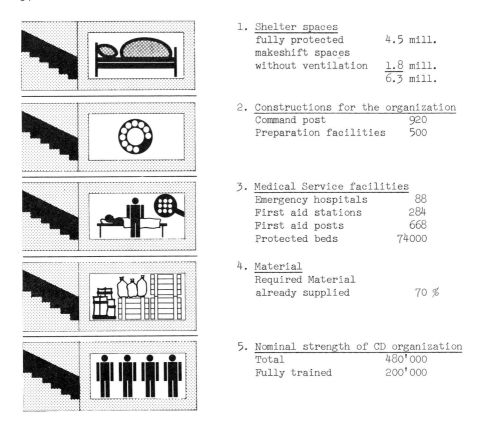

1. Shelter spaces
   fully protected   4.5 mill.
   makeshift spaces
   without ventilation   1.8 mill.
                          ‾‾‾‾‾‾‾‾‾
                          6.3 mill.

2. Constructions for the organization
   Command post   920
   Preparation facilities   500

3. Medical Service facilities
   Emergency hospitals   88
   First aid stations   284
   First aid posts   668
   Protected beds   74000

4. Material
   Required Material
   already supplied   70 %

5. Nominal strength of CD organization
   Total   480'000
   Fully trained   200'000

ASSUMPTIONS ON THREAT, MEANS OF PROTECTION

Possible War Scenarios

Unfortunately, the possibility of war in Central Europe cannot be excluded without else. This does not only ensue from ideological contrasts between East and West, but just as much from the weapon systems kept ready. Never before in history such large annihilation potentials, conventional and nuclear, faced each other in Europe. It is very difficult to foresee the course of a possible war. Therefore, it is necessary to take as a basis for civil defence, different conceivable impacts of war. Existing military means and operational doctrines lead to the following thinkable war scenarios:
- Blackmail, for instance by the threat to use arms of massive destruction in order to achieve unhindered passage or simply capitulation.
- Limited use of means of mass destruction, either during attempted blackmail or in an attack, assisted by tactical nuclear weapons and chemical warfare.
- Use of massive conventional means of attack terrestrially and by air.
- Strategic blow of annihilation. A nuclear blow in order to eliminate the military and industrial potential does not seem likely, however, it cannot be completely excluded.
- War in neighboring countries and in such a case, joint affection through large-scale effects by mass destruction weapons.

Should one of the described war scenarios become real, then civil defence, especially adequate constructional measures, can considerably diminish human losses.

It is obvious that war scenarios as outlined above for Central Europe need not apply
without else to countries or continents more distant from a possible scene of war,
such as for instance the USA.

However, even the US are not out of reach of nuclear warheads delivered by long
range missiles. Nuclear attacks on missile shelters and strategically important installations cannot be excluded. It may, therefore, be wise to consider the implementation of protective structures and the necessary equipment into all new buildings. Particularly the newly, for reasons of fuel economy propagated earth sheltered homes and other underground constructions lend themselves to such measures.

Weapon Effects and Principal Means of Protection

It can be assumed that the civilian population is primarily threatened by nuclear
weapons of medium to large size but also by chemical and to a lesser extent by conventional and possibly biological weapons. Owing to the prevailing nuclear threat
it may be appropriate to briefly recall process and effects of a unclear explosion:

Development of the explosion. At the moment of the explosion of a nuclear weapon, a
strong radiation of heat and light is emitted for seconds. At the same time primary
nuclear radiation begins. The blast wave (pressure wave) strikes objects a few seconds after the explosion. With its arrival a wind having the multiple force of a
hurricane sets in. It lasts as long as the overpressure from the blast wave, that
is generally a few tenths of a second up to several seconds in case of large bombs.
Buildings collapse and large quantities of debris are hurled through the air. Flammable material will be ignited by the heat rays. Explosions on or near ground cause
radioactive fallout which may become effective after some quarters of an hour and
which may last for hours or even days.

Mechanical effects. The mechanical effect of a nuclear weapon consists mainly of
the air blast. It's intensity depends on the distance from the centre of the explosion, on the yield of the weapon and on the height of the explosion. The air
blast causes a very rapid increase of the air pressure up to a maximum, with an initially steep, later slow decrease, which lasts for some tenths of a second. This
phase of overpressure is followed by a prolonged phase of a relatively slight under-
pressure. The entire pressure development is shown in Fig. 4.

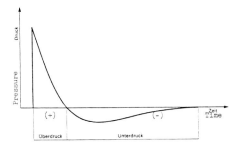

Fig. 4   Pressure time characteristic
         over plane ground

The approximate distances where, with different yields and an explosion near the
ground, a maximum incident pressure of 14 psi or 42 psi overpressure occurs, are
shown in Fig. 5. For comparison, the 1945 explosions over Hiroshima and Nagasaki
were 12 and 22 KT respectively.

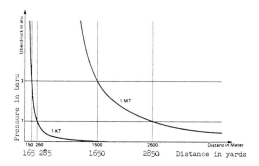

Fig. 5 Max. overpressure as a function of distance from
explosion center. Explosion on ground level

The blast is reflected on buildings above ground and on exposed parts of the shelter
structure, which leads to multiple increase of the peak pressure. In addition, the
blast produces a pressure wave in the ground similar to that of an earthquake, the
air-in duced ground shock. This air-induced ground shock causes a pressure load on
the side walls and a shock to the whole shelter. The intensity of this shock largely depends on the structure of the ground, but also on the installation itself. In
small size shelters and soft ground the shocks are felt very much stronger than in
large shelters in hard grounds.

Optimum protection against such mechanical effects may be achieved by
- a shelter arranged underground
- a simple and solid construction of re-inforced concrete
- a limitation of equipment and furnishing to what is required for survival.

Primary nuclear radiation. The primary or initial nuclear radiation propagates - invisible for a human being - from the rapidly increasing fireball of the explosion.
This radiation is the sum of all directly or indirectly ionizing rays, which emanate
within a minute from the fireball and the atomic cloud (mushroom). Only two components of this radiation are important for endangering the shelters and/or their
occupants: the gamma and the neutron rays.

The protection against the effects of the initial nuclear radiation (down to the
practically harmless dose of maximum 100 rem) is achieved by the underground arrange
ment of the shelter, covered by soil or the building above and the concrete shelter
shell itself.

Secondary nuclear radiation (fallout). Secondary nuclear radiation can practically
only develop in explosions near the ground, i.e. by falling dust particles with a
high content of radioactivity. Through the wind-transport of the dust clouds large
surfaces on the ground may be contaminated (the fallout is visible). With increasing
distance from zero ground and time after explosion, the radiation intensity decrease
Radioactive contamination can spread many hundreds of miles.

A solid protective screening of concrete, soil etc. can reduce secondary radiation to fractions of its initial value. 1 bar (14 psi) shelters or even normal underground closed rooms (cellars without windows), enable in the case of fallout a survival. However, in such a case we must allow for a stay of days or even weeks in such a protected room until the radiation has decreased so much as to allow a stay in the open again.

Thermal radiation. The thermal radiation emitted from a unclear explosion is of great danger to unprotected people. For shelters arranged partly or totally underground it is, however, of little to no consequence. However, secondary effects may develop by buildings set a fire above a shelter and ceiling or exposed walls may heat up to an extent that such danger may become a determining factor for the design of a shelter.

Shelters

Requirements on planning and construction. In order to obtain optimum protection against mechanical effects of weapons as well as nuclear radiation and heat it is best to place a shelter entirely underground (Fig. 6).

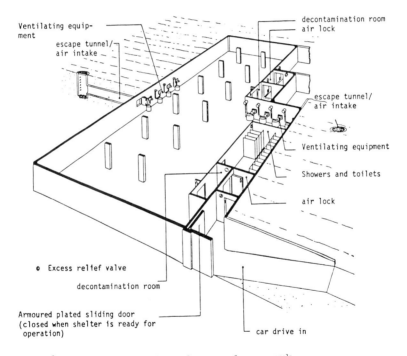

Fig. 6 Public shelter in underground car park

It should be of simple, cubicle shape and consist entirely of re-inforced concrete. Openings should be limited to a minimum. Entrances should be provided with armoured doors and air lock. Emergency exits with escape shaft or escape tunnel should be provided and be furnished with armored covers. Air intakes and exhausts should be protected with explosion protection valves. Particularly air intakes and escape openings should be arranged so as to not be covered by debris of collapsed buildings.

Basic design considerations:
- Resistance against heat (surface fires)
- Resistance against nuclear radiation
    → Largely determines thickness of concrete walls
- Mechanical stress from weapon effects
- Mechanical stress from peacetime use
    → Largely determines extent of steel re-inforcement

It has further to be considered that nuclear weapons produce a dynamic stress on the shelter which is characterized by a very short period of load increase and a relatively long load duration.

Minimum requirements for space, ventilation and climate

| Space: | Floor area approx. | 11 ft^2/person |
| | Room volume approx. | 90 ft^3/person |

Climate:	Upper limit	85 °F at	100 % RH
	or	86 °F at	90 % RH
	or	88 °F at	80 % RH

Ventilation: Air volumes required to maintain above climatic conditions without mechanical refrigeration in moderate outside climate:

Fresh air operation	3.5	cfm/person
Gas filter operation	1.75	cfm/person
in shelters with up to	200	persons

Fresh air operation	5.3	cfm/person
Gas filter operation	2.65	cfm/person
in shelters with more than	200	persons

Figures apply to life support shelters only.

Degree of protection. A degree of protection of for instance 1 bar (14 psi) applied to a shelter means that this shelter can withstand a 1 bar (14 psi) incident air blast and all other effects of a nuclear explosion which may be expected in a distance from the detonation center at which such pressure prevails.

Bomb size and ground distances for which an overpressure of 1 bar (14 psi) and 3 bar (42 psi) occur at low level explosion are listed overleaf.

| Yield | Approx. distance for pressures of | |
	1 bar (14 psi)	3 bar (42 psi)
1 KT	0.26 km (0.16 miles)	0.15 km (0.09 miles)
100 KT	1.20 km (0.75 miles)	0.70 km (0.43 miles)
1 KT	2.60 km (1.60 miles)	1.50 km (0.93 miles)
100 KT	12.00 km (7.50 miles)	7.00 km (4.30 miles)

The objective of shelter construction now is to reduce the distance of survival from the center of explosion as much as possible relative to the distance of survival of unprotected man. The practical limit of such reduction of distance, however, will always be governed by a comparison of cost and use respectively by the relation of cost and life saving.

Studies on the basis of an <u>equally dense population</u> in a target area have shown that the life saving efficiency of structural measures rapidly decreases at degrees of protection smaller than 1 bar (14 psi) but only marginally increases at degrees of protection larger than 3 bar (42 psi) as shown in Fig. 7. If cost of construction is taken into consideration as well, a curve showing the ratio of the above structural efficiency versus cost of a shelter place would clearly show a maximum in the region of 1 to 3 bar (14 to 42 psi).[1]

In consideration of the above, the protection degrees for shelters and shelter construction parts for civil defence in Switzerland have been laid down as follows:
1 bar (14 psi) for private and public shelters, first aid posts
3 bar (42 psi) for command posts, first aid stations and emergency hospitals.

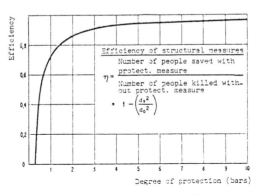

Fig. 7 Efficiency of structural measures

[1] It must be made clear that above considerations on degree of protection are valid for an equal distribution of population and in this sense also for military troops or installations. However, they may not apply without else to military point targets of which the coordinates are known to a possible agressor.

STANDARD SHELTER CONSTRUCTION PARTS

Closing Devices

<u>Armoured doors and covers</u>. Proper shelter doors and covers are of great importance since they have
- to provide the same degree of protection as the shelter structure itself against air blasts and their reflection, radioactive radiation, bomb fragments, dust, gas and fire
- to remain operable even under extreme conditions after an attack

<u>Standard door sizes</u> (interior dimensions)

Armoured door size	1	33½ x 73	in
	2	39¼ x 73	in
	3	55 x 86½	in
	4	23½ x 47	in
Armoured cover		23½ x 31½	in

Armoured doors and covers consist of door frame with anchoring rods, integrated locking device and special hinge anchors. Door leaf made of pressed steel frame with joint groove and rubber gasket, concret re-inforcement bars, locking levers and hinges (Fig. 8).

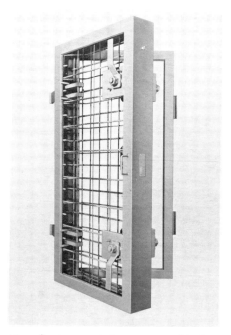

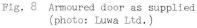

Fig. 8 Armoured door as supplied (photo: Luwa Ltd.)

Fig. 9 Preparation for casting of armoured doors (photo: Luwa)

<u>Installation</u>. The door, whilst being in closed position is fitted to the shuttering of the shelter wall (Fig. 9) and cast in in one go with the casting of the shelter wall proper. After the concrete has sufficiently set, the door leaf itself, still in closed position and properly wedged, can be filled with concrete.

The armoured doors (Fig. 10) and covers discribed above are designed to withstand air blasts of 160 psi (reflected pressure).

Fig. 10   Armoured doors installed in shelter
          (photo: Luwa Ltd.)

Explosion protection valves. Explosion protection valves are to prevent blast waves
from entering a shelter through air intake and exhaust openings. They are preferably
to be of the selfactivated type and provide very short closing times, thus reducing
blow by pressure and impulse to a minimum.

The explosion protection valves described
below provide closing times of 1 to 2 milli-
seconds at a reflected pressure of 160 psi.

Figure 11 shows the LUWA explosion protection valve type F. The valve consists of a
valve body from cast light alloy with leaf
springs from specially treated steel as the
only moving parts. Individual elements can
be combined in a wall frame for air volumes
up to 7000 cfm per valve assembly at 1 ins
WG pressure drop. The operating principle
is shown in Fig. 12.

Figure 13 shows the LUWA combined explosion
protection/pressure relief valve type K.
The valve consists of a valve body from cast
light alloy with a light weight, high
strength aluminium bar suspended from guiding levers. Individual elements can be combined in a wall frame for air volumes up to
5000 cfm per valve assembly at 1 ins WG
pressure drop. The operating principle and
selection of operating mode is shown in
Fig. 14.

Fig. 11   Luwa spring type explosion valve type F

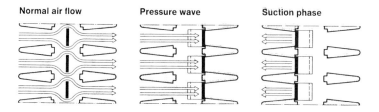

Fig. 12   Operating principle of Luwa valve type F

Fig. 13   Luwa explosion protection/pressure relief
valve type K

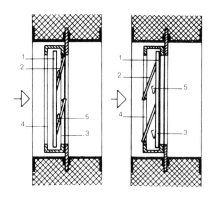

→▷ direction of blast
1   valve bar
2   guiding lever
3   internal valve seat
4   external valve seat
5   spring

Fig. 14   Operating principle and
operating mode of Luwa
valve type K

Ventilation Systems

Principle requirements. All shelters must be provided with a ventilating system which maintains bearable air and climatic conditions. This even more so as modern warfare may require continuous shelter occupation over long periods of time.

Air intake

Explosion protection valve

Primary air filter

Ventilation rate meter

Gas filter

Ventilator

Pressure relief valve

Main purpose of the ventilation system is:
- to supply oxygen and keep carbon and carbon dioxide concentration below a critical level
- to maintain temperature and humidity within bearable limits
- to build up positive pressure within the shelter to prevent ingress of contaminated air (war gas, radioactive dust).

Ventilating appliances

Fig. 15 Ventilating appliance in
         private shelter
         (photo: Luwa Ltd.)

Ventilating Appliances as shown in Fig. 15 conform with the requirements described above.
The assembly consists of:
- Explosion protection valve
- Air flow meter
- Gas filter (Hepa and activated carbon)
- Ventilator with manual and motor drive
- Flexible hose connections

The units are available in 5 sizes with capacities as follows:

Unit size (nominal m3/h	20	40	75	150	180
Air flow					
with gas filter (cfm)	12,5	25	45	90	105
without gas filter (cfm)	25	50	90	180	210
Nos. of persons served	7	13	25	50	60

Multiple arrangements of the larger units are used for ventilation of larger life
support shelters for up to several hundred people.

Standardized ventilating units for filter air capacities from approx. 700 to 2800
cfm together with gasfilters for air volumes of 300 and 700 cfm are also available.
Units are equipped with electric fan drive motor and emergency hand crank. These
unit combinations (Fig. 16) are used for very large shelters or for shelters of the
civil defence organization where, owing to the activities involved, air volumes mus
be larger than for pure life support shelters.

Fig. 16 Ventilating appliance in emergency hospital
(photo: Luwa Ltd.)

REFERENCES

Sager F., Federal Office of Civil Defence, Informative meetings 1980
Basler E. Dr., Zurich, Schweiz. Bauzeitung, volume 28, 1965
Heierli W. Dr., Zurich, Schweiz. Bauzeitung, volume 46, 1976

USING THE KANSAS CITY UNDERGROUND FOR CIVIL DEFENSE: AN
ALTERNATIVE TO EVACUATION OF AN URBAN POPULATION

D. Ward

Dept. of Geosciences, University of Missouri-Kansas City
5100 Rockhill Rd., Kansas City, Missouri 64110

ABSTRACT

Greater Kansas City possesses over thirty underground space sites, of which fifteen
have been identified as potential civil defense shelters. A few are being used for
civil defense supply storage at the present time. It is estimated that 20-30
million square feet could immediately be used, given minor modifications and stockage. Planning is underway with area civil defense officials and underground operators to give the population of Greater Kansas City another option for survival.

KEYWORDS

Underground shelters; civil defense; mined space; disaster planning; urban population protection; emergency preparedness.

INTRODUCTION

The Kansas City metropolitan population of 1.3 million is spread over an area in
excess of 1000 square miles. The area encompasses seven counties in northwest
Missouri and northeast Kansas. The entire region is underlain by sedimentary rocks
composed primarily of limestones and shales. One layer of limestone, found from
50-200 feet beneath the surface, is the Bethany Falls. This massively bedded limestone, with an average thickness of 22 feet, is exposed along valley sides throughout the metropolitan area where erosion has stripped away the overburden and rock
strata. The Bethany Falls is nearly horizontal, with only a gentle dip. Thus,
mining the limestone has always been relatively simple and cost efficient.

Since the late 1800's the easily accessible limestone has been mined. For many
decades the limestone was mined for the rock alone. However, since the early 1950's
the primary purpose of mining has changed from one of extracting rock to one providing space. Rapidly increasing demands on urban space, together with skyrocket-
ing energy costs, have resulted in increased emphasis on the use of underground
space. Today, additional useable space is being added as the advantages inherent
in the underground become more apparent. Thus, it is the secondary use of mined
out space -- space of high value in an urban area -- that encourages further expansion.

For several years area civil defense directors have considered using the underground facilities as emergency shelters. In fact, some emergency supplies have
been stored in a few mines since the late 1950's. However, not until recently
have more concerted efforts been initiated to plan for the emergency protection of
large numbers of people.

Only in recent years has it become readily apparent that Kansas City has a valuable
resource beneath the surface capable of insuring the survivability of a large percentage of the total population in a natural or man-made disaster.

In recent years many civil defense experts have noted the potential of the Greater
Kansas City underground (Goure, 1979; Kilpatrick, 1979) for use as emergency shelters. On December 13, 1978, in a speech given in Kansas City, Major General George
J. Keegan (U.S.A.F., Ret.) stated that the Kansas City area has the greatest potential civil defense facilities in the United States because of the underground sites.
Even a well-known Soviet analyst (Kamenetskiy, 1978) implies in the Russian publication Voyennyye Znaniya that the Kansas City area possesses some of the best
potential civil defense sites in the world (Fig. 1).

Fig. 1. Modern available underground space.

PRESENT UNDERGROUND FACILITIES

At the beginning of 1981 there were 34 underground sites in or near the Kansas
City metropolitan area. Owing to poor mining methods in the past, roughly half of

these sites are deemed inappropriate for the secondary use of underground space
as emergency shelters. At the present time sixteen underground sites are developed to the degree where large numbers of the Kansas City metropolitan population of
approximately 1.3 million could be safely protected with minimal preparation time
and cost.

The sites are currently utilizing over 22 million square feet of space for manufacturing, warehousing, and services. An additional 30 million square feet is available for use -- with roads, lights, water, and sewer installed. Up to 25 million
more square feet of space could be made available for emergency use with the addition of portable generators, medicine, food, emergency and sanitation stations.
Thus, for civil defense shelter purposes, roughly 75 million square feet is available, given varying degrees of preparation. With proper planning the inhabitants
of Greater Kansas City could have ample footage for an extended period of shelter
time.

The greatest expense incurred in shelter development is the shelter itself. This
is especially true for an underground shelter where the rock must be extracted to
provide the needed space. In Kansas City the shelters are already in place. The
situation is comparable to having a house already built with only some interior
work and stockage left to be done.

Expansion of current developed facilities is progressing at the rate of 6 million
square feet per year. Yearly square footage increases more than compensate for
the population increase of Greater Kansas City. Future projections indicate annual added space will increase as the appeal of underground space -- with its
60-70% energy savings, minimal physical plant overhead, and security -- continues
to grow.

The Kansas City facilities are located in the lower portion of the previously
named Bethany Falls limestone. Modern mining employs the "room and pillar" method
of leaving aligned 20-25 pillars of limestone with a spacing between pillars of
approximately 40 feet. Ceiling heights average 13 feet.

The pillars are aligned so as to allow roads and other construction with minimal
development costs, while at the same time providing a safe working environment.

The sixteen sites currently under consideration are clean, large, and easily acess-
ible to the urban population. The mines are not the dark, damp, inhospitable
places that may come to mind when one thinks of going underground. In fact, over
2000 people now work daily in the Kansas City underground facilities.

The underground facilities are ideally located along major transportation routes.
As previously noted, the sites are located along valley floors where erosion has
dissected the valleys to expose the limestone. It is along the level valley floors
where transportation networks -- both rail and highway -- are naturally built.
Because of the dispersed facilities, large numbers could reach a shelter in a short
period of time. No one in the Greater Kansas City area is more than 10 miles from
a underground space development and the majority are less than 5 miles from a
potential shelter. If one were to initially plan the location and building of
shelters and connecting highway links in Kansas City to insure the protection of
its population, it would be difficult to improve upon what Kansas City has today.
And the situation continues to improve as new sites are opened and older sites are
expanded.

The facilities vary greatly in size; from under 100,000 square feet to over 22
million square feet. The size of the average mine is approximately 5 million
square feet with one-third of the mines continuing to expand their facilities.

Fig. 2. Underground roadway - Commercial
Distribution Center

Fig. 3. Emergency supplies in modern underground facility

Where expansion is occuring it is done with the idea of eventually using the secondary space as a valuable addition to the available urban real estate.

Current uses for mined space focus on warehousing, factories, and offices. Warehousing is by far the number one use for the Kansas City underground with food storage the main warehousing operation. The temperature is constant 57°F-58°F. and the humidity is low. Factories and offices are being drawn to the underground as energy costs continue to increase above ground. In the underground there is no requirement for heating or cooling. From a shelter standpoint it is estimated that the influx of a large number of people, along with lights and other generating equipment, would raise the temperature from 5°-20°F, depending on the size of the area and the numbers involved. Those working in the underground facilities today find conditions excellent. It is for all of the above reasons that the sites continue to expand.

PROTECTION AFFORDED BY THE UNDERGROUND

Existing underground facilities provide protection not available in most existing shelters. The majority of community shelters are referred to as "fallout" shelters for a reason--they provide partial fallout protection. The underground puts mass --via layers of sedimentary rock and overburden-- between the nuclear detonation and the sheltered population.

A nuclear device destroys by (1) blast effects, (2) thermal effects, and (3) fallout radiation. Nuclear weapons are similar to conventional weapons in that most of their destructive effect is due to blast. The effect of blast is dependent on the yield of the weapon, whether it was a surface or air blast, and the distance from ground zero. Although it cannot be stated that the underground will completely protect against blast, the facilities are definitely blast-resistant and far superior to existing shelters on the surface.

Thermal effects are greatly reduced. The rock mass, the small entrances in relation to the mined areas (which are below the dominant surface level), and the fireproof nature of the underground, all greatly reduce potential thermal destruction.

The civil defense shelter program was designed to protect the population against fallout--the third and potentially the deadliest long-term hazard of a nuclear device. The underground provides the shielding necessary to virtually eliminate fallout dose rates high enough to kill.

Three types of nuclear radiation may be emitted by fallout. Alpha radiation is effectively stopped by the outer layers of the skin and presents no major problem. Beta radiation penetrates the skin, but not the internal organs. Any existing surface fallout shelters protect against alpha and beta hazards. The greatest fallout threat is from the emission of gamma rays; the extremely penetrating rays that virtually destroy body cells with X-ray like intensity.

Shielding is the key to protecting against gamma radiation. The greater the mass of the shield, the greater the reduction in the dose rate--the more shielding, the more protection. Earth and rock are extremely effective at stopping the rays. Thirty inches of earth give the same comparative protection as 2-1/2 inches of lead or 24 inches of concrete. Mass quickly reduces the dangers inherent in gamma rays. Just 2 inches of concrete will stop 50% of the gamma radiation. In general, the denser the material the less thickness is required. The protection afforded by up to 200 feet of dense rock and earth above the underground is obvious. However, where the entrance tunnels and walls are exposed and thin additional

Fig. 4. Water and Sanitation Supplies

Fig. 5. Underground office and potential E.O.C.

efforts must be made in the future.

CURRENT SHELTER USE AND IMPROVEMENT PLANS

It has long been recognized that the Greater Kansas City underground has potential shelter use. Since the late 1950's some civil defense supplies have been stored by area civil defense directors in selected sites. Food, water drums, sanitation kits, medical kits, geiger counters, survey meters, and dosimeters are representative of what has been stored. At one site in the City of Independence, a 250 bed hospital is stored for emergency use.

The Greater Kansas City region has been fortunate to have a strong organization composed of county and city civil defense directors, area planning organizations, representations of other federal, state, county, and city agencies and departments, as well as an important nucleus of concerned citizens. Working together and individually within their jurisdictions with underground operators, the planning and shelter implementation phase is continuing.

Perhaps the greatest progress is being made today in Clay County, Missouri, at the Great Midwest Corporation's underground site. The site is rapidly developing as a prototype for the area. Properly stocked, this site will be capable of protecting many thousands of citizens in an emergency situation. 600 KW generators, a 200 bed hospital (with 2 additional ones planned) food and water supplies and other emergency supplies are already located. Blast doors are being planned and ventilation requirements surveyed. The county has laid out a complete evacuation plan to include emergency routing of people and supplies. In addition, the Clay County E.O.C. (Emergency Operations Center) will locate its nerve center within the Great Midwest complex.

In summary, the Greater Kansas City area has the potential to become a showcase for civil defense. Given the vast underground facilities now available, the population of the metropolitan area has the resource base to protect itself in time of emergency in a minimal period of time and at extremely low cost per person sheltered. However, much planning and work remains to be accomplished. Sites vary considerably with respect to improvements already carried out by private operators to service the private sector now in place. Each site deemed appropriate for development as an emergency shelter requires a detailed survey and analysis with respect to resistance to blast and thermal effects, numbers of people to be sheltered, ventilation requirements, and emergency equipment and supply stockage. Finally, the problem of outside evacuation plans to each site as well as shelter management at each site remain to be addressed. Given that all of the above are accomplished, the Kansas City area will have the best emergency shelters for the largest number of people anyplace in the world.

ACKNOWLEDGEMENT

The author would like to express his gratitude to the area civil defense directors and to the underground operators who have been so instrumental in planning for the protection of the Kansas City population, and who assisted me in my initial site surveys and mapping. A special thanks is extended to Dr. Truman Stauffer, Director of the Center for Underground Space Studies at the University of Missouri - Kansas City.

REFERENCES

Disaster Planning Guide For Business and Industry (1978). Defense Civil Prepared-
    Agency CP62-5, 6-11.

Goure, L. (1979). Comments made before the 1979 Seminar-Conference of the American Civil Defense Association, Kansas City, Missouri. Sept. 28.

Keegan, G. J. (1979). Speech made before the February meeting of the Greater Kansas City Emergency Preparedness Group.

Kilpatrick, K. (1979). Way to Go: Down. Journal of Civil Defense, XII, 5-8.

Sisson, G. N. (1975). Mining More Protection. Foresight, May-June, 18-19.

Stauffer, T. S. ed. (1978). Underground Utilization: A Reference Manual of Selected Works, 8 volumes. Dept. of Geosciences, University of Missouri-Kansas City.

CHAPTER II

EARTH SHELTERED BUILDINGS

SESSION DEVELOPER: Raymond Sterling

EARTH SHELTERED BUILDINGS COUPLED WITH THE SUN

OPPORTUNITIES AND CONTRAINTS IN DESIGN

David J. Bennett, AIA
Principal, BRW Architects
2829 University Avenue S.E.
Minneapolis, Minnesota 55414

ABSTRACT

The Civil/Mineral Engineering Building at the University of Minnesota U.S.A. is a 2,064,770 cu. ft. building, 95% of which is underground, 35% in mined space 110 feet below the surface. The building is programmed as an energy conservation demonstration, linking underground construction with a number of solar responsive components including solar shading with plant material, passive solar storage with water tubes and a system for illumination of interior spaces by tracking the sun with mirrors and optical lenses.

The design of the Civil/Mineral Engineering Building was preceded by the design and construction of several other projects. Each of these have addressed similar issues relating to energy utilization and may be considered as antecedant work.

The Civil/Mineral Engineering Building is under construction. Completion is anticipated in the Winter of 1982/1983.

KEYWORDS

Earth Sheltering/Underground Space; Mined Space; Passive/Hybrid Solar Heating; Landscape Microclimatology; Solar Optics/Remote View Optics.

INTRODUCTION

In 1978, the University of Minnesota commissioned BRW Architects (formerly Myers and Bennett Architects/BRW) to design a new Civil/Mineral Engineering Building on its Minneapolis campus. The building was mandated by the University as a demonstration design in energy conservation and innovation in active and passive solar energy applications. It is currently under construction.

The building has been designed 95% below ground, with approximately 35% of its 76,423 cubic yards of volume in mined space 110 feet below ground in sandstone, below a 30 foot thick limestone layer. The remaining 60% of underground space is in the 53 feet of earth above the limestone layer.

The 5% of the building which projects above ground houses a number of passive solar energy systems: supplementary heating with a trombe wall system, solar shading and wind control with plant material and a unique system for optically projecting sunlight and exterior views into deep interior space, including sunlighting of the mined space 110 feet below ground.

The design of the C/ME Building has its antecedents in several earlier buildings. Each of these represent contributory steps in the identification of issues and the development of design principles for earth sheltered buildings. A description of each, and its particular characteristics follows:

WILLIAMSON HALL

In 1973 Williamson Hall, a combined Bookstore and Admissions and Records facility, was designed for the Minneapolis Campus of the University of Minnesota. The building of 83,000 square feet is 95 percent underground.

The reasons for constructing the building largely underground included concerns with Urban Design, Historic Preservation and Energy Conservation. Williamson Hall was completed and occupied in the summer of 1977.

The design problem involved the integration of pedestrian circulation patterns, above grade, on-grade and below-grade as these relate several university service functions to both campus and public access.

In order to preserve an open space adjacent to Folwell Hall, most of the facility occupies two levels below grade. An open sub-grade court provides daylight exposure to the building's occupants.

As well as meeting urban design and energy conservation objectives, the design of Williamson Hall demonstrates that earth sheltered buildings can provide a humane and pleasant living/working environment. Completed and occupied, Williamson Hall is described as the "sunniest building on the campus." Numerous articles published about the building have quoted occupants as expressing their pleasure with the interior spaces.

A unique planter system was designed for Williamson Hall using Engleman Ivy as a solar control device to screen out summer sun and permit winter solar collection. This may be among the first contemporary deliberate applications of landscaping for passive solar control in a building.

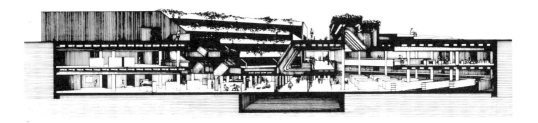

The mechanical and electrical systems were developed to extend the energy conserving characteristics of the basic concept, while maintaining normal comfort levels. The building employs a low pressure air distribution system and an electrical buss duct system for plug-in ceiling lights. This duct system allows the application of a "task lighting" concept which helped reduce the number of required fixtures by about 20%.

The construction cost of the building was .6% below its pre-established budget, which was originally determined for a conventional on-grade building.

The University of Minnesota received a grant of approximately $350,000 from the National Science Foundation to monitor the building for its energy conservation characteristics as an earth sheltered building. Subsequent to the first grant, the University received a second grant of $430,000 from the Energy Research and Development Administration to construct a high-temperature solar collector in conjunction with the building. The performance of this active solar system will be monitored to further our understanding of the cost-effectiveness of the high-temperature solar collector systems when used in conjunction with energy conserving architectural design.

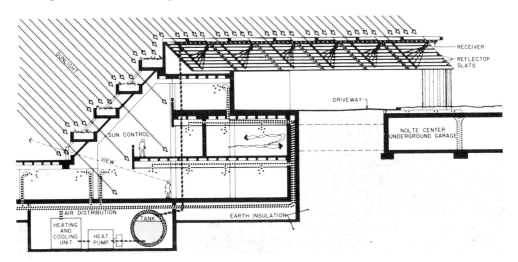

ST. PAUL CAMPUS STUDENT CENTER EXPANSION, UNIVERSITY OF MINNESOTA

The design of the Student Center Expansion was a comprehensive response to the goals of the St. Paul Campus Masterplan, the needs of the St. Paul Campus community, and passive energy conservation.

A significant segment of the building has been extended under an existing major street which bisects the Campus. This building element provides a protected pedestrian concourse – an underground "street" – which helps complete the Campus circulation system and binds the two halves of the Campus together.

The major program elements included remodeling and expansion of the main dining facility, the lounges, and the recreational spaces, as well as the addition of a new bookstore, and a new art and music lounge.

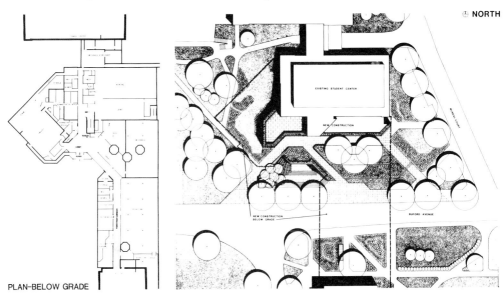

⊕ NORTH

PLAN-BELOW GRADE

71% of the new construction was built underground. This segment of the building demands only 20% as much energy per square feet as the existing above-ground areas. New insulated building elements were added to the present above grade structure to improve its energy conservation performance.

WALKER COMMUNITY LIBRARY, MINNEAPOLIS MINNESOTA

The Walker Community Library is located on a small site at the juncture of an open space greenway and the edge of an urban commercial node in a larger residential neighborhood. In this pivotal location, the library site represented a complicated urban design problem.

The design solution was an underground building which allowed the roof to be used as a community open space area. This concept was particularly appropriate since the neighborhood has a history of active public events, including an annual regional art fair.

Constructing the building underground not only provided a satisfactoy physical solution, but also provided the Library Board with an immediate economic benefit. By reducing the required land area by 15,000 sq. ft., land acquisition costs were reduced by $195,000. Set against this savings were the additional building construction costs, not of building underground itself, but of preparing the roof surface to accommodate automobile parking and landscaping. These costs totalled approximately $70,000. Therefore, the Library Board enjoyed an immediate net saving of about $125,000 - roughly 9% of the construction cost of the project.

The energy benefit accruing to the underground construction will result in a conservatively estimated 40% reduction in demand for heating and cooling. In a public facility that will remain in service for many decades, the ultimate economic benefit of this dramatically reduced demand is worth many times the initial site and building costs.

In addition to the physical and economic benefits already discussed, several other social, environmental and aesthetic benefits are evident. The surrounding neighborhood has a history of active public events, including an annual art fair in the streets, which draws exhibitors from a five-state region. Because of its strategic location in this setting, the library building has been conceived as an urban "place," connecting the commercial node to the Mall, and the boulevard to the lake system. Its underground location provides an opportunity to develop part of the roof area as a public plaza, suitable for community activities.

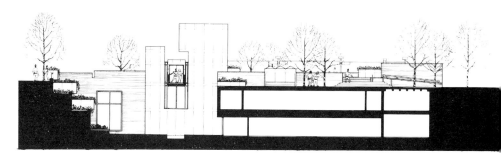

Despite the high activity and noise level created by vehicular traffic along arterial streets, relatively quiet exterior spaces can be created with sunken courtyards. By placing the main reading room of the library below grade, a sunken outdoor reading court directly adjacent to it has been made available for outdoor activities related to library programs.

FORT SNELLING I, DESIGN STUDY FOR ENERGY INDEPENDENCE

A conceptual design study of the Fort Snelling Center was conducted to help the Minnesota Historical Society gain funding approval from the legislature. The Fort Snelling Center located on the site of Fort Snelling State Park near the Fort, is designed to serve as an interpretive center, Historical Society Archaeological Laboratory and central office facility for the Society's Historic Division.

An important objective of the Fort Snelling Master Plan is to reconstruct the area as it was between 1820 and 1848. Therefore, in the study we recommended that the center be conceived as an unobtrusive "non-building."

In the conceptual plan, the Center is placed into the edge of the cliff along the Mississippi River gorge with its roof surface approximately a half-story level above natural grade. The roof surface is developed into a terrace, the perimeters of which are gently sloped to natural ground level away from the cliff edge. This provides a vantage point from which the Fort Camp Coldwater upstream, and the sweep of the Mississippi River may be seen. This concept minimizes the impact of the Center on the historic setting and the natural environment. As seen from the river, the geometry of the building is integrated into the natural configuration of the cliff face.

The concept of energy independence was explored in the Fort Snelling Center study. To accomplish this, the architectural design has been integrated with a constellation of mechanical systems designed to maximize energy conservation and replace reliance on fossil fuels with energy drawn from the natural conditions of climate and environment:

Solar Collection/Heating

A solar collection system of the high temperature concentrator type is proposed as a part of the site/building design. In order not to intrude on the historic setting, the collectors are located at the edge of the site. The collectors would be used to apply solar energy to space heating, domestic hot water and electricity generation.

Solar Collection/Electrical Generation

The proposed solar collector would generate steam at 400-650° F, sufficient to drive a turbine generator supplying electricity to the building. During part of the spring and fall and all of the summer, the generator would produce electricity in excess of the building's demands. This excess would be diverted to the local utility power grid (NSP) at times when electrical demand for cooling is normally highest. This concept has profound implications for the design and operation of utility power plants on a national scale.

Natural Cold Storage/Cooling

The Fort Snelling Center would be cooled by a new application of an ancient idea — cold drawn from stored ice. During the winter months a heat exchanger, in the form of liquid flowing through a continuous pipe exposed to the atmosphere, is

used to freeze water contained in a storage tank buried in the earth. During the
summer, by a simple valve system, the liquid continues to be pumped through the
stored ice and then into the cooling coils of the air distribution system.
Instead of cold from coils in a conventional absorber, the cold would be taken
from the coils connected to the stored ice system. From that point the cooling is
taken from the coils connected just like any other air conditioning system.

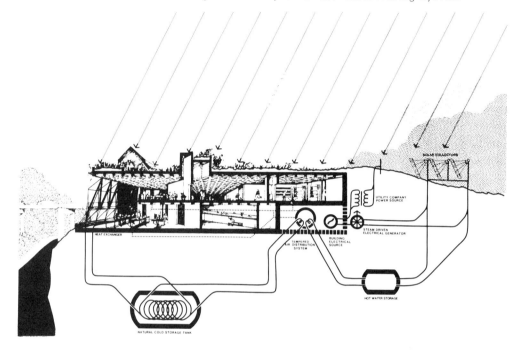

FORT SNELLING II, BUILDING DESIGN AND CONSTRUCTION

The design solution for the Fort Snelling Visitor Center was greatly influenced by
the Master Plan and by State legislation controlling development along the edge of
the Mississippi River. Important considerations including urban design, natural
area preservation, river ecology, historic restoration and energy conservation led
to the development of a "non-building" - an underground, earth sheltered, unobtrusive facility.

The parking and bus drop-off areas are arranged to bring visitors into almost
immediate visual contact with the Mississippi River. The Visitor Center is placed
between the parking areas and the Fort, where it is best situated to provide its
interpretive function and act as a control center. By being depressed into the
site, it allows an unobstructed view of the Fort and minimizes the impact of contemporary development on the historic setting.

The design solution for the Visitor Center includes several innovations in earth
sheltered and solar-passive technology. These include natural thermal control of
entries with unusual entry concourse design, refinement of the use of planting
material for solar screening, and the application of a continuous strip
"periscope" within a mechanical plenum to bring light and view to spaces two
levels below grade.

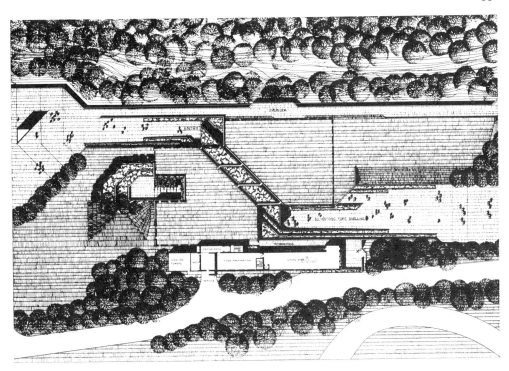

Solar Passive Design/Landscape Microclimatology

All the major spaces in the Fort Snelling Center are earth sheltered. The office areas, one level below grade, are organized around three sides of a small depressed courtyard. The fourth side is dished out to the south and west to provide along horizontal view and to permit passive solar gain from the low winter sun. Solar screening of this area during the cooling season will be accomplished by the use of plant material. Engleman Ivy, which drapes verticaly, will be placed along the west edge. Another vine material will be grown horizontally on a tension-wire trellis along the south side of the court.

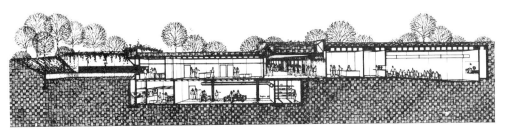

Concourse Design

The Fort Snelling Center design will employ a central concourse as an interstitial space between the entry vestibules and the activity spaces. This solution is directed toward both servicing user comfort demands and improving energy performance. The mechanical system will be zoned to maintain the concourse as a separate transitional space, at a lower temperature in the winter and higher tem-

84

perature in the summer than the surrounding areas, from which it is physically
separated. In this way it will serve as a buffer between outside and interior
conditions.

CONCOURSE PERSPECTIVE   PERISCOPE SECTION

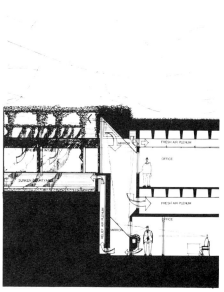

The Periscope Principle

The office areas in the Fort Snelling Center provided an opportunity to develop a
design solution for light and view access for areas two stories underground which
will satisfy user needs and provide a more energy-efficient solution than conventional skylighting. The design applies the principle of the periscope. By using
a view plenum with reflective angled surfaces at the top and bottom, the totally
underground area will receive a view of the sunken courtyard visually similar to
that of the area with windows one level above.

CIVIL MINERAL ENGINEERING BUILDING

The C/ME building has been shaped by a combination of program requirements, site
forces and energy conservation/energy demonstration considerations. These have
combined to give the building its unique configuration and to determine its site
location and orientation.

Building Program

A fundamental purpose for the construction of the building is to replace the Civil
and Mineral Engineering Department's outmoded facility, built in 1912, with a contemporary facility providing laboratory, instructional and administrative space.
By doing this, the Department expects to enhance its research and teaching
programs, provide its faculty with an improved working environment, and attract
talented new faculty to the University by the research opportunities which only a
new laboratory facility can provide.

The basic program components include the following: Environmental Engineering Laboratories, Soil and Rock Mechanics Laboratories, Water Resources Laboratories, Mineral Engineering Laboratories, Transportation Engineering Department, Surveying Laboratories, and The Underground Space Center. Departmental services, faculty offices, auditorium lecture rooms, computer and support spaces, and lounges are also included. A pedestrian tunnel to the Institute of Technology Complex via the Architecture Building is an additional part of the program.

Site Forces

The site is located at the northern end of the existing building complex housing most of the University's Institute of Technology. This location was selected by the University prior to the beginning of the design process. Several unique site conditions had to be addressed in the design process.

In a preceding and independent activity, the University developed a planning document, the "University Area Short Range Transportation Program," which recommends the site area as the new location of the Campus bus terminal. In the definition of the Study, the "terminal" is not a structure, but the waiting area for up to twelve sixty-passenger buses with destinations within all of the University's three Twin City campuses.

The location of the buses at the building site area, combined with the relatively close proximity of the majority of the off-campus parking areas (a city block or so to the north), serves to make the building's location the nexus of a pedestrian circulation pattern throughout the Minneapolis East Bank Campus.

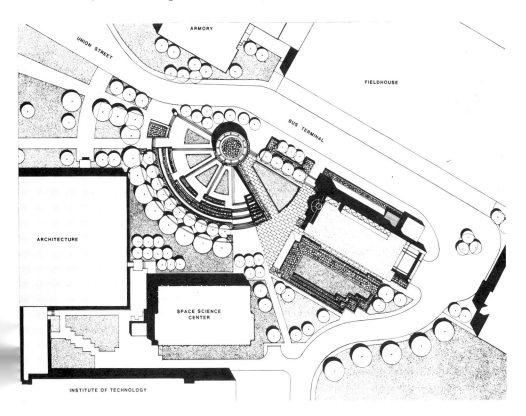

Site Planning/Urban Design

The site planning/urban design approach has been shaped by the site forces
described above and has resulted in the following organizational strategies:

The building is divided into three major areas: classrooms, laboratories and
offices. The classrooms, actually lecture/demonstration spaces, have been placed
underground. Since they require absolute•light control, and close proximity to
the classrooms in the School of Architecture basement, they are located at the
west end of the site, in the cool shadow of the Space Science Center, and are connected to the School of Architecture by an underground tunnel.

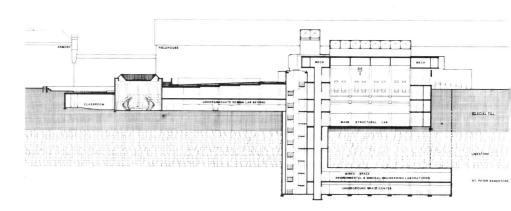

The laboratories, which also may require light control, are underground along the
northern edge of the building. Most of the laboratories are in cut-and-cover
space in the top 53 feet of earth cover. The exceptions are the Environmental and
Mining Laboratories, which along with the offices of the Underground Space Center,
are in space which is about 110 feet below ground, mined out of the soft sandstone
below a 30 foot deep limestone layer. This temperature-stable, vibration-free
environment, with a natural clear-span of at least 50 feet, has proven to be less
expensive to create than equivalent space in either cut-and-cover or conventional
above-ground construction, once the access to it is created. As a result of this,
the 8500 square foot Mining Laboratory has been shifted down to the mined space
from the cut-and-cover space above as a construction economy after bids were taken
demonstrating the cost-saving opportunity such a move would provide.

The other exception is the main structural laboratory, which, though its floor is
some 30 feet below ground, has its roof well above ground. This laboratory is the
most prominent part of the building which projects above ground level. The extension of this great space into the atmosphere is both to allow it to receive the
large structural sections which will be tested within it and to serve as the
housing for the solar-related demonstration components of the building. For these
reasons it is located at the east end of the site, as much out of the shadow of
the Space Science Center as could be accomplished.

The faculty offices have been placed along the southern flank of the building,
also at the east end of the site, where they can enjoy access to the sunlight. In
order to not interfere with the solar access required for the demonstration components housed in the Structural Laboratory to the north of them, and to optimize
their energy conservation opportunity, the offices are stepped down into a sunken
courtyard.

The "Bus Terminal" is laid along the northern edge of the site, between the underground laboratories and the Field House. Here the buses will create the least environmental disturbance, their noise trapped between spaces which are not noise sensitive, their fumes flushed out through a corridor open to the prevailing northwest winds and the sight of them shielded from the campus by the above ground C/ME Building mass. At the same time, by strategic location of entries into the C/ME Building and access to the I.T. tunnel system, this location of the bus terminal maximizes the opportunity for pedestrian circulation from the buses to a significant amount of the campus through protected passageways. The location of the bus terminal at this edge of the site also facilitates excellent on-grade access to the remainder of the Campus to the southwest, without conflict with the I.T. pedestrian movement patterns.

The underground classrooms are arranged in a series of fan shaped segments in 30° intervals, stepped down successively so that their roofs form a great 150° spiral entry court which begins at grade level and terminates in the main entry to the building. This courtyard not only provides an entry into the C/ME Building itself, but, because of its site location in relationship to the bus terminal and the Campus entry, as well as its access to the classrooms and tunnel to the Architecture Building, the courtyard also becomes the northern gateway to the I.T. Complex. In so doing the courtyard creates a symbolic entry to the Institute of Technology as a distinct unit of University.

Energy Demonstration Design

The energy demonstration design represents a synergetic relationship between a variety of energy conserving systems which, in an additive way, are each intended to contribute toward making the building as nearly independent as possible.

The energy demonstration program for the building has been divided into two categories:

Active Systems, including solar heating, solar electricity generation and ice energy cooling. These all depend on mechanical subsystems and are conceived to

act as an integrated unit. The basic principle of the system is that a high temperature solar collector will be used to heat the building and to partially
electrify it through a Rankin-cycle generator, while summer cooling is achieved by
using ice stored in an underground tank, formed and employed by a system of heat
exchangers at the surface and connected to the building's cooling system. The
design and installation of this system is dependent on the approval of a U.S.
Department of Energy Research/Field Application Grant applied for by a University
research team.

Passive Systems, including earth sheltering and underground space, Landscape
Microclimatology, passive solar heating employing hybrid systems, Solar Optics and
Remote View Optics. Each of these systems deserves a brief description:

Earth Sheltering/Underground Space: The C/ME Building is designed 95% below
grade. Of the 151,000 square feet in the building, 44,000 square feet is in mined
space about 110 feet below the surface. Earth sheltering, the baseline of the
energy conservation system, will result in a total building energy demand of less
than half of that required for a similar above ground building.

Landscape Microclimatology: The application of landscape material as an integral
part of the building design provides an opportunity to employ it as a working component of the Environmental Control System. In the C/ME Building this includes
the use of draping deciduous vines. These provide solar screening of south facing
surfaces in the summer time and, by shedding their leaves, allow passive solar
heat gain during the winter months. Measurements taken at an earlier
installation, Williamson Hall, indicate that vines can reduce insolation 50% with
moderate growth and 75% with heavy growth. Other applications of the use of
landscaping to effect the microclimate around the C/ME Building include the use of
masses of conifers at strategic locations which will direct the winds to flush out
the bus terminal corridor.

Passive/Hybrid Solar Heating: A trombe wall system of water-filled fiberglass
tubes in an enclosed glass chamber has been included along the south wall of the
main structural laboratory. This system is a hybrid design which operates in two
primary modes, depending on demand conditions. In one mode, it acts as an
instantaneous pre-heater, with duct-fed outside air introduced at the bottom of
the chamber and heated air carried off by duct at the top and distributed as makeup air. In its other mode, the ducted air is sealed off and vents at the bottom
and top of the chamber's interior wall are opened to allow the gravity heating of
air in the main structural laboratory. The heated air is collected at the ceiling
and recirculated through the space. During the night, when no fresh air is
introduced, the heated air from the trombe wall chamber will be used to maintain
the temperature level of the 72,000 cubic yards in the main structural laboratory.

Solar Optics/Remote View Optics: The solar optical system is a means of
collecting sunlight, concentrating it and directing it through the interior of the
building by an assembly of lenses and mirrors. As environmental design, it has
the potential of enchancing the habitability of deep interior spaces, whether
underground or not, by providing sunlight where there was none before. As energy
design, it promises to be among the most cost effective of active solar systems.

Beginning with the sun, a light source which provides about 10,000 ft. candles of
light on the earth's surface, solar lighting can be optically compressed into
narrow light streams, passed through small apertures and and corrridors, and
spread out over target spaces using no transmission energy but its own.
Furthermore, the light does not have to be converted into another form of energy.
When combined with conventional artificial lighting under controlled conditions,
it has the potential to substantially reduce the electrical energy demanded for
lighting interior spaces. Eventually, the introduction of very bright, high

lumen/watt artificial lighting into the same optical stream promises the potential of providing all interior lighting from a single source, replacing conventional electric lighting distribution systems altogether.

In the C/ME Building, the Solar Optics system will be a limited installation solely for demonstration purposes. It will, however, include an installation to bring sunlight into the mined space, about 34 meters below the surface.

The Remote View Optics is an optical viewing system essentially based on the principle of the periscope, but in a much more refined form. A purely optical system, it offers the immediacy of looking through a window. Its function is not directly energy related, but is primarily intended to enhance the habitability of underground space. The demonstration application in the C/ME Building will provide a view of the site area (looking westward over the campus, as seen from a second story window) to the reception area of the Underground Space Center offices in the mined space 110 feet below grade.

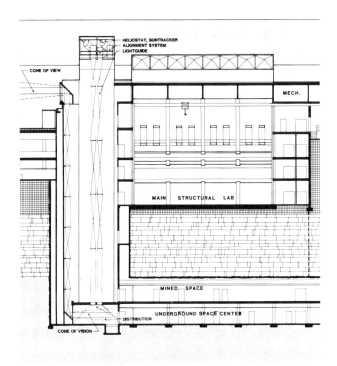

This integrated system of energy demonstration components is intended to establish a level of single building energy independence not achievable with conventional technology. Several of the innovative energy components will be used for the first time in this building. It is intended that the demonstration of these energy conserving systems will encourage the development of further technological innovations to conserve energy resources.

Building Systems and Materials

The structural system includes cast in place concrete walls, columns and waffle floor slabs. A structural steel truss roof supports a 15 ton capacity traveling crane over the main structural lab floor. The exterior will be clad with face brick, glass, and metal panelling.

The mechanical systems make use of traditional tempered air distribution, integrated with systems which make use of captured solar energy.

The electrical system incorporates high efficiency motors, electrical supply systems and high lumen/watt lighting which will be used in conjunction with solar lighting systems.

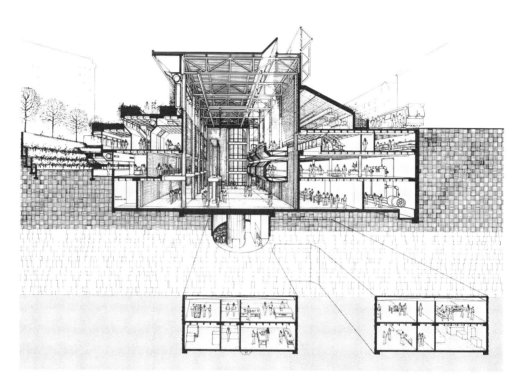

SQUARE FEET SUMMARY

Level	Class rooms	Laboratories	Offices	Underground Space Center	Mechan.	Tunnel To Arch.	Gross Square Feet	Elevation
Level A					o		11,000	Grade Level
Level B		o	o		o		30,000	– 15 Ft.
Level C	o	o	o			o	42,142	– 37 Ft.
Level D		o			o		21,873	– 51.5 Ft.
Level E					o		2,164	– 63 Ft.
Level F		o			o		22,067	– 95 Ft.
Level G		o		o			22,067	–112.5 Ft.
TOTAL							151,313	

A COMPUTER SIMULATION OF THE THERMAL PERFORMANCE OF EARTH COVERED ROOFS

Jerome J. Speltz
Setter, Leach and Lindstrom, Inc.
Minneapolis, Minnesota

and

George D. Meixel, Jr.
Underground Space Center, Univ. of Minn.
Minneapolis, Minnesota

ABSTRACT

A comprehensive computer model of the transient one dimensional heat flow through earth covered roofs has been developed and utilized to study thermal performance for both winter heating and summer cooling periods. This model couples a proven conduction transfer function calculation scheme with current models of surface convection, radiation and evapotranspiration to generate the capability for evaluating the influence on roof heat transfer corresponding to changes in the following roof characteristics:

(1) thermal mass,
(2) insulation,
(3) evapotranspiration, and
(4) convective and radiative heat exchange.

The details of the model are described. Results of applying the computer simulation to the assessment of heat flow through simple roof structures and more complicated earth covered roofs during summer cooling periods are presented.

KEYWORDS

Earth covered roof, computer model, earth-contact heat transfer, earth surface boundary conditions, thermal performance.

INTRODUCTION

Earth covered roofs have unique properties that offer energy benefits in addition to the aesthetic appeal, shelter from the wind, and the solar shading possibilities of having plants on the roof. The expected advantages associated with having the large thermal mass of an earth covered roof have been qualitatively discussed by several authors [1, 2, 3]. The additional potentials of the plant cover to reflect summer sun, to counteract radiant heat gain with latent energy losses due to evapotranspiration, to lose energy to the night sky through radiation, and to reduce winter heat loss by holding an insulating cover of winter snow may all contribute to a highly energy efficient roof structure. However, the key questions concerning the quantitative level of the summertime cooling available from evapotranspiration, the winter reductions in heat loss due to the increased presence of snow, or the improved performance because of the large thermal mass, are largely unanswered. There is little laboratory or field data relevant to earth covered roofs, and only a few computer simulations have been described. The purpose of this paper is to present initial details of a comprehensive computer model that is being developed to predict the thermal performance of earth covered roofs. Application of this model to the analysis of several roofs will demonstrate its utility. Kusuda and Achenbach [4], Shipp [5], Szydlowski [6] and Blick [7] have utilized finite difference techniques to make computer estimates of the heat transfer through earth covered roofs. Except for Blick, these authors have been concerned with the overall performance of an entire earth-sheltered building and, consequently, they have not focussed on the details of the roof component. Features, such as evapotranspiration, snow cover and high thermal mass, that distinguish the performance of earth covered roofs from traditional above-grade roofs have received very limited attention. Generally the model of heat transfer that has been used at the ground-air interface has consisted of a heat flux determined solely by the temperature difference between the air and the ground surface. None of the models referenced previously can easily predict the changes in the surface transfer of sensible heat that are associated with different types of ground cover or the latent exchange due to evapotranspiration.

The comprehensive model of heat transfer described in this paper:

(1) accurately calculates the influence of the thermal mass of the earth covered roof on the heat transfer,

(2) appropriately modifies the surface convection coefficient at the ground surface to account for changes in ground cover,

(3) performs a complete radiation balance calculation at the ground surface including both shortwave solar and longwave infrared radiation, and

(4) includes an experimentally validated model of potential evapotranspiration.

Following an outline of the elements of this detailed model as presented in the next section, simulation results for several roof compositions with varied thermal mass and surface covers will be presented.

COMPUTER MODEL

The computer program for predicting the dynamic thermal performance of earth covered roofs has been developed by integrating subroutines from the National Bureau of Standards Load Determination Program (NBSLD) [8] and algorithms from the American Society of Heating, Refrigeration and Air-Conditioning Engineers (ASHRAE) 1972 ASHRAE Handbook of Fundamentals. These procedures were then supplemented by additional calculation procedures for predicting heat transfer at the roof surfaces and are a direct extension of the work by Speltz [9]. Transient heat flow through the roof is calculated with an hourly time step using SOLMET solar-meteorological data inputs [10]. This section of the paper describes the general calculation procedures for each component of the estimation of the surface boundary conditions and reviews the overall process for calculating the roof heat transfer.

Figure 1 presents a hypothetical schematic of an earth covered roof section indicating the heat flow for a sunny summer morning when the roof is cooling the

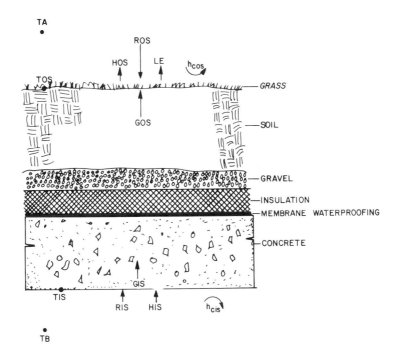

Fig. I. Schematic of earth covered roof cross-section defining the major components of heat flow, temperatures and surface coefficients.

building. The major components of the heat balance at the inside and outside surfaces are shown with arrows indicating the assumed directions of the one dimensional heat flow. The additional temperatures and heat flux coefficients will be
referenced in later discussions. At the upper surface, where the outside air
and the ground cover meet, the incoming energy must be balanced by the energy
leaving. Net incident radiation, ROS, combined with the heat conducted up from
the inside of the roof mass to the ground surface, GOS, must equal the sensible
energy convected into the outside air, HOS, plus the latent energy lost through
evapotranspiration, LE, i.e.

$$ROS + GOS = HOS + LE. \tag{1}$$

Note that the net incident radiation, ROS, is the algebraic sum of the incident
shortwave solar radiation, the outgoing shortwave radiation reflected from the
ground surface, the incident infrared radiation, the outgoing infrared radiation
reflected at the ground surface, and the infrared radiation emitted from the ground
surface. At the inside surface of the roof a similar heat balance occurs but with
fewer terms in the radiation component, and the latent exchange at this surface is
assumed to be zero. The following paragraphs in this section discuss the major
components of the heat balance.

Heat Conduction Through the Roof Mass

Transient heat conduction through the roof mass is calculated by use of conduction
transfer functions. One dimensional heat flow is assumed; edge and corner effects
are not considered. The methodology employed follows closely that of Kusuda [8,
11]. A thorough discussion of this method of solution is given in the cited
references. The major difference in the model developed for this study is
that the conduction transfer function computation does not include the surface
thermal resistance in order to allow the ground surface temperature to be computed
from a more sophisticated heat balance of all of the heat fluxes at this surface.
As explained in the references, and using the notation of Fig. 1,

$$GIS_t = \sum_{j=0}^{N} X_j \star TIS_{t-j} - \sum_{j=0}^{N} Y_j \star TOS_{t-j} + CR \star GIS_{t-1} \tag{2}$$

where GIS_t = heat flux into the roof at the inside surface at time t

TIS_{t-j} = inside surface temperature at time t-j

TOS_{t-j} = outside surface temperature at time t-j

GIS_{t-1} = heat flux into the roof at the inside surface for the
previous hour (i.e. time t-1)

N = number of the significant terms to be used for the conduction
heat transfer calculation

CR = common ratio of the conduction transfer functions, and

X_j and Y_j are conduction transfer functions, there being N X_j's and Y_j's

It should be noted that as the thermal mass of the roof is increased, the number N
of each type of conduction transfer function increases also. For an 8 inch concrete

slab covered by 2 inches of polystyrene with 12 inches of soil on top N = 75 conduction transfer functions.

Similarly at the outside of the roof

$$GOS_t = \sum_{j=0}^{N} Y_j \star TIS_{t-j} - \sum_{j=0}^{N} Z_j \star TOS_{t-j} + CR \star GOS_{t-1} \tag{3}$$

where the new variables are

GOS_t = heat flux arriving at the outer surface from the inside of the roof at time t

and Z_j is also a conduction transfer function.

Beginning with equation (1), the heat balance at outside surface, then by rearranging equation (1)

$$HOS_t = ROS_t + GOS_t - LE_t \tag{4}$$

and when the sensible heat transfer, HOS, is written in terms of the outside surface convection coefficient and the temperature difference,

$$HOS_t = h_{cos_t} \star (TOS_t - TA_t), \tag{5}$$

equations (4) and (5) may be solved with equation (3) to obtain

$$TOS_t = \left[ROS_t + h_{cos_t} \star TA_t + \sum_{j=0}^{N} Y_j \star TIS_{t-j} + CR \star GOS_{t-1} \right.$$

$$\left. - \sum_{j=1}^{N} Z_j \star OS_{t-j} - LE \right] / \left[h_{cos_t} + Z_o \right] \tag{6}$$

At the inside surface the heat balance is

$$HIS_t + RIS_t = GIS_t, \tag{7}$$

and since

$$HIS_t = h_{cis_t}*(TB_t - TIS_t) \, ,$$
(8)

solving equations (7), (8) and (2) for the inside surface temperature, TIS, gives

$$TIS_t = \left[h_{cis_t}*TB_t + \sum_{j=0}^{N} Y_j*TOS_{t-j} - CR*GIS_{t-1} \right.$$

$$\left. - \sum_{j=1}^{N} X_j*TIS_{t-j} \right] / \left[h_{cis_t} + X_0 \right]$$
(9)

Calcuation of the net radiation balance, ROS, the surface convection coefficients h_{cos} and h_{cis}, and the evapotranspiration are discussed in succeeding paragraphs.

Radiation Balance at the Outside Surface

Using the notation of Sellers [12] the net radiation at the surface of the ground can be written

$$ROS = (Q + q) (1 - \alpha) - I$$
(10)

where $Q + q$ is the sum of the direct and diffuse solar radiation incident on the surface of the ground, α is the surface albedo, and I is the effective outgoing infrared radiation from the ground surface. In our case, assuming a horizontal surface, $Q + q$ is the global solar radiation, and hourly values may be read directly from SOLMET weather tapes. The surface albedo, α, or solar reflectivity is a paramater characteristic of the different ground cover. For example, approximate values for α are: $\alpha = 0.3$ for grass, $\alpha = 0.8$ for snow and $\alpha = 0.2$ for asphalt.

Again following Sellers

"The effective outgoing radiation from the earth's surface consists of two basic components, the total longwave energy sent out from the surface I↑, which is a function of the surface emissivity ε and temperature T_s, and the counter radiation from the atmosphere I↓ which is a function primarily of the air temperature T_a, the precipitable water vapor w_a, and the cloud cover n. Thus,

$$I = I↑ - I↓ = f (T_a, T_s, \varepsilon, W_a, n). \text{"} ‡$$
(11)

In the model developed for this study, infrared radiation heat exchange between the earth covered roof and the sky was estimated using the statistical method developed by Clark and Allan [13]. This analysis includes the effects of clouds and humidity on the apparent emissivity of the sky. Their equation for the effective outgoing radiation is

$$I = \varepsilon [\sigma TOS^4 - S]$$
(12)

The outgoing infrared radiation is $I↑ = \varepsilon\sigma TOS^4$, and the incoming infrared radiation is $I↓ = \varepsilon S$, where ε is the surface infrared emissivity (approximately equal to 0.9) and σ is the Stephan-Boltzman constant. The downcoming infrared

‡Sellers, William D., Physical Climatology, the University of Chicago Press, Chicago & London, 1965, p. 47

irradiance was calculated from commonly available meteorological parameters using the following results from Clark and Allan:

$$I\downarrow = \varepsilon(1.0 + 0.0224*n - 0.0035*n^2 + 0.00028*n^3) *$$

$$\left[0.787 + 0.764*\ln\frac{TDP}{273}\right] * \sigma *TAK^4*0.31695 \tag{13}$$

where   n = opaque sky cover (units from 0 to 10, 10 being fully overcast)

TDP = the dew point temperature (°K)

TAK = the ambient air temperature (°K)

and   ln is the natural logarithm.

Surface Convection

The sensible heat flux between the ground surface and the air due to surface convection is estimated with the equation

$$HOS = h_{cos} (TOS - TA) \tag{14}$$

Using the empirical model developed by Kreith and Sellers [14] the surface convection heat transfer coefficient is expressed as

$$h_{cos} = \rho\ C_p\ D_h \tag{15}$$

where ρ = air density

C_p = specific heat

and D_h = nebulous transfer coefficient.

The nebulous transfer coefficient, D_h, is estimated by the relationship

$$D_h = D_{ni} = 0.16\ u\ (\ln\frac{L}{Lo})^{-2} \tag{16}$$

for stable conditions, i.e. where the air temperature does not vary greatly with height due to an overcast sky. In equation (16) u is the wind speed, L is the height at which the wind speed is measured, and Lo is the roughness length, a parameters characterizing the texture of the surface. When the air temperature does vary with height due, for example, to the heating of the air near the ground by solar radiation on a clear day, D_h takes the form

$$D_h = D_m (1 + 14\ \frac{TOS - TA}{u^2})^{1/3} \tag{17}$$

under unstable conditions when TG - TA > 0. For stable conditions (TG - TA < 0)

$$D_h = D_m \left(1 - 14 \frac{TOS - TA}{u^2}\right)^{-1/3} \qquad (18)$$

The convective heat transfer coefficient at the outside surface of the roof, h_{cos}, is calcuated each hour as a function of the horizontal wind speed, u, and the roughness length of the surface. For example, an earth covered roof may be characterized by a roughness length of $.32 \leq Lo \leq .75$ in the summer months depending on the length of the grass which during winter the snow cover would be characterized by a roughness length $Lo = .03$.

Surface convection at the inside surface of the roof is modeled by

$$HIS = h_{cis} (TB - TIS) \qquad (19)$$

where h_{cis} is the combined surface convection coefficient and linearized radiation coefficient on the inside surface. Values of h_{cis} were obtained from a thorough investigation by Buchberg [15]

Evapotranspiration

To do more than direct the interested reader to the relevant literature on the complex subject of evapotranspiration is beyond the scope of this paper. A helpful paper by Kreith and Sellers [14] and a book edited by Jensen [16] provide many references and Jensen's book has useful data.

For the purposes of this paper only the theoretical maximum possible evapotranspiration, termed potential evapotranspiration, is examined. This assumes that there is sufficient moisture available at the ground surface so that evapotranspiration can proceed at the maximum rate. Such a condition would occur, for example, at the surface of a typical lawn or a well irrigated field.

Following Sellers [12], the potential evapotranspiration, LE, may be written

$$LE = \frac{(ROS + GOS)\Delta}{\Delta + \gamma} \qquad \rho\, C_p\, D_h\, (TA - TW) \qquad (20)$$

The new variables are:

$$\Delta = \frac{Rd \star TA^2}{0.622 \star LW \star e_{sa}}$$

where   Rd is the gas constant for dry air,

   e_{sa} is the saturation vapor pressure at the air temperature,

   LW is the latent heat of vaporization for water,

$$\gamma = \frac{C_p \star P}{0.622 \star L}$$

with P the atmospheric pressure,

TW is the wet-bulb temperature,

and the other variables have been previously defined. (The book edited by Jensen has table of $\frac{\Delta}{\Delta + \gamma}$.)

From equation (20) it can be seen that the first term generates a latent energy exchange that reduces the warming of the ground by the net radiation, ROS, and the heat conducted to the ground surface from below, GOS. The second term convects latent energy away from the surface until the wet-bulb temperature is approached. In the next section, results for earth covered roofs with this level of potential evapotranspiration will demonstrate the efficiency of this mechanism in lowering the surface temperature of the roof and thereby reducing the required summer cooling associated with the roof component.

RESULTS

Prior to presenting the relatively complex results of the computer analysis of earth covered roofs, the results of the application of the computer model to a simple homogeneous roof with no plant cover are discussed. Examination of the roof heat transfer in this situation, where the analysis has been uncomplicated through elimination of the evapotranspiration from the plant cover at the upper surface of the roof, facilitates presentation of the thermal mass effects as modeled by this computer program.

Homogenous Roof - No Evapotranspiration

On Fig. 2 the hourly heat fluxes entering the inside surface of three different roof slabs, as predicted by the computer simulation, are plotted for three August days in Minneapolis, Minnesota. The hourly outside air dry-bulb temperatures are plotted above the values of heat flux. Each roof slab has the same steady-state thermal resistance (thermal resistance equals the thickness of the slab multiplied by the thermal resistivity of the material.) However, each roof slab has a different density and, consequently, a different volumetric heat capacity (volumetric heat capacity equals density times specific heat) and a different thermal diffusivity (thermal diffusivity equals conductivity divided by volumetric heat capacity). By examining the temporal response of the heat flux for each roof, the impact of changing the density of the material (hence changing the thermal mass and also the thermal diffusivity) while holding all other independent variables constant is readily observed. Increasing the density of the material in the roof slab increases the thermal mass and decreases the thermal diffusivity. The thermal diffusivity governs the rate at which changes in heat flux travel through the roof.

The blackened circles on the heat flux plot of Fig. 2 are computed results for the response of a standard density, 8 inch thick concrete slab. The thermal diffusivity for this material is 0.0268 ft^2/hr. The unblackened circles on this figure are the computed response for a slab of a fictitious material that is ten times less dense than the standard concrete, but with the same thickness, conductivity and specific heat. The thermal diffusivity for this less dense material is 0.2678 ft^2/hr.; it is ten times greater than for the standard concrete. Comparing these two materials, note how closely the low density material (thermal diffusivity = 0.2678) mirrors the difference between the outside dry-bulb temperature and the constant inside air temperature. The standard concrete, which has a ten times higher density (and a correspondingly lower thermal diffusivity = 0.0268) responds at a slower rate. There is a noticeable phase shift between the responses of the two roofs. Where the heat flux through the low density (high thermal diffusivity) roof slab follows the difference between the outside air temperature and the constant inside air temperature very closely, the relatively higher density (lower thermal diffusivity) roof slab responds at a slightly later time. The peak amplitude of the heat flux is also reduced for the higher density slab.

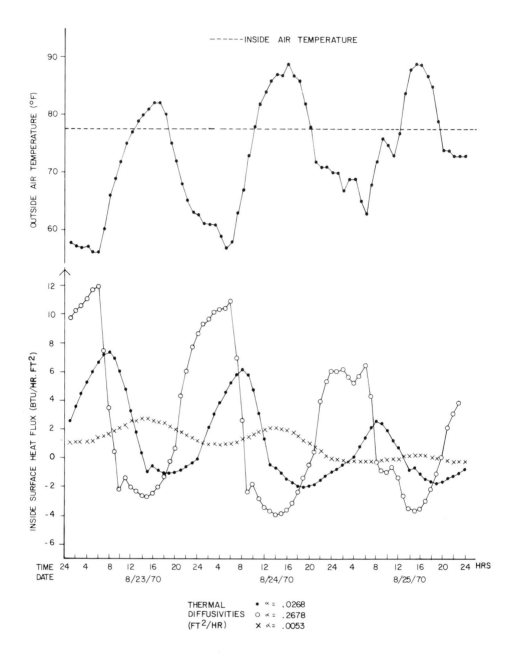

Fig. 2. Summer heat flow through three uninsulated slabs with constant thermal resistance and different thermal diffusivities. Thickness = 8 in., Conductivity = 0.75 BTU/hr. ft. °F, specific heat = 0.20 BTU/lb., density = 14, 140, or 700 lbs/ft³, surface absorptivity = 0.85, and infrared emissivity = 0.90.

Heat flux results for the highest density slab shown in Fig. 2 corresponds to the case with the smallest thermal diffusivity. The response of this roof slab exhibits a marked phrase shift of nearly four hours compared to the response of the roof slab with the lowest density. This high density roof has a thermal diffusivity of 0.0053 ft^2/hr which is one fiftieth of the lowest density roof plotted in Fig. 2.

These changes in phase of the thermal response due to the changes in thermal mass and concomitant changes in thermal diffusivity illustrate the capability of the computer program to effectively model these characteristics. Noting that the steady state thermal resistance, commonly designated as R value, is the same R = 0.89 hr*ft^2*°F/BTU for each roof slab, the reductions in peak heat flux out of the building as well as the potential benefits associated with reductions in the peak heat gains to the building that are predicted by the computer model must result from the differences in thermal mass. For the standard 8 inch concrete roof slab modeled to obtain the data for Fig. 2, the heat flux is out of the building for roughly the same number of hours as heat is flowing into the building during this three day summer period. By increasing the thermal mass to the highest value plotted in Fig. 2 (thermal diffusivity = .0053), the heat flux is nearly always out through the roof of the building. This computer model could be used to determine the appropriate roof configuration to maximize this beneficial summertime cooling response.

Figure 3 further defines the response of these three roof slabs by presenting their response to the lower frequency transients characteristic of Minnesota winters. During the first two days plotted in this figure, a diurnal outside dry-bulb temperature swing does not occur. Instead a 24 hour period of relatively constant outside air temperature is followed by a frigid period from the afternoon of January 4 to mid-morning on January 5. Note how the different slabs respond to this change at different rates. Note also that they reach different peak heat flux values. The lowest density slab reaches its peak loss nearly in phase with the lowest outside temperature while the standard density concrete slab (thermal diffusivity = 0.0268) exhibits a definite phase lag and a somewhat reduced peak load. However, examination of the heat flux response of the roof slab with the highest thermal mass reveals no peak at all.

Earth Covered Roofs

Figures 4 and 5 present computer predictions of the thermal response of four different roofs during three summer days in Minneapolis, Minnesota. Each of the roofs has a steady state thermal resistance of R = 24. The four roofs studied are a traditional low mass built-up roof, an insulated 8 inch poured concrete roof, an earth covered roof with no evapotranspiration, and an earth covered roof with evapotranspiration occurring at the maximum potential rate. The heat flux at the inside surface of each roof, the outside surface temperatures for all but the insulated 8 inch concrete slab roof, the global solar insolation and the cloud cover are plotted on Fig. 4. On an expanded temperature scale, the ground surface temperature data for the earth covered roofs operating with and without potential evapotranspiration are repeated on Fig. 5. Outside air dry-bulb temperature is also included on Fig. 5.

The roof heat loss data on Fig. 4 clearly demonstrates one of the effects of increasing the thermal mass of the roof. Comparing the three roofs without evapotranspiration, the expected reductions in diurnal peak heat flow are readily observed. The different phase shifts expected with increased thermal mass are, however, not so apparent because the thermal diffusivity of each of the different composite roofs is already very small due to the insulation. Note that the phase shift between the minimum surface temperature (or minimum outside air temperature)

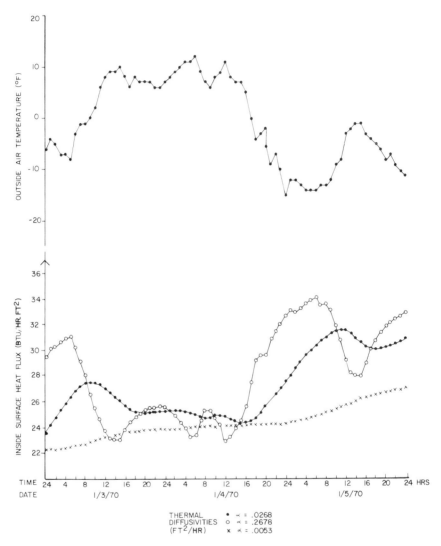

Fig. 3. Winter heat flow through three uninsulated slabs with constant thermal resistance and different thermal diffusivities. Thickness = 8 in., Conductivity = 0.75 BTU/hr.ft.°F, specific heat = 0.20 BTU/lb., density = 14,140,or 700 lbs/ft³, surface absortivity = 0.85, and infrared emissivity = 0.90.

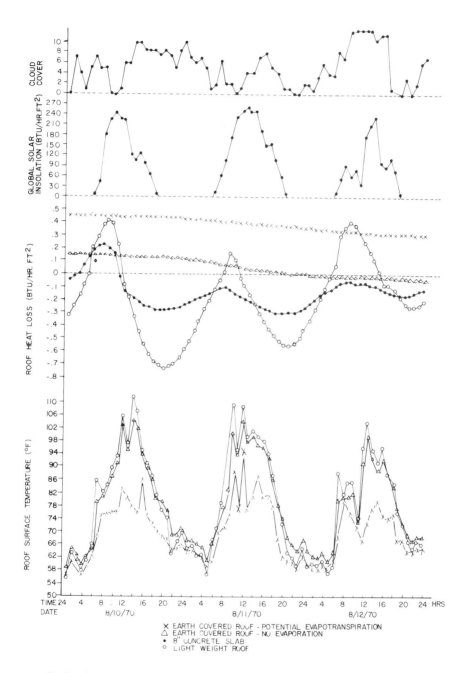

Fig. 4.   Characteristics of earth covered roofs compared to traditional roofs for a summer period in Minneapolis, Minnesota.

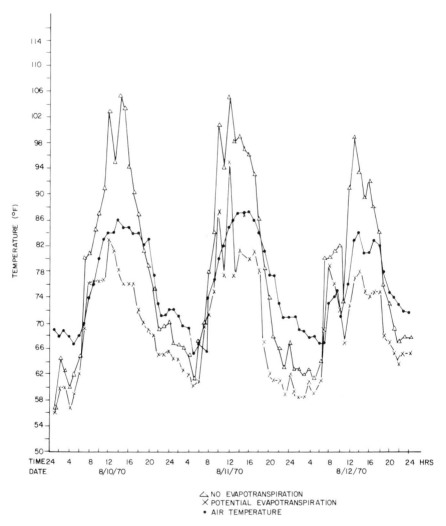

Fig. 5. Ground surface temperature for earth covered roofs with and without evapotranspiration, Minneapolis, Minnesota.

and the peak heat loss for the low mass roof (and for the insulated concrete roof) is already several hours. The large thermal mass of the earth covered roofs (12 inches of soil, 2 inches of rigid polystyrene insulation, and 8 inches of concrete) completely averages away the diurnal peaks.

Comparing the heat loss for the two earth covered roofs, the computer predicted positive benefit of evapotranspiration in reducing summer cooling load is seen. Because the evapotranspiration at the surface of the ground often lowers the surface temperature below the dry-bulb air temperature, and, for most of the three days plotted in Fig. 5, often below the inside air temperature, heat may be continually lost from the inside of the building to the earth covered roof. Occasionaly increases in outside surface temperature due to relatively high air temperature or incoming radiation are averaged out by the thermal mass of the roof.

CONCLUSIONS

A comprehensive computer model for predicting the thermal performance of earth covered roofs has been developed. The physically reasonable results obtained from applying this simulation procedure to several test cases indicates the possible validity of the approach. Thermal mass of the roof, insulating properties of the roof materials, surface heat convection, surface radiation exchange and potential evapotranspiration from the plant cover have been included in the model. In addition to the thermophysical properties of the roof materials, only readily available hourly weather data is needed for input to the computer model. Comparisons of this model with experimental results on earth covered roofs, and use of this model to estimate the annual thermal performance of earth covered roofs, including heating and cooling benefits, can now be performed.

ACKNOWLEDGEMENTS

The authors would like to thank Charles Fairhurst, Anthony Starfield and Alan Wassyng of the Department of Civil and Mineral Engineering for the use of the departments PDP 11/60 computer on which these computer studies were performed. Donald Slack of the Department of Agricultural Engineering and Donald Baker of the Department of Soil Science at the University of Minnesota were helpful in clarifying details of the anlaytical models of evapotranspiration.

George Meixel's work on this project has been supported by the United States Department of Energy under contract DE-AC03-80SF11508. The United States Navy, Northern Division, and the Legislative Committee on Minnesota Resources under a grant to the Underground Space Center for underground building design have supported the work of Jerome Speltz.

NOMENCLATURE

n   cloud cover, 0-10

C_p specific heat

CR   common ratio of transfer functions

D_h nebulous transfer coefficient

G1S   heat flux into the roof at the inside surface (BTU/hr*ft^2)

GOS   heat flux into the roof at the outside surface (BTU/hr*ft^2)

h_{cis} combined surface convection coefficient and linearized radiation coefficient (BTU/hr*ft^2*°F)

h_{cos} convection heat transfer coefficient at the outside surface (BTU/hr*ft^2*°F)

HIS   sensible heat flux between the inside surface and air (BTU/hr*ft^2)

HOS   sensible heat flux between the outside surface and air (BTU/hr*ft^2)

I   effective outgoing infrared radiation from the ground surface (BTU/hr*ft^2)

$I\uparrow$ outgoing infrared radiation from the ground surface (BTU/hr*ft^2)

$I\downarrow$ incoming infrared radiation from the sky (BTU/hr*ft^2)

L   height at which the horizontal wind speed is measured

Lo   roughness length

LE   latent heat of evapotranspiration (BTU/hr*ft^2)

LW   latent heat of vaporization for water

N   number of each type of conduction transfer functions

P   atmospheric pressure

Q   direct solar radiation incident on the surface of the ground (BTU/hr*ft^2)

q   diffuse solar radiation incident on the surface of the ground (BTU/hr*ft^2)

Rd   gas constant for dry air

RIS   net incident radiation at the inside surface (BTU/hr*ft^2)

ROS   net incident radiation at the outside surface (BTU/hr*ft^2)

TA   ambient dry-bulb air temperature, °F

TAK   ambient dry-bulb air temperature, °K

TB   inside dry-bulb air temperature, °F

TDP   dewpoint temperature of the ambient air, °F

TIS   inside surface temperature, °F

TOS   outside surface temperature, °F

TW   ambient thermodynamic wet-bulb temperature, °F

u   horizontal wind speed

X_j, Y_j, Z_j conduction transfer functions

ε surface infrared emissivity

σ Stephan - Boltzman constant

ρ air density

e_{sa} saturation vapor pressure at the air temperature

REFERENCES

1. Underground Space Center, Earth Sheltered Housing Design, Van Nostrand Reinhold Company, New York, NY (1978).

2. Labs, Kenneth, Earth Tempering as a Passive Design Strategy, Proceedings of the 1st Earth Sheltered Building Design Innovations Conference at the Skirvin Plaza Hotel, Oklahoma City, Oklahoma, April 18-19, 1980.

3. Metz, Don, The Latest - Not the Last - Word in Underground House Design, Solar Age, October, 1980.

4. Kusuda, T. and Achenbach, P.R., Numerical Analysis of the Thermal Environment of Occupied Underground Spaces with Finite Cover Using a Digital Computer, ASHRAE Transactions, Vol. 69 (1963).

5. Shipp, P.H., Thermal Characteristics of Large Earth-Sheltered Structures, Ph.D. Thesis, University of Minnesota (1979).

6. Szydlowski, R.F., Analysis of Transient Heat Loss in Earth Sheltered Structures, Ph.D. Thesis, Iowa State University (1980).

7. Blick, Edward F., A Simple Method for Determining Heat Flow Through Earth Covered Roofs, Proceedings of the 1st Earth Sheltered Building Design Innovations Conference at the Skirvin Plaza Hotel, Oklahoma City, Oklahoma, April 18-19, 1980.

8. Kusuda, T., NBSLD, The Computer Program for Heating and Cooling Loads in Buildings, NBS Building Science Series No. 69, Washington, D.C. (1969).

9. Speltz, J., A Numerical Simulation of Transient Heat Flow in Earth Sheltered Buildings for Seven Selected U.S. Cities, M.S. Thesis, Trinity University (1980).

10. SOLMET - User's Manual, Vol. I, TD-9724, Hourly Solar Radiation - Surface Meteorological Observations, prepared by the U.S. Department of Commerce National Oceanic and Atmospheric Administration Environmental Data Service, National Climatic Center, Asheville, N.C., December, 1977.

11. Kusuda, T., Thermal Response Factors for Multi-Layer Structures of Various Heat Conduction Systems, ASHRAE Transactions, pp. 250-269, Part I, 1969.

12. Sellers, W.D., Physical Climatology, University of Chicago Press, Chicago, (1965).

13. Clark, G. and Allan, C.P., The Estimation of Atmospheric Radiation for Clear and Cloudy Skies, Proceedings of the 2nd National Passive Solar Conference, Vol. 2, Philadelphia, PA, 1978, pp. 676-679.

14. Kreith, F. and Sellers, W.D., General Principles of Natural Evaporation, Heat and Mass Transfer in the Biosphere, Part I Transfer Processes in the Plant Environment, Scripta Book Company (1975).

108

15. Buchberg, H., Sensitivity of Room Thermal Response to Inside Radiation Exchange and Surface Conductance, Building Science, Vol. 6, 1971, pp. 133-149.

16. Jensen, M.E. (ed), Consumptive Use of Water and Irrigation Water Requirements, American Society of Civil Engineers, New York, (1973).

EARTH SHELTERING IN THE EASTERN UNITED STATES

John Barnard
Architect
Osterville, MA

When late last fall I received a call from Dr. Ray Sterling, Director
of the Underground Space Center, University of Minnesota requesting
my participation in a June U.S.C.E. Space Conference, I was delighted
because it would not only give me a valid excuse to get out of the
office and meet with others who share my interest, or rather passion,
for the underground, but it would provide an opportunity to leave my
own narrow sphere and take a long hard look at the triumphs, trials,
and tribulations of others. Also, I'm afraid I must admit that back
in November, June was a lifetime away so that I would of course have
more than enough time to leisurely compile the required information
for a relaxed presentation.

Now with only about 10 days to my "Photo Ready" copy deadline for "Construction Activity and Innovations of Earth Sheltered Buildings in
Eastern U.S.", I sit at my desk surrounded by photos, brochures, and
miscellaneous bits of information trying to run down last minute leads
and separate proposed and shelved projects from those completed or
under construction.

I estimate that there may be as many as 500 underground houses either
occupied or under construction in the Eastern portion of the United
States. As it is impossible and basically meaningless to even attempt
a compilation of individual homes, I am limiting this paper to the work
of professionals, architects, builders, and authors which may be considered a resource for those looking to the better use of underground
space.   The range of the projects - Civil Defense, Telecommunications,
Commercial, Educational and Residential is intentional.

Time, space, and my limited research facilities will make omissions
inevitable. To those I've missed, my sincere apologies. Unfortunately
photographs could not be obtained for all projects.

DON METZ

Don Metz, a New Hampshire Architect/Builder is a pioneer in the earth shelter residential field. Starting with his award-winning "Winston House", built in 1972 and continuing to this date.

He has developed a stock house design of 2,000 sq. ft. which he calls Earthtech. These plans are available at nominal costs. In the past two years alone, he has eight earth sheltered houses to his credit. His most recent houses feature a double floor heat storage air distribution system, composed of two concrete slabs with 8" block spacers. His roof systems are of heavy timber with approximately 12" of earth covering.

For waterproofing, Mr. Metz uses Bituthene either by itself or in conjunction with Volclay Panels or Carlisle Butyl Rubber.

Mr. Metz has conducted seminars on earth sheltering at Dartmouth, Yale, Columbia University, University of Montreal and many other locations.

He is currently engaged in writing a book, which will be published by Gardenway, "High Performance Houses".

Fig. 1

Interior view
Living Room &
Dining Room
Earthtech 6
looking south.

Don Metz, Architect, Pinnacle Road, Lyme, New Hampshire 03768

THE DEWOLFF PARTNERSHIP

This firm has two underground houses; one has been occupied for two years, and construction has just started on the other. These are some-what unusual in that they are both north-facing and do not utilize passive solar.

The completed house has been monitored for two years and a saving of 68% energy is reported over a comparable, fully insulated 2800 sq. ft. above-ground house. The architects feel that there is so little sun in the Buffalo area that spending a great deal of money to utilize pas-sive solar could not be justified.

The cost of the first house was 18% more than a comparable above-grade house with the second one running about 12% more.

This firm has two more underground houses on the board at present.

The DeWolff Partnership, A.I.A., 530 Fairport Office Park, Fairport, N.Y. 14450.

HAVEN EISENBERG ASSOCIATES, INC.

Haven Eisenberg, a Boston architect was responsible for one of the very early underground structures on the East coast. That was the Emergency Operations Center for the Massachusetts Civil Defense Agency in Framing-ham, Massachusetts. This was completed in 1963 at a cost of $2,500,000. The building is 220' x 160' in size and covered with 4 to 5 feet of com-pacted earth. The concrete roof is three feet thick and the walls are twenty inches thick. It was designed to withstand the equivalent sta-tic pressure of 30 lbs. per sq. in. (blast affect of a 20 megaton bomb, three miles distance). This self-contained disaster command center provides for the needs of a staff of 300 for a minimum of two weeks. The building is an early example of the use of a well water to air heat pump for economical heating of the building.

All air intake and exhaust openings are fitted with valves that auto-matically close when the atmospheric pressure increases by 2 lbs. per sq. in., so that these will be protected when the main shock wave reaches the shelter. The shock wave following a nuclear blast is expected to expose a building to an acceleration of 12 times gravity. Provision has been made throughout the building to absorb this tremendous shock, so that all mechanical and electrical equipment are mounted on vibra-tion insulators.

Haven Eisenberg Associates, Inc., 29 Temple Place, Boston, MA. 02100

WILLIAM MORGAN, ARCHITECTS

In my opinion, the most dramatic use of underground space has been made by William Morgan F.A.I.A., an architect from Jacksonville, FL. His well-published seaside duplex, built in 1975 at a cost of $25,000 per unit, is both practical and imaginative. Although small in area (750 sq. ft. for each side) these apartments contain commodious and luxurious living quarters consisting of a living room, kitchen, sleeping balcony, bath, entry, and terrace. Constructed on two levels, the house is as nearly concealed into the dune as it is humanly possible. The only exterior clue of the living inside is the entrance way on the land side and a porthole for each unit which opens onto a terrace on the beach side.

Structurally, the walls and ceiling are vaulted and are constructed of gunnite concrete sprayed to a thickness of 2 1/2" over a base of reinforcing steel and post tensioned by the backfill of sand. The interior of the concrete is painted, the exterior is waterproofed and there is no other insulation other than that naturally provided by the earth cover.

Mr. Morgan's second house, no less dramatic in appearance, consists of a truncated earth covered pyramid containing study, reflecting pool, dining/kitchen and sleeping areas at the lower level, with an elevator and stairs leading to an observation living room which sits on the top of the truncated pyramid. The pitch of the roof carries out the pyradimal effect of the structure. Located in central Florida on one of the few high points, this building looks down 230 feet to the citrus groves below.

Fig. 2

Night view,
Dune House
from Patio.

William Morgan, Architects, Professional Association, 20 E. Forsyth Street, Jacksonville, FL   32202.

KENNETH LABS, ARCHITECT

No discussion of the underground movement would be complete without
giving full credit to Ken Labs, lecturer, author, researcher, and consultant, whose considerable contributions should never be under estimated. I, for one, am constantly indebted to Ken for help with some
of my more vexing problems.

Ken Labs is a design consultant and researcher specializing in problems of climate, site planning, and underground building. He is the
author of over 20 articles and professional papers dealing with various
issues in environmental design and planning. Among these are suggested
land use regulations for underground housing published by the American
Planning Association.

Mr. Labs is co-author with Donald Watson, A.I.A., of Climatic Design
for Home Building, a book prepared for the National Association of
Home Builders Research Foundation, with the support of the Department
of Housing and Urban Development. He has also contributed chapters to
the books, Energy Efficient Buildings, edited by Walter F. Wagner, Jr.
(Architectural Record Books, New York), and Housing in Arid Lands:
Design and Planning, edited by Dr. Gideon S. Golany (The Architectural
Press, Ltd., London).

Mr. Labs is the recipient of a 1981 Individual Project Fellowship from
the National Endowment for the Arts, which is intended to support work
toward a manual of climate design methods and data for architects. He
is presently completing work on Architecture Underground, a design survey and desk manual for design professionals to be published by the
McGraw-Hill Book Company, Architectural Record Books Division.

Ken Labs may be contacted through UNDERCURRENT, 147 Livingston Street,
New Haven, CT.   06511

EARL R. FLANSBURGH

This Boston Architect is the designer of Cornell University's underground campus store. This, in common with the Pusey Library and Northeastern's new law school, utilizes the space between existing buildings
without interfering with the visual openness of the area. When it became necessary to build a larger campus store, the idea of a walking
campus was in the architect's mind. The store would have to be cental-
ly located, but again no central space was available. Thus a most happy use of underground space came into being.

The building which is cast in place concrete set on bedrock and with
two feet of earth covering was completed in 1976 at a cost of $1,702,000
including equipment, finishing fees, and landscaping.

Earl R. Flansburgh & Associates Inc., 77 N. Washington Street, Boston,
MA.   02100

HUGH STUBBINS AND ASSOCIATES

Their contribution to the earth shelter field is the prestigious Nathan
Marsh Pusey Library at Harvard University in Cambridge, Massachusetts.
This building containing 87,000 sq. ft. of floor area was completed in
1976 at a cost of $5,653,000.

The roof of this building is covered with plantings, grass and paths,
and features a central garden which serves as a light court to introduce natural lighting for the building. Side windows are concealed
in the sloping berms of two sides.

Originally the proposed library was to be completely underground, but
due to new concepts of landscaping the building was allowed to project
slightly above ground. The resulting design provides a bright daylight
interior and opened up new vistas in the yard.

Fig. 3   Nathan Marsh Pusey Library, Harvard University

Hugh Stubbins and Associates, Inc. Architects, 1033 Massachusetts
Avenue, Cambridge, MA.   02138

HERBERT S. NEWMAN ASSOCIATES

This is an architectural firm in New Haven, Connecticut which is currently involved in the building of a new school of law at Northeastern
University in Boston, Massachusetts. When completed, this will increase
the sq. ft. available to the law school from 27,800 to 47,000 sq. ft.
The concept, in many respects, has similarities to Hugh Stubbins' Pusey
Library at Harvard, making use of two large central courts to provide
light to the space below as well as perimeter windows. The roof of
this structure being raised slightly above the existing grade will
provide an attractively landscaped plaza or park, with trees, planting, and seating areas for conversation and contemplation. One of
the two courtyards is encompassed by the Administrative Office and
the other by the Faculty Office and Lounge. Both introduce natural
light into the areas surrounding them.

To quote Judith O. Brown, professor of law, "What an exciting prospect
it will be, to teach and study in a civilized urban place where people
will congregate because it is aesthetically appealing and academically
satisfying. How can we escape being communicative in a place so inviting for thinking and thinking out loud?"

Fig. 4   Model of New Law School, Northeastern University now under
           construction. NOTE: Landscaped Courtyard light wells.

Herbert S. Newman, Associates, 300 York Street, New Haven, CT. 06511

DOUGLAS ORR, WINDER AND ASSOCIATES, ARCHITECTS

This firm designed the earth sheltered structure which houses the
first Emperor Tandem Van de Graff accelerator on the campus of Yale
University. This building, completed in 1964, provides laboratory as
well as office facilities. To protect the staff from radiation, the
walls between the accelerator vault and the rest of the lab are five
feet thick and boron-loaded for additional shielding. Also for the
protection of the personnel, the vault roof is 3 feet thick on concrete, which together with the earth berms, provides the shielding
for the exterior walls. The moving of 40,000 cubic yards of earth
took three months and accounted for approximately 7% of the total
cost of the job. It was indeed a mamouth undertaking.

Because the steepness of the sides would make a grass cover impractical, the ground was covered with honeysuckle which not only prevent
the erosion of the soil cover but enhances the appearance of the stru●
ture. While on the subject of appearance, this attractive yet utili-
tarian building, to me, is most reminiscent of some of the early Maya
structures.

Fig. 5

Douglas Orr, Winder & Associates, Architects, 85 Willow Street, New
Haven, CT.   06511.

PHILIP JOHNSON, ARCHITECT

A discussion of underground buildings of the East Coast would not be
complete without the inclusion of Mr. Johnson's famous art gallery
built on the grounds of his home in New Canaan, CT. The use of underground space was decided on for two reasons. Mr. Johnson just didn't
want to have another building in his backyard, and the insulation provided by building underground made it simple to provide for nearly constant 50% humidity and 70° year-round temperature for the protection
of his valuable collection.

The gallery is as closed and free from contact with the outside world
as his famous glass house is open. The architect was interested in
the aesthetic effect of a mound as an art form, the psychological effect
of a bermed building, and the dramatic effect of this most elegant cave.

Fig. 6

Underground Art
Gallery
New Canaan, CT.

Johnson/Burge, Architects, 375 Park Avenue, New York, N.Y.   10152

J.L. HARTER, ASSOCIATES

A Pennsylvania Architectural firm which are designers for the operation center for the Rodale Press Inc. of Emmaus, PA. This complex, comprising a gross area of 46,000 sq. ft. and a new office area of 36,000 sq. ft. is the operation center's function for all magazine and book subscriptions for this international publishing firm.

This building is earth bermed to the roof line and is oriented for maximum use of passive solar, which is augmented by solar air collectors charging a 5000 cu. ft. thermal heat sink. Another feature is natural building ventilation developed by a landscaped cold sink, also utilizing the envelope. Windows and skylights are shuttered against the cold with insulating panels.

The architect has been honored with the 1980 Interior Magazine trophy, for the best in energy efficient designs.

Fig. 7

Operations Center
Rodale Press Inc.
Emmaus, PA.

J.L. Harter Associates, 41 S. Tenth Street, Allentown, PA. 18109

DAVIS, SMITH, AND CARTER, ARCHITECTS

This architectural firm of Reston, Virginia is responsible for five
underground elementary schools in the Eastern United States. The first
being the "Terraset" Elementary School in their home town of Reston.
The basic plan consists of classrooms with three feet of earth cover
and glass walls opening onto a central courtyard or patio. The patio
is partially shaded by a steel network which supports a battery of
solar collectors. In addition to forming an interesting pattern of
light and shade in the courtyard, these collectors contributed $14,400
worth of heating and cooling in one year alone.

(Note: Ironically enough, funds for the solar system were provided by
the Al Dyr Iyyah Institute and University of Petroleum and Minerals at
Dharahn, under the auspices of the Saudi Arabian government.) The
school is operating, at the projected energy savings of about 75%;
Enough savings to quickly pay back the 3 or 4% construction penalty
for going underground.

Construction on this school was started in 1975. Since its completion
there have been three more built in Fairfax County, Virginia and one
in Kissimmee, Florida. All these schools utilize the same basic design
concept.

Davis, Smith, and Carter, Architects, Reston, VA. 22070

ENERGY EFFICIENT ENVIRONMENTS, INC.

Atlanta based, this architectural/building firm is the most prolific
in the underground field of any of the groups that I have surveyed to
date.

They report the completion through last year of 40 underground homes
and that they had, at the present, 20 under construction.

The single largest problem which confronts the earth sheltered architect is that of procuring qualified as well as enthusiastic builders
for their projects. I have seen several very good jobs go down the
tube because of the lack of skilled contractors. Energy Efficient
Environments have solved this problem by the use of an in-house construction crew.

The principals of the firm are Architect Jack Schupp, and the founder/
builder Dick Lang whose experience in underground goes back for 6 years.

In size, their houses range from 1200 to 2500 sq. ft. with a cost of
$32 to $45 per sq. ft., not including land.

For further information contact: Energy Efficient Environments, Inc.,
2531 Briarcliff Road, Suite 212, Atlanta, GA. 30329

120

JOHN E. BARNARD, JR., A.I.A.

Mr. Barnard has long been recognized as the third of the triumvirate of East Coast pioneers in the underground housing field. His "Ecology House" in Marstons Mills, MA. has achieved world-wide acclaims since its completion in 1973. His designs include both atria and elevationa styles with many being a combination of both.

Mr. Barnard has sold approximately 175 sets of one, two, and three bec room stock and/or custom design house plans to date. Of these, more than a quarter are either under construction or have been completed.

Fig. 8   Exterior View of Elevational House, combining covered Atrium design.

Fig. 9   Interior of covered Atrium, maintains min. 40°F in 0° weather, due to greenhouse effect.

John E. Barnard, Jr., A.I.A., Race Lane, Marstons Mills, MA.   02648

SWANKE, HAYDEN, CONNELL & PARTNERS

This New York Architectural firm has been involved in underground buildings since 1959. Theirs is a specialized business and they have completed 69 hardened underground projects in the U.S. and abroad.

The majority of these were for telecommunications facilities. There were various proto-type structures developed which then became a series of building types which were constantly upgraded and became more sophisticated as the art and science of nuclear protection advanced. The knowledge of materials for below ground usage became more complete; and, success with their application, more accepted.

These buildings were reinforced concrete, founded some 40 to 60 feet below grade with an earth cover of two to four feet. All ventilation and egress openings were protected by special blast-resistant enclosures. Siting is critical for sub-surface drainage and a perimeter drainage system of porous backfill with porous wall piping directed to an outfall downgrade and away from the building was always included.

Waterproofing consisted of tar pitch and saturated felt tops and sides; five plied top, three plied sides. Sub-surface water was avoided at all costs to a point where special deep work slabs or coffer dams of clay, bentonite, or other impervious soils were used.

In addition to their home office at 400 Park Avenue, New York City, they maintain branches at 1333 New Hampshire Ave., N.W., Washington D.C., and Two Illinois Center, 233 N. Michigan Ave., Chicago, Ill.

SHORT AND FORD, ARCHITECTS

Under the direction of William H. Short, A.I.A., Short and Ford, completed in 1979 an Undergraduate 3374 sq. ft. addition to the Woodrow Wilson School of Public and International Affairs at Princeton University.

Two main considerations influenced the selection of underground construction: The original building by Mineru Ymasaki is so architecturally symetrical and self-contained that it was impossible to provide compatible above-ground space for needed library expansion. Furthermore, local zoning setback restrictions do not apply to underground construction and since the library occupies an end of the building near a street corner, this allowed maximum use of the site for a contiguous addition.

Short and Ford, Architects, RD 4, Box 864, Mapleton Road, Princeton, N.J. 08540.

GIATTINA & PARTNERS

This is an Alabama architectural firm which has under construction at this time a three-story 460,000 sq. ft. office building. The complex includes a three-level parking area for 1460 cars. When completed, this building will house the operations center for the South Central Bell Telephone Company.

One of the more interesting features of the project is an earthen dam which was installed to retain storm water. This creates a 3.2 million cu. ft. lake which prevents the overloading of the downstream areas. The water of this six acre pond is used for a source of irregation for the planting.

Mechanically, the pond water is used in three different ways. When the water temperature is below 85°F, it is used as a chiller for air conditioning condensers. It is used as a heat source when the pond water temperature is above 45°F and directly for building cooling when the pond water temperature is below 50°F. The engineers estimated that using the pond in this manner will provide a total energy savings of $31,000 in its first year of operation.

Since this building is being built in the sun belt, and energy conservation is not the prime consideration, the earth cover on the roof is limited to 18". The East and West elevations of the project are buried into natural slopes on the site to reduce the solar exposure of these elevations. The budget cost of this project is $39,000,000.

Fig. 10   Southern Central Bell, Alabama Operations Center

Giattina & Partners, Architects, Inc., 2031 11th Avenue South, P.O. Box 3381-A, Birmingham, AL.   35255 Attn: Orin C. Smith, A.I.A.

CALCO, INC.

One of the real innovators in the Earth Sheltered field is CALCO, Inc.,
St. Johnsbury, VT. Since 1977, they have been custom designing and
erecting shells for earth sheltered homes -- 11 to date in the New
Hampshire and Vermont area. These range anywhere from their smallest
480 sq. ft. "Hut" to two-level above and below grade four bedroom designs. These buildings are constructed with precast Waffle-Crete,
concrete panels which they manufacture under license from Van Dorn
Industries Inc., of Hays, Kansas. These panels are used for the wall,
floor, and roof construction. They are all 8" thick, reinforced and
cast of 5000 p.s.i. concrete. Their standard sizes are 12'-0" x 8'-0",
8'-0" x 12'-0", and 8'-0" x 16'-0". Steel pins are used to lock the
prefabricated wall panels to the cast in place footings and steel bolts
tie the panels together at their joints and corners. Seams between
the panels are filled with strips of butyl tape which are compressed
when the panels are bolted together to form a water-tight seal.

CALCO, Inc. erects the shell on the owner's property for $18 to $20
per sq. ft. This price includes waterproofing with Bentonize 80 and
styrofoam insulation consisting of 4" on the roof, 3" for sidewalls,
and 2" for frost wall and under slab.

Among the advantages claimed for this type of construction are economy,
speed of construction, and they can be erected without regard to freezing weather and less moisture remaining in the structure after construction than in conventional poured in place concrete.

Fig. 11   Two Bedroom CALCO home in St. Johnsbury, VT.

Further information can be obtained by writing Robert Pelkey, at
CALCO, Inc., Box 28, St. Johnsbury, VT.   05819

MALCOLM WELLS, ARCHITECT

Malcolm Wells, pioneer in the East Coast underground movement, started
some 11 years ago by constructing his underground office in Cherry Hil
New Jersey. This was a time when the public thought of underground
space to be suitable only for subway trains and moles. During the yea
Malcolm has done some 8 or 9 buildings under his own name. The second
being "Solaria", also in New Jersey, in 1974.

Probably his most note-worthy contribution is the Cary Arboretum of
the New York Botanical Garden. Also viewed with much interest is his
current home and office experiment - an Easterly facing passive solar
bermed complex.

Although he spends the major part of his time writing and lecturing,
he is still very active as a consultant and designer for other architects. The eighteen such designs currently on the boards, swells the
number of his designs close to 100. A list of his books includes:

Underground Designs, 1977. Self-published 9th printing. Edited versi
to be distributed by Brick House this spring.

Notes from the Energy Underground, 1980. Van Nostrand Reinhold.

Underground Plans Book 1, 1980. Self-published with co-author, Sam
Glenn-Wells, his son.

Gentle Architecture, 1981. McGraw-Hill.

Passive Solar Energy, 1981. Brick House. Bruce Anderson, co-author.

Fig. 12

Malcolm Wells' new
Home/Office Complex,
Brewster, MA.

Mr. Wells can be contacted by writing him at P.O. Box 1149, Brewster,
MA. 02631.

M.S. MILINER CONSTRUCTION INC.

This company, located in Maryland is the builder of Terra Vista (HUD award-winning, passive solar earth sheltered house). Although not a small house, 2300 sq. ft. including the solar greenhouse, Terra Vista is a most happy and logical marriage of passive solar and earth sheltering. It contains such innovative features as an earth pipe to temper the make-up air summer and winter, and features the new quadrupal glazing system as patented by 3 M. Passive solar is expected to provide 75% of the total heating needs, the remainder to be furnished by a wood stove and heat pump. Domestic hot water is also provided by a new heat-pump system which should cut water heating costs by at least 60%. In all, a thoroughly imaginative demonstrative project which requires a minimum addition of non-renewable energy.

Fig. 13

Terra Vista house as viewed from the Southwest.

M.S. MILINER CONSTRUCTION INC., 302A E. Patrick Street, Frederick, MD 21701

ROB ROY

Rob Roy, a Builder/Designer in upper New York State, is an advocate of the do-it-yourself school of low-cost home building. His present home is a Log End Cave. This is basically an Elevational type house with 12" concrete block sidewalls, a wood-frame roof, and 4" of topsoil and sod.

He is currently involved in a much more ambitious project of building a 36' diameter two-story earth sheltered structure utilizing the same log-end technique on the exposed walls.

Mr. Roy conducts hands-on seminars on his construction technique at his home site; and, is the author of a book entitled "Underground Houses" published by Sterling Publishing Company, Inc., New York.

Contact Robert L. Roy, RR #1, Box 105, West Chazy, N.Y. 12992

SHELTER DESIGN GROUP

This is an Eastern Pennsylvania architect/builder team which specializes in the construction of earth sheltered houses. To date they have
completed 13 homes.

Most of these houses are of the Elevational type using reinforced concrete block walls and roofs of precast panels. The average construction
price in 1980 of their earth sheltered homes was $74,000 complete.

For the most part, their buildings have been insulated with extruded
polystyrene; but they have been experimenting successfully with the use
of sprayed-in-place urethene as both insulating and waterproofing agents
This technique is currently being carefully monitored and results appear
to be most encouraging. Although the urethene does lose a certain amoun
of insulating value, due to moisture absorbtion, it appears to be considerable more economical to just add a little more thickness to make
up for the loss.

Fig. 14   Kohout Residence, Oley, PA. 1580 sq. ft. house, represent-
           ative of Sheltered Design Group work.

Shelter Design Group, Box 66, Stony Run, PA.   19557

TERRA DOMUS CORP.

This central New York architectural firm has under construction a most ambitious earth sheltered condominium complex, consisting of 54 units in three, three-story 45' x 204' buildings. Each floor sets back 15' from the story below, providing on-grade terraces for each unit. Completion date is scheduled for mid May.

Prior to this project, the architect had designed four earth sheltered homes.

Terra Domus Corp., 3939 Sholtz Road, Oneida, N.Y.   13421 Attn: Lynn Elliott.

The compilation of the data for this project has led me to the following conclusions:

• The buildings constructed prior to the early nineteen seventies, were primarily designed as protection from nuclear attack. Non-residential structures were designed to be bomb resistant, whereas underground additions to be used as fall-out shelters were added to conventional dwellings.

• Non-residential earth-sheltered buildings are being designed for ecological reasons. Educational Systems, both private and public, are making use of underground space more than any in the non-residential field.

• The residential designers show a strong preference for the bermed elevational type with varying degrees of passive solar use. Active solar appears to be largely relegated to the heating of domestic water. The houses that do not include wood or coal stoves are by far in the minority.

EARTH SHELTERING IN THE MID-WEST

Brent D. Anderson
Underground Space Center
University of Minnesota
Minneapolis, Minnesota

ABSTRACT

Discussion of systems used in earth sheltering in the Upper Midwest.

KEYWORDS

Structural Systems; passive techniques.

INTRODUCTION

As more people have come to realize that earth sheltered housing will be a part of
our housing market, questions have arisen as to the general construction techniques
and costs. Most of the homes built in the upper midwest have been constructed in
rural areas. A large percentage of the finished homes have been constructed by the
home owners at very low costs. The fact that the costs have been low and
publicized extensively, the general public has been in many cases led to believe
that earth sheltered homes are less expensive than conventional above ground homes.
Usually when a house is built at a low cost, it will reflect in its general
appearence. Although we must give much credit to the do-it-yourself builders for
their pioneering efforts, many of their designs have left poor impressions of earth
sheltered homes on the general public, especially the banking society. There are
many companies now in existence that are producing excellent designs and builders
with much practical experience and knowledge. This degree of professionalism has
been helping the general impression of earth sheltered housing. Large scale
builders and developers are now starting to get involved with the concepts and some
excellent projects are emerging. An example of this is the Muir Woods project in
Indianapolis, Indiana. This project is being built by Bay Development Corporation
under the supervision of Bruce Sklare. The project will consist of both earth
sheltered and conventional above ground homes.

THE SYSTEMS

As more and more people get involved with earth sheltered construction new systems
are being developed. A review of the general systems used shows that building an
earth sheltered home is similar to light commercial construction. The following

list reveals most of the major systems.

> A. Cast in Place Concrete
> B. Post-tensioned Concrete
> C. Reinforced Block
> D. Precast, Prestressed Plank
> E. Precast Reinforced Plank
> F. Dry Stacking Block
> G. Treated Wood
> H. Fiberglass Coating Over Wood
> I. Shotcrete Domes
> J. Steel Culverts

To describe how people are building earth sheltered homes in the upper midwest by state by state basis would be difficult. A discussion of the systems will follow along with a description of where they are commonly built. Also remember that mos homes actually use a combination of the above mentioned systems.

Most earth sheltered homes are built with poured concrete walls and precast, prestressed concrete plank roofs. There are many reasons for this type of construction. Because of the high soil loads concrete has been the traditional material used to support these loads. Many of the first homes were built by poure wall contractors. Since most poured wall contractors had been pouring basements for years and never had the necessary forms to pour roofs, the precast prestressec concrete plank made it easy for the poured wall contractor to adapt to this type construction. The plank roofs also were very economical to construct as compairec to poured roofs.

Costs will vary from one location to the next in the midwest. The following is an examination of many of the different systems.

A. Cast in Place Concrete

Completely cast in place concrete homes became more numerous as homes were built long distances from metropolitian areas where the planks were produced. The completely cast in place home is quite common in northern Wisconsin and Minnesota Indiana and southern Illinois. The completely cast in place system offers some unique advantages in that the entire system is tied together. We have found that waterproofing problems are far less frequent on cast in place systems compared to precast, prestressed plank systems.

B. Post-tensioned Concrete

This system probably offers more advantages than any other system. The process is such that all concrete is in compression. The structure is tied together very tightly. Waterproofing a post-tensioned system should offer excellent results. The costs may be slightly higher than cast in place systems and inspection is very important when building with this system. Most post-tensioned earth sheltered homes are in Kansas, Missouri, and Nebraska.

C. Reinforced Concrete Block

Reinforced concrete block walls are very common in Minnesota. Block walls are mo common in large metropoliton areas. Most earth sheltered homes will have 12 inch exterior walls with 8 inch interior walls. Generally precast, prestressed plank used in conjunction with block walls. Waterproofing block walls can be more combersome than cast in place walls.

D. Precast, Prestressed Concrete Plank (PPC)

Hollow core plank as they are commonly called have been the most common roof material used to date. Typically plank run about $3.00 per square foot installed and grouted. This has been very competive except where transportation is over great distances. More success has been achieved when plank are applied with a two inch structural topping than when not. The topping offers a means to slope the roof, adds strength to the roof, and allows for a much more uniform surface in which to waterproof. Topping will increase the cost of the roof by about $1.00 per square foot.

Hollow core and solid plank have also been used successfully in wall design. The costs have been very competive when comparing precast systems to cast in place systems. Much precast work has been done in Minnesota and Michigan. A few homes have used prestressed double tee sections in place of flat slab plank. Usually double tee sections are more expensive.

E. Precast Reinforced Plank

A system called waffle-crete has been used extensively for precast construction. The panel uses standard reinforcing steel. Most of the earth sheltered work has been done in the Dakatos. Spans up to 16 feet are commonly achieved. Bids on homes using the panel for walls and roof have been very competive.

F. Dry Stacking Concrete Block

A fiberglass reinforced mortar is troweled or sprayed on the surface of the block. The skin on each face has high strength. Walls still must be reinforced vertically, but lesser amounts of steel may be used. Many homes of this nature have been built in northwestern Illinois. Hollow core planks are usually used on the roofs. The troweled finish can be left exposed on the inside surface of the wall.

G. Treated Wood

Treated wood offers the advantage that most builders are accustomed to its use. The treated wood earth sheltered home uses the soil mass below the floor system to its advantage. Most treated wood structures have plenum spaces below the floor. Wood walls and roofs must be carefully designed to transmit all shear loads safely to the ground. Roof systems are usually glued beams. Spans up to 16 feet can be achieved. Most treated wood earth shelters are located in Minnesota and Iowa. Although wood structures should be founded on poured concrete footings, some builders are experimenting with gravel pads and finding them successful. Wood shells seem to be running about the same costs as concrete shells.

H. Fiberglass Coating over Wood

This system is a four foot by twelve foot panel that is of wood construction with a fiberglass skin on each face. The precast system is erected on concrete footings. The same type of panel is used on the roof. Costs are very competative. Only a few homes have been constructed this way. They are located in Michigan, Iowa, and Illinois. Some question has arisen as to the fire safety of these homes and that must be further researched.

I. Shotcrete Domes

The dome and arch shape has intrigued builders for some time. Most domed structures have been constructed above ground. The dome and arch offer great economies for enclosing space. There major problem is insulating them. Waterproofing should always go directly on the structure and extruded polystyrene should be placed over that. Most domes are insulated with urethane foams which should not be used underground.

J. Steel Culverts

Steel culverts generated much interest when used on the Clark-Nelson house in River Falls, Wisconsin. Since then, only a few structures have been built with culverts. Culverts in their pure form have and can be economical. When the design calls for excessive openings in the shell, structural problems can be excessive. Some question still remains as to the long term effects of the steel system underground.

The layout of floor plans has not changed considerably over the past few years. Most designs are of the elevational design with the east wall partially exposed. Because of the basic code requirements that all habital spaces must have direct access to the outside, floor plans will retain some basic formats. The long east-west format is still quite common and probably will remain because it is the most efficient shape for passive solar gain.

A number of new ideas have been introduced into earth sheltered homes of recent. These include deep heat heating, cooling tubes, underslab plenums, wood parapets, trombe wall fireplaces to name a few.

Deep heat systems are a series of electrical resistance lines which are installed in the concrete floor. The system is turned on during off-peak power and the floor is heated and the heat radiates into the space. Floor coverings should not be carpet or covered excessively with rugs. The system by using off peak power has been found to be very economical.

The most recent application was installed in northern Wisconsin. Wisconsin Power and Electric supervised the job.

Earth tube cooling and heating has also raised the curiousity of many builders. The success of earth tubes has been quite random. Systems must be about 400 feet long and buried at least 6 feet in the ground. More success has been found with winter time application than summer. Cooling warm moist air has lead to humidity buildup in the pipes. Mold and mildew can also be a problem in summer time use. During winter time use air at -10°F has been brought up to +35°F. One must realiz that under high volume use the efficiency of the system will become less because you will gradually freeze the soil around the pipes. It is difficult to assess the true performance of earth cooling pipes.

Underslab plenum spaces have proven to be very successful. A number of the treate wood systems contain these systems. Air is blown under the slab over a large rock bed of washed cobbles. The variation of temperature of the cobble bed maintained relatively constant range of temperatures from 60-64°F. The system contains a series of blowers which draws air warmed by passive solar process and blows it under the floor slab. Air is then circulated around the structure. Plenum spaces vary from 12 to 36 inches. The most recent ones built are around 12 inches deep.

Fireplaces that also serve as trombe walls can also be very effective. The fireplace is placed about 8 inches behind the glass facade. During the day the sun shinning through the glass strikes the verticle walls of the brick fireplace. The large mass of the fireplace is heated up and then will radiate heat back into the space.

Some work has been done with using large swimming pools as heat sinks hooked up to a heat pump. As of yet, no one seems to have an effective system. Many systems like this seem to work in theory but getting them to work in practice has been more of a problem.

The double envelope type of earth sheltered home has also been designed in the upper midwest. These type of homes have performed well but the costs have been excessive. It will be difficult to justify the additional cost and lost space in double wall earth sheltered homes.

Costs of homes has bothered many potential home owners for some time. We are finding that the costs of earth shelters are about the same as custom built above ground homes. We have found that once earth shelters go below 1600 square feet the costs increase quickly.

Based on current research as to the type of buyer for earth sheltered homes, we find that over 50% of the people are 50 years or older looking for low maintenance and a way to keep their energy costs down. Most people going into retirement will be on fixed incomes and want a retirement home with low operating costs. In conclusion, I think it would be fair to say that earth sheltered homes could be about 5 percent of the housing market in the next few years.

EARTH SHELTER TRENDS IN THE SOUTH CENTRAL PLAINS

Dr. Lester L. Boyer, P.E.

School of Architecture, Division of Engineering
Oklahoma State University, Stillwater, OK 74078

ABSTRACT

Identified earth sheltered houses in a 9-state region of the south central United States were examined via detailed questionnaires, extended telephone communications, and on-site visits. The work was initially supported by a Presidential Challenge Grant from Oklahoma State University and subsequently by a research grant from the Control Data Corporation. Reported results of energy and habitability aspects are presented herein.

Habitability and passive energy design of earth sheltered structures are key focal elements being investigated at Oklahoma State University. Habitability aspects have received little treatment elsewhere, and existing passive energy design strategies have generally not considered the passive cooling benefits of earth sheltered construction.

Initial results indicate that occupants are generally satisfied with such attributes as structural safety, thermal stability, and acoustical environment; but some occupants have reservations concerning site design, daylighting, thermal radiation control, and energy design and performance.

Sizable energy savings are, in fact, being realized by owners of current generation earth shelters. However, further significant savings could be anticipated with optimized passive systems design coupled with selected modifications in lifestyle. Further, an appropriate design balance between passive heating and passive cooling needs must be sensitive to changing latitudes and geography.

KEYWORDS

Earth shelter; dwellings; questionnaires; habitability; construction; energy performance; passive cooling; Oklahoma.

INTRODUCTION

Nature of Inquiry

Earth sheltered structures are becoming more numerous and are gaining increasing importance as a contemporary building concept. This increased activity is especially prominent in the south central plains of the United States, where traditional storm protection concerns have recently been coupled with desires for reduced heating and cooling needs. Almost the entire stock of earth sheltered dwellings in this region has been constructed since the mid-1970's.

What specifically is the earth shelter idiom of this high plains region which has caused such widespread interest? What type of people are building them, how are they built, what about habitability concerns, and how do the structures perform energy-wise? Finally, what is the prognosis for the future for the serious informed enthusiast?

Background

In June 1979, a Presidential Challenge Grant was awarded to an interdisciplinary team at Oklahoma State University in order to evaluate and assess habitability and energy performance aspects of earth sheltered dwellings in Oklahoma (Boyer, 1980a, 1980b; Weber, 1980). A considerable wealth of data had already been collected by that time as a result of Architectural Extension teaching activities for the public and practicing design professionals which began in early 1977 (Boyer, 1979; Grondzik, 1979). In December 1980, the formalized survey project was extended to nine (9) states with funding support from the Control Data Corporation of Minneapolis. Some preliminary findings have subsequently been published (Bice, 1980; Boyer, 1980c, 1981; Grondzik, 1980).

Objectives

This paper provides an overview of the findings thus far developed from the data obtained in the projects described above. Brief profiles are presented on occupant and construction characteristics, habitability aspects, and energy performance. In addition, recommendations for the future development of the earth shelter idiom are provided.

Procedure

The data have been collected through the use of detailed questionnaires, extended telephone communications, and in many cases with on-site interviews and instrumentation measurements. The computer coded data were then statistically analyzed via item-by-item frequency distributions, including the calculation of means and standard deviations. Further analyses, such as correlational studies and comparison of group differences on various response items, have been only partially completed at present, and will be published fully at a later time.

OCCUPANT DATA

Number of Projects

The approximate density of earth sheltered structures throughout the 9-state region ranging from Iowa to Texas and Colorado to Missouri is illustrated in Fig. 1. Preliminary studies in August of 1980 indicate that nearly 200 earth integrated structures may exist in Oklahoma alone. The expected total inventory of buildings in the 9-state region is anticipated to be in the neighborhood of 1000 projects. It is believed that these earth shelter projects far exceed the number of active solar or deliberately designed passive solar buildings constructed in the region during the same time period. Table 1 indicates the number of presently identified earth shelters in each of the nine states.

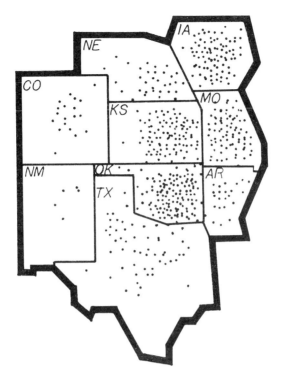

Fig. 1. Density of earth sheltered dwellings in
south central United States.

TABLE 1 Verified and Projected Earth Shelters in 9-state Region - 1980

State	Verified Sites	Estimated Percent Identified	Projected Total
Arkansas	23	43%	54
Colorado	19	37%	50
Iowa	85	59%	144
Kansas	96	52%	186
Missouri	79	44%	180
Nebraska	30	50%	60
New Mexico	5	40%	12
Oklahoma	120	60%	200
Texas	49	45%	108
Totals	506	--	994

Decision Factors

The purpose for building contemporary earth shelters in this region was examined from occupant ranking responses to a series of twelve possible decision factors. Four of the items dealing with energy reduction and storm protection were all ranked very high, while such factors as environmental noise reduction, lifestyle improvements, and preservation of roof-top land were of no particular significance. Intermediate ranks were listed for reduced maintenance, security from vandalism, insurance rate modification, personal privacy, and personal demonstration or experimentation interests. The beginning of the extensive construction activity seems to have been overwhelmingly due to the energy concerns, since almost all projects analyzed were built since 1975.

Occupant Profile

The "typical" occupant might be identified by the use of median values obtained in the statistical analysis. For example, the median age of questionnaire respondents was about 45, which means that half of the owners were older and half were younger than this median age. The "typical" number of occupants is about three (3), including many couples and many multiple-child families. The median family income was about $22,000, and the construction cost was something in excess of $55,000, or just over $25 per sq. ft. ($275 per m^2).

Site Preference

The "typical" gross floor area, less garage, is close to 2200 sq. ft. (about 200 m^2), which is substantially larger than the median contemporary dwelling built during the same time period. The median lot size is 12 acres (about 49 000 m^2), which is rather non-typical for contemporary above ground housing. The absence of urban examples in the plains area is probably due to the preferences of the owners, who find subterranean living to be a rather viable housing alternative for a country setting.

CONSTRUCTION CHARACTERISTICS

Design Formats

An earth covered dwelling in this 9-state region is basically either located at grade and bermed, or integrated into a partially or completely excavated site. Terratectural constraints such as a high water table, expansive clay soils, rock strata, or flat rural sites have often been overcome with the use of a bermed configuration. The hillside site utilizing a partial excavation is by far the most common site application. Also, nearly all recent examples utilize the thermal advantages of a substantial sod covered roof.

A variety of architectural design priorities and constraints must be considered during the selection of a basic plan type. Such priorities and constraints deal with factors which include orientation for climate, opportunities for views, site access routes, external noise sources, proximity of neighbors, and building egress requirements. Regardless of whether a bermed or excavated format is indicated, the resulting plan could incorporate the elevation, penetration, or atrium motif. An example plan of the typical hillside earth shelter format with one exposed elevation is illustrated in Fig. 2.

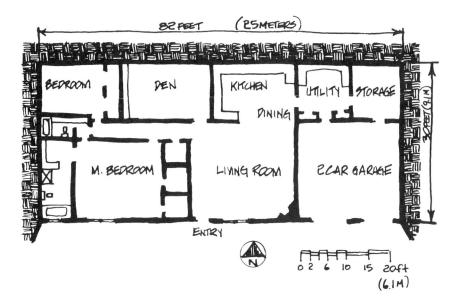

Fig. 2. Example plan of hillside earth sheltered
house.

Building Structure

Poured concrete is by far the most popular construction material being used for
earth shelters in the south central United States. More than two-thirds of all projects utilize poured-in-place concrete for both roof and earth contact walls. Concrete block is also used quite often for wall systems, and the tendency to use this
material seems to increase with constructions in the northern latitudes.

Roof systems are predominantly of the flat slab variety; however, barrel vault,
waffle slab, inverted beam, and post-tensioned slab techniques have also been
utilized. Post-tensioned structures seem to be most prevalent in Missouri and Texas.
Also, the use of bar joists with metal deck and concrete topping is somewhat common
throughout the entire region.

Professional Input

The owner of such current generation earth shelters is also apt to be the designer
and builder as well. Although the owners are typically well educated, with about
one-third having post-high school education at the college level, many of them have
not had professional structural advice on their projects. Exactly how their satisfactorily performing structures were derived is not entirely clear. Selective in-
quiries and visits regarding pre-existing successful projects was undoubtedly a
useful planning resource to many. At any rate, a grassroots approach to design
and construction has been very much in evidence. Correspondingly, it is also quite
evident that a strong design/build industry for earth sheltered housing has been
conspicuously lacking in the region.

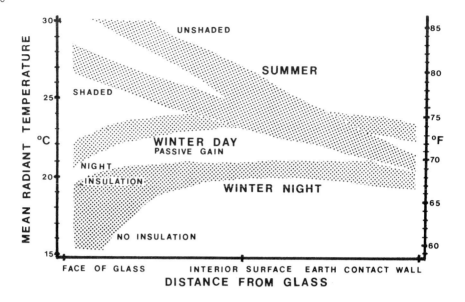

Fig. 3. Estimated variation in the radiant thermal environment for earth sheltered interiors.

HABITABILITY ASPECTS

Earth Shelter Interiors

Five (5) parameters were studied with regard to habitation in earth sheltered spaces including thermal comfort, acoustical qualities, spatial attributes, lighting aspects, and lifestyle conditions. An average of about six questions were addressed for each of these categories.

Thermal. Overall, the group of thermal conditions studied received the highest rating of all groups. However, a question in this group which did cause a limited concern dealt with condensation on the walls. Evidently, conditions exist in some projects, mostly in the northern latitudes, which cause some condensation to occur at certain times during the year. Possible omission of insulation in the northern portion of the analyzed region could be associated with such effects. A further explanation might be that the use of dehumidification equipment in the north is not as widespread as in the southern areas, where condensation has not been reported as a problem. A general tendency to undersize air circulation systems as a result of reduced space conditioning loads may also contribute to the problem.

Mean radiant temperature (MRT) is a major factor contributing to the achievement of thermal comfort in passively conditioned buildings. Figure 3 shows an example of the estimated variation in the radiant thermal environment for earth sheltered spaces. For locations remote from the exposed facade, the beneficial passive cooling effect of the earth heat sink becomes apparent. These MRT impacts can be strongly affected by both explicit and implicit insulation treatments, which will be discussed later. At exposed perimeter locations, the fenestration treatments have the most pronounced MRT effects.

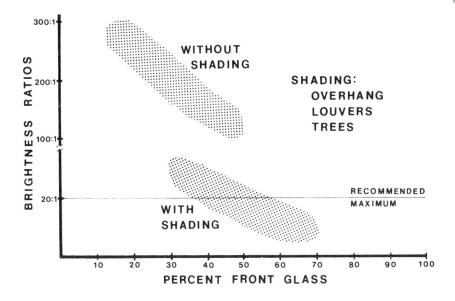

Fig. 4. Variation in brightness ratios typically encountered
in earth sheltered interiors.

Acoustic. The acoustic conditions were evaluated as the second highest group of interior parameters. Only the concern of some for living in a "too quiet" environment
diminished the otherwise rather high average rating. These occupants often preferred to have a radio or a fan operating to produce a background sound level,
especially at night.

Spatial. The third highest group of interior parameters was that of the spatial environment. Except for minor reservations about amount of inside storage space and
overall space flexibility, the respondents generally viewed their homes as being more
than adequate with regard to spatial requirements.

Lighting. As a group, the lighting parameters were not rated as highly as the previous groups of interior conditions, although acceptable ratings were generally still
maintained. It should be noted that lighting design seems to take on more significance when contemporary dwellings are placed below ground. Thus, substantial
improvement opportunities seem to be present with regard to lighting design for
earth shelters. For example, brightness ratios to which occupants are exposed when
gaining visual access to exterior views are often quite excessive. The measured
variations of such quantities are shown in Fig. 4. As can be noted, values often
exceed typical recommended ratios by more than a factor of ten. More attention
should be given to control of these brightness levels in the field of view.

Lifestyle. The final interiors assessment area dealt with the initiation of lifestyle
modifications on the part of the occupants in order to further enhance already substantial energy reductions due to earth sheltering. In general, occupants indicated
that their total energy expenses were so much lower than in their previous residence,
on the order of 30 to 40 percent lower, that they made little effort toward major conservation practices through lifestyle modifications. In fact, their space conditioning
systems typically operate continuously at the mid-point of the comfort range. Except for attention to turning off lights when not in use, occupants do not seem to

be concerned with lifestyle modification factors such as selective thermostat settings, modulated use of various rooms, use of additional clothing, or any other type of energy-sensitive adjustments.

Exterior Environs

Three (3) parameters were addressed with regard to exterior planning within the context of the exterior surroundings of the project. These included property maintenance aspects, community setting, and site planning and design.

Maintenance. The maintenance parameter assessments revealed some interesting results. Major maintenance requirements for insulation, waterproofing, and structural components have not been statistically significant, but there have been isolated cases of major concern. However, general concerns of a serious nature have been expressed with regard to maintenance of ground cover and erosion control. Also, dehumidification is often being accepted as a necessary component of comfort conditioning and energy conservation, especially in the southern part of the region.

Community Setting. The group of community setting items did not reveal any particular interests or concerns on the part of the occupants. They acknowledged that the exterior appearance from the street or highway could be enhanced, but this fact is not very important for projects located on large sites.

Site. As a group, site parameters were not very highly rated in general. However, the attribute of storm protection was very highly rated in comparison to other suggested attributes. The general absence of appropriately located shade trees seemed to be an important adverse factor with regard to solar and daylighting control.

ENERGY PERFORMANCE

Energy Systems

About one-third of the earth shelter projects have incorporated the use of a special energy system such as attached greenhouses, solar assisted domestic water heaters, or windows deliberately designed for passive solar space heating. In general, such enhancements were not strongly considered in view of the substantial expected benefits from the earth shelter effect. About 70 percent of the occupants, in fact, indicate that they have obtained energy reductions at least as good or better than they originally expected. Overexpectations, larger living areas, and possibly unanticipated design problems probably account for most of the unfavored energy results.

Conventional heating systems are installed in about one-half of the projects, and the remainder utilize heat pump systems. Roughly two-thirds have either a wood stove or a fireplace as an additional heat source. Mechanical space cooling systems are provided in 80 percent of the projects, when heat pump systems are included. In general, the button-up mode of building operation is rather continuous throughout the year. However, a full 20 percent have no mechanical cooling whatsoever, and these would rely mostly on earth cooling and natural ventilation effects.

Figure 5 shows occupant evaluations of the various energy source contributions to their heating and cooling needs. In general, mechanical space conditioning is considered to account for about 60 percent of the total energy requirement in both the heating and cooling seasons. Thus, a full 40 percent is attributable to the use of passive or renewable systems such as wood heat in winter and earth cooling in summer. Unfortunately, as previously mentioned, few sites have incorporated the

Heating
Season

Cooling
Season

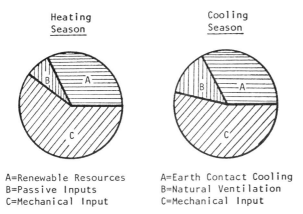

A=Renewable Resources
B=Passive Inputs
C=Mechanical Input

A=Earth Contact Cooling
B=Natural Ventilation
C=Mechanical Input

Fig. 5. Occupant evaluations of environmental control
energy sources.

use of planned passive or active solar heating. If this condition were rectified,
the dependence on mechanical space conditioning could be still further reduced.

Measured Consumption

Actual energy consumption information was obtained for a number of projects in
Oklahoma with the use of utility billing records. Figure 6 shows monthly total
energy usage in earth sheltered homes compared with conventional above ground
homes. The earth shelter data reflect usage in five typical all-electric houses of
recent vintage, and the above ground consumption data represent the mean for 20
randomly selected all-electric homes in the same area. This limited comparison clearly demonstrates the performance potential of earth shelter designs. In addition to
the dramatic peak-to-valley reductions achieved, the total annual energy usage of
the earth sheltered dwellings is substantially decreased by about 40 percent.

These major reductions shown above occur in spite of the fact that appliance and
hot water usage may be unchanged. Still further energy reductions would be expected as building envelope designs are improved with regard to insulation and
thermal mass treatments. Additional reductions due to modest lifestyle and comfort
modifications could also be expected with the adoption of a systems approach to
solar design.

Design Load Analysis

A final energy analysis, shown in Fig. 7, illustrates a comparison of load calculations for the previous five earth shelters with those for equivalent above ground
dwellings utilizing several widely accepted energy-sensitive constructions. Specifically, projected design heating and cooling loads are shown for several equivalent
above ground dwelling types meeting the construction requirements of the HUD
Minimum Property Standards, ASHRAE Standard 90-75, and the widely publicized
"Arkansas House". Each of the eight vertical levels on the figure represents an
average value for five individual earth shelters or above ground equivalent houses.

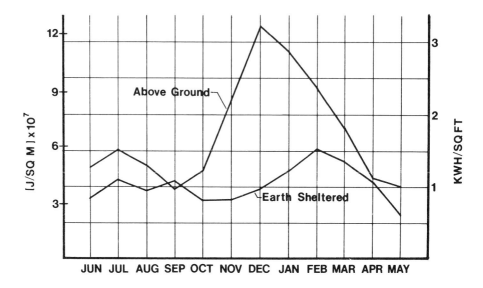

Fig. 6. Monthly total energy usage in conventional
above ground and earth sheltered homes.

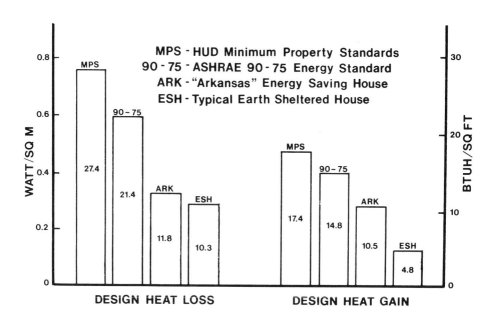

Fig. 7. Design heating and cooling loads for earth
shelter and energy standard equivalents.

The earth sheltered structures fare two to three times better than equivalent above ground dwellings meeting MPS or ASHRAE 90-75 requirements. The super-insulated "Arkansas House" compares quite favorably with the earth shelter design heating load, but requires more than twice the design cooling capacity in summer. As these comparisons illustrate, the earth shelter alternative is a viable response to the energy imperative for residential-scale structures.

Performance vs. Potential

Designers of earth sheltered dwellings must be cognizant of the differing, and some-times dichotomous, requirements imposed by the dual potentials of passive design strategies to meet both cooling and heating needs. The thermal mass concept is an important element that can be utilized to illustrate this point. For passive heating, this mass should be accessible to direct solar gain and be thermally isolated from the external environment. On the other hand, passive cooling requires thermal mass to be in direct contact with the heat sink effect of the earth. Likewise, the use of buffer spaces such as closets requires careful placement to satisfy both heating and cooling energy design strategies.

One of the reasons why earth cooling capacities may not be fully utilized could be the widespread use of implicit insulation treatments which reduce direct earth con-tact potential. Oklahoma earth shelter experiences have shown that, in fact, ex-tensive decoupling effects are inherent in roughly one-half of the projects surveyed. Not only have the houses been decoupled from the earth, in many cases the interior environment has also been decoupled from the substantial structural mass of the building envelope. This inadvertent decoupling effect can result with the use of furred or suspended interior surface treatments, as well as with the direct use of interior insulation materials.

The extent of the decoupling problem has been previously documented (Boyer, 1980a). In many cases, the effective thermal mass is reduced to less than one-half of the actual installed structural mass; the large potential benefits of the earth mass are also not fully realized. In addition to the drastic effects on energy con-sumption, the efficient utilization of the installed thermal mass is also expected to have a dramatic impact on habitability responses for the occupants. Further de-tailed examination is needed to determine the actual impacts on overall energy usage for comfort conditioning for such projects.

SUMMARY

Occupants. Owners of earth sheltered dwellings are generally between 30 and 50 years old and are well educated, with about one-third having post-high school edu-cation at the college level. About half are employed in the technical/professional and management/administration categories. Better than half of all respondents had incomes in excess of $20,000.

Construction. Earth shelter construction systems in the Oklahoma region consist predominantly of poured-in-place concrete components. Steel bar joist construction is a significant alternate technique for roof systems. A variety of other individual roof constructions have also been successfully used, but appear to be in a minority. Reinforced concrete block construction is an important alternate wall system. Many projects are occupant designed and built.

Habitability. Significant energy savings have been achieved in these dwellings with no compromise in the normal expectation of a contemporary living environment. In fact, many respondents indicate enhanced occupant comfort and habitability quotients. Most projects are located on large rural sites with considerably larger than average floor areas. Construction costs are generally similar to comparable above ground dwellings.

Expectations. Over 40 percent of the occupants are saving energy to the extent expected. However, 30 percent indicate their energy consumption is much higher than expected. Initial studies indicate that in many cases the potential thermal mass of the structure has been decoupled from the interior by insulation and furred interior surface treatments. Significant reductions in free earth cooling can thus accrue. Other factors now under study also undoubtedly contribute additional adverse effects. In addition, several published up-beat articles have probably led to over-expectations in many cases.

Passive solar heating. Earth sheltered housing represents a perfect vehicle for vigorous application of alternative heating strategies specifically including passive and active solar systems. Unfortunately, few earth shelter sites in the Oklahoma region incorporate planned participation of alternative energy sources other than wood heat. On the other hand, many occupants have experienced unwanted passive solar heating in the summer.

Energy savings. Total energy consumption in Oklahoma earth sheltered structures averages 30 to 40 percent below usage in comparable above ground dwellings. Research studies at Oklahoma State University indicate that these savings could be more than doubled, if well considered dwelling construction and lifestyle adjustments are jointly implemented.

Systems design. In order to obtain a highly effective optimized solution, a sensitive systems design approach is essential. Key energy design parameters include effective thermal mass, earth contact potential, interiors and spatial design impacts, and systems coordination considerations among others. For example, items as diverse as insulation, wall finishes, waterproofing, and room location may all have an impact on earth cooling potential and thermal storage optimization.

Overall assessment. Earth sheltering represents a viable energy conservation/energy advocacy strategy, especially for the south central plains region of the United States. Energy savings several magnitudes greater than that possible with other much publicized strategies appear to be not only possible, but probable. Also, the personal discomfort typically associated with many conservation schemes need not be associated with earth sheltered designs, which in many cases represent exemplary contemporary habitats.

RECOMMENDATIONS

Expand data base. The process of identifying new sites and questionnaire distribution should continue and be expanded regionally and nationally to enlarge the data base and permit identification of regional similarities and differences. An emphasis on habitability and energy performance aspects seems warranted.

Design development. Recommendations for the improvement of the design development process for earth sheltered buildings should be formulated. The image presented by the external appearance, circulation problems addressed by the entrance design, and the impacts of energy engineering of the interiors will all be strongly influenced.

Habitability index. Interior environmental design requirements for earth sheltered and passive structures should be more thoroughly examined relative to contemporary equipment intensive non-passive buildings. Environmental comfort parameters should be developed into an index format to enable architects and engineers to modify their design practices and capitalize on energy saving potentials.

Daylighting investigation. Daylighting in earth sheltered buildings requires special emphasis on such features as brightness and glare control, shading design, back-lighting, and control of the field of view. Energy savings due to daylighting in both small and large buildings should be investigated in addition to the habitability aspects which are normally ignored.

Energy assessment. Regional and nationwide survey and energy metering studies should be instituted in order to determine specific energy requirements for space heating and cooling, domestic water heating, and appliance usage. Such studies would also investigate the impacts of selected design parameters such as vegetation, insulation, thermal mass and associated decoupling effects.

Energy monitoring. Thermal transactions across the building envelope should be examined with the use of selected case studies in various regions as well as by para-metric studies in well controlled test cells. These investigations would provide veri-fication of energy metering study results and provide input for heat transfer algo-rithm development.

Solar augmentation. The existing situation of minimal passive solar inputs to earth sheltered dwellings should be reversed. Potential benefits should be quantified and simplified methods of analysis should be made available. Potentially adverse effects of unwanted passive solar impacts during the summer need to be identified in a quantitative manner. Recognition of the above effects should substantially influence the physical appearance of earth sheltered buildings.

Back-up systems design. The concept of systems optimization for earth sheltered buildings should be refined and quantified. Equipment needs and performance should be delineated. Alternative equipment augmentation schemes such as earth tubes and ground coupled heat pumps should be evaluated in a total systems context.

Validation research. Validation studies of energy performance models and habitability indices should proceed at several levels. A first level validation effort should in-volve application of public domain thermal analysis computer programs such as DOE-2 or BLAST. Comfort and energy concerns could be simulated in a full-scale mock-up situation in a facility such as the Environmental Control Complex at Okla-homa State University.

ACKNOWLEDGEMENTS

The support provided by a grant from the Office of the President of the Oklahoma State University to assess earth sheltered housing in Oklahoma is hereby acknowl-edged. Credit is also extended to the Control Data Corporation of Minneapolis, Minnesota for continued funding to extend and expand initial research efforts to include a regional analysis and in-depth studies relative to the engineering of earth sheltered interiors. The Division of Buildings and Community Systems of the U.S. Department of Energy in Washington, D.C., and the Oak Ridge National Laboratory in Tennessee have contributed to an extension of information dissemination efforts.

REFERENCES

Bice, T.N. (1980). Energy Analysis of Earth Sheltered Dwellings. Thesis in Architectural Engineering. Oklahoma State University, Stillwater.

Boyer, L.L. (1979). The Oklahoma experience cited. Earth Shelter Digest, 1(2), 32-35.

Boyer, L.L., W.T. Grondzik, and M.J. Weber (1980a). Passive energy design and habitability aspects of earth sheltered housing in Oklahoma. Underground Space, 4(6), 333-339.

Boyer, L.L., M.J. Weber, and W.T. Grondzik (1980b). Energy and Habitability Aspects of Earth Sheltered Housing in Oklahoma. Presidential Challenge Grant. Oklahoma State University, Stillwater.

Boyer, L.L. and W.T. Grondzik (1980c). Habitability and energy performance of earth sheltered dwellings. In T.N. Veziroglu (Ed.), Proc. 3rd Intl. Conf. on Alternative Energy Sources. Miami Beach.

Boyer, L.L., W.T. Grondzik, and T.N. Bice (1981). Energy usage in earth covered dwellings in Oklahoma. Underground Space, 5(4), 227-236.

Grondzik, W.T. and L.L. Boyer (1979). Oklahoma earth shelters: a state-of-the-art review. In R. Sterling and N. Larson (Eds.), Going Under to Stay on Top - Housing Conf. Proc. University of Minnesota, Minneapolis.

Grondzik, W.T. and L.L. Boyer (1980). Performance evaluation of earth sheltered housing in Oklahoma. In G.E. Franta and B.H. Glenn (Eds.), Proc. Intl. Solar Energy Society 25th Annual Meeting, Vol. 3.2. Phoenix. pp. 729-733.

Weber, M.J., L.L. Boyer, and W.T. Grondzik (1980). Implications for habitability design and energy savings in earth sheltered housing. In L.L. Boyer (Ed.), Proc. Earth Shelter Design Innovations Conf. Oklahoma State University, Stillwater. pp. VI-19 to 27.

Earth Sheltered Design and Construction Activity
in the Western States including
Arizona, California, Nevada, Utah, Wyoming
Montana, Idaho, Oregon and Washington

David M. Scott, FAIA
Department of Architecture, College of Engineering
Washington State University
Pullman, Washington 99164

ABSTRACT

This paper is limited to work accomplished since 1970. Through a telephone survey an
attempt was made to identify contact persons including university faculty as well
as architects and builders who are providing leadership in earth sheltered design.
While limited, it does demonstrate that a new family of buildings is being spawned
that includes institutional, commercial and residential building types.

KEYWORDS

Buildings types; faculty members; architects; designers; builders; extension agents;
Universities.

INTRODUCTION

The 9 Western states represent approximately 1/3 of the land area of the United
States excluding Alaska and Hawaii. The urban areas are concentrated on the Pacific
Coast where the climate tends to be mild, both summer and winter, and extends from
Southern California to Northern Washington and the Puget Sound area. Earth sheltered
buildings such as University Libraries and office structures have been generated as
a response to saving open space, such as the Sedgwick Library at the University of
British Columbia, Vancouver, B.C.

In Arizona, Paolo Soleri created an earth sheltered house in the desert near Phoenix.
One or more houses have been constructed on the Pacific Coast near Monterey as a way
of subordinating the house to the landscape.

In 1968, Architect Donald R. Heil developed a house using a 2 story free standing
wood platform in a concrete cylinder cut into a hillside and top lighted, to create
a pavilion in a garden experience for a Japanese American family living in Pullman,
WA.

In 1972, Architect John Morse, created an olympic size swimming pool for the city of
Seattle on a tight site that uses the roof surface of the building as a park.

The earth sheltered concept has been used on buildings of all types, in California,
Oregon and Washington, not so much for saving energy but as a way of achieving an

appropriate design expression or gaining two uses from the land. Because the 9 Western states are all agricultural states, earth sheltered facilities have long been a part of a way of life on the farms and in the agribusiness community whether it was for wine cellars in California, potato cellars in Washington and Idaho, or just plain storage cellars, or ice houses for the family on the farm, in the suburban areas or the urban areas.

Because of the missile sites and other military bases that are located in the 9 Western States people have become accustomed to the notion of storing ammunition in earth sheltered enclosures or missiles in deep silos.

Throughout the west there are many of survival type structures built by groups and/or individuals as protection again fallout and were built during a time when our concern for civil defense was being emphasized. There are many survival structures being built even today. One designer-consultant from Spokane, Washington has consulted on some 300 such structures.

While there has been some concern for a number of years on the part of a few individuals for the use of soil as a tempering or sheltering medium, it was not until 1973 that the public interest was sparked.

Generally speaking, the West has always been blessed with low cost natural gas and electrical power. The Pacific Northwest has a very favorable hydroelectric rate, in fact, the cheapest rates in the U.S.A. Because of this abundance, the cost of energy did not become a significant issue until about 1979. The power planners, however, had long known that the Pacific Northwest was reaching the time when restrictions would have to be placed on large industrial users, such as the aluminum industry.

There seems to be 2 general areas where interest in earth sheltered buildings is focused in the 9 Western states. One area is Arizona. The primary interest in that state has been on solar heated housing of both passive and active systems. They have had a long tradition of above grade structures that are energy efficient. The other area in the West where there is a concentration of interest in the north, in the large agricultural area in Washington between the Cascade Mountains and the Rocky Mountains and extending into Montana. The climate is hot and dry in the summer, and cold and dry in the winter. During the years of 1977/78 the public interest was ignited and by 1979 in Spokane, Washington through the office of the Washington State Energy Extension Service a number of seminars was held over a 6 month period of time. The attendance was over 5,000. The lay public is leading the banks, builders, and even the architects in their interest in an alternative form of housing.

1980 was the year that Oklahoma State University Extension Service offered programs on earth sheltered buildings in Arizona. The Underground Space Center held programs in Portland, Oregon and San Francisco, California. All were well attended.

1980 saw the development of the first commercial project or office building in Spokane, WA for the Central Pre-Mix Co.; a project that won a national award for design excellence. In Boise, Idaho an earth sheltered solar assisted elementary school was built. In California under the leadership of Syn Van der Ryn a large institutional building was developed that utilized earth shelter concepts. In Oregon, Multonomah County has developed a large earth sheltered garage facility. In Arizona a private developer is about to start construction on a 4 storey commercial facility. The building will have one storey above grade and 3 stories before grade using the atrium concept. The project is in a residential neighborhood. The primary motivation was to find a way to relate a large facility to the scale of the neighborhood.

The issue of solar access, solar rights and solar easments is another motivating factor in bringing the concept of earth sheltering to the forefront. In Seattle, a lawsuit has been brought against a hotel owner because the reflective nature of the glass is creating a hazard on the freeway. This kind of publicity is focusing attention on alternative forms of building.

On March 18, 1980 Mt. St. Helens was a pristine snow capped volcano in the Cascade chain located in Southwestern Washington. It erupted and covered a substantial portion of the state of Washington, Idaho and Montana with up to 6" of volcanic ash. Those people who had earth sheltered houses were able to clean up their house or their yards in a matter of a few hours, while those who did not, like this author, spent 32 hours cleaning 1200 sq. ft. of a shingle roof. He is still spending time on recleaning the roof and the wall surfaces due to the dust that accumulated on trees.

Figure 1. Location of Construction Activity

Further, the construction industry has had a moment to reflect due to the high cost of money. It has caused everyone to re-evaluate what they have been doing, and it has given them time to think about alternative products and projects. We have begun to see action by manufacturers, suppliers, architects, and real estate people, who are promoting alternative forms of buildings, both at the residential and commercial scale as a response to both the economic and the energy realities of our time. Had we not had the energy crunch; had we not had the fiscal crunch, people would have gone on doing exactly what they have always been doing with gay abandon. I believe that it is fair to state that on the Pacific Coast that 1979 and 1980 were the years of awakening and that we are now in a phase where in we are beginning to see the development of products, systems, plans, services, legal instruments and even legislation to encourage earth sheltered buildings of all types.

ACCREDITED SCHOOLS OF ARCHITECTURE IN THE NINE WESTERN STATES	No Information Available	Random Student Interest	Random Faculty Interest	Consistant Student Interest	Consistant Faculty Interest	Research Activity	Extension Activity
1. Arizona State University Tempe					●[1]	●	
2. California State Polytechnic Pomona					●[2]		
3. California Polytechnic State San Luis Obispo		●	●				
4. Montana State University Bozeman		●	●				
5. Southern Calif. Inst. of Arch. Los Angeles					●[3]		
6. University of Arizona Tuscon				●	●[4]	●	
7. University of California Berkeley	●[5]						
8. University of California Los Angeles	●						
9. University of Idaho Moscow				●	●[6]		
10. University of Oregon Eugene		●			●[7]	●	
11. University of Southern Calif. Los Angeles						●[8]	
12. University of Utah Salt Lake City		●	●				
13. University of Washington Seattle		●			●	●[9]	
14. Washington State University Pullman					●	●[10]	●

CONTACT PERSONS: 1. James Scalise, 2. John Halldane, 3. Roland Coate, 4. Kenneth Clark, 5. Sym VanderRyn, 6. Kip Eder, 7. G. Z. Brown, 8. Ralph Knowles, 9. Charles Kippenham, 10. David M. Scott

Figure 2. Activity in Accredited Architecture Programs

Figure 1 is a map of the nine western states showing the locations of some of the significant instituional structures and some housing. The map in no way represents all of the activity. The architects, builders, and projects are listed under each state.

Figure 2 represents a listing of the accredited schools of architecture in the nine state area. An indication of the degree of involvement is expressed.

A contact person is listed for each shool where the information was available. Interest and involvement in the schools varied from no interest to very intensive interest including a significant amount of extension activity.

No attempt was made to contact the engineering schools. There is reason to believe that interest is developing in most civil and agricultural engineering departments.

A check with the home office of the International Conference of Building Officials in Whittier, California revealed that while there has been a great deal of interest in earth-sheltered buildings reported by various code officials, who are members of the ICBO, the primary area of concern seems to be with the direct exit requirements and the 44" height from the floor to the window sill. It appears as though recommendations are in the process of being forumalted that allow exits to be made into light wells and other secondary forms of the egress.

Arizona

There has and is much activity in the State of Arizona related to active and passive solar housing. Earth sheltering is a concept which very closely relates to the indigenous building of the state. It has recently become more popular in the University as well as with the building community. At the University of Arizona, Tucson there is a strong interest on the part of 2 faculty members and a graduate student. Two houses have been built in the last year. The names of the people are Les Wallach, Kenneth Clark and Mike Stanley. They can be contacted at the School of Architecture at the University of Arizona.

At Arizona State University the students have been involved in the production of a book on earth sheltered buildings under the direction of Professor John Scalise. The Department of Architecture and the College of Engineering have a joint research project.

A local architect/engineer, Herman J. Von Franhoffer, has started a company called Concept 2000, Inc., with the express purpose of promoting the construction of earth sheltered houses. To date the company has built three houses.

California

California has a long history of earth sheltered structures, particularly in the agricultural community.

Because of the mild to warm climate of most of the state, in addition to the agricultural connection to earth sheltered buildings, there has been less activity in the field of housing than one might imagine.

Perhaps one of the most significant houses to be designed as an earth sheltered concept has been developed by Roland Coate, Architect, in Los Angeles. The house approaches sculpture and fits into the land in a most elegant and graceful way.

154

At Cal State Polytechnic University in Pomona two faculty members, John Halldane and Brooks Cavin have demonstrated an interest in research activities related to earth sheltered buildings for California.

Perhaps the most significant work relating to energy, solar access, solar rights, solar easements, and zoning has been accomplished at USC by Professor Ralph Knowles. Professor Knowles is the author of a book entitled Energy and Form, published by MIT Press. This book establishes a theoretical foundation for the development of energy conserving communities, energy conserving land use, and building forms.

At the University of California, Berkeley, Professor Sym Van der Ryn, Architect and former consultant to Governor Jerry Brown, has been involved in the development of large scale state institutional buildings that use the earth sheltered concept in a very sophisticated urban context.

The interest in the development of earth sheltering as a concept is only now beginning to emerge and it is anticipated that there will be rapid development over the next 5 years.

One of the most significant buildings in the United States in an urban environment is the Oakland Museum. While the Oakland Museum is not technically an earth sheltered building it has the characteristics of a earth sheltered building in that the exterior expression of the building is subordinate to the development of the roof surfaces for gardens and exhibition spaces. It is a good example of a new type of urban building.

Nevada

I was unable to find a contact person in the State of Nevada. It appears from a search of the literature that there is some interest. The MX program will have shelters for missiles and support areas for operation crews and maintenance and family situations. It is my understanding that earth sheltering concept is being considered for the military facilities. The MX program will affect not only the State of Nevada but Utah and perhaps other states as well.

Utah

In conversations with the School of Architecture it seems that there is only modest interest or awareness of interest at this time, in the State of Utah. Western Sun sponsored a solar conference in the fall of 1980 where Ray Sterling and others were invited to make presentations. Emphasis at the conference was on active and passive concepts. It is my judgement that a strong interest in earth sheltering will emerge within the next two years.

Wyoming

The State of Wyoming contains some of the most elegant and majestic natural geological formations and natural scenic events in the United States, particularly in the Northwest corner where the Grand Teton and the Yellow Stone Park areas are located. The majority of the state, however, is rolling natural land, that is less majestic but nevertheless has a beauty of its own. The population of the state is small, but it is becoming developed because the available oil and coal.

At the present time there is interest in earth sheltering and there are a number of projects already complete. In Jackson Hole is a Forest Service Information building, a wood structure that has a sod roof. It stands in sharp constrast to some of the more "modern" buildings built since WWII. South of Jackson Hole is one earth sheltered house. There are two houses in Laramie designed and built by contractors. It is my understanding that there are a number of survival type homes scattered throughout the western edge of the state. A contact person in the State of Wyoming is Mr. Michael McNamee c/o of Agri Extension Service, University Station Box 3354, Laramie, WY 82071, phone (305) 766-4396.

Montana

Montana has been a very active state in the last several years with the majority of the work focused in the communities of Missoula, Hamilton and Helena. Perhaps the most visible designer/builder is Mr. Joe Frechette of Hamilton, Montana. Joe has completed 9 houses as of this date, is working on 3 others and has consulted on 12 owner/built, owner/designed houses. Approximately 18 houses have been built by Paul Bessler in the area of Helena. The winter climate in Montana is very harsh and can be very severe. The landscape is majestic. The majority of the work being accomplished in the state of Montana is by designer builders or owner/builders. At this time neither the university nor the architecture profession is actively engaged in the development of earth sheltered concepts. It would appear, however, that there is an emerging interest by the profession as well as the students and faculty in the university.

Idaho

Idaho is another predominately agricultural, mining and timber state. In 1979 an earth bermed school was constructed in Boise, Idaho. It is an elementary school that houses 780 students. It is fully occupied and the community appears to be pleased with the thermal performance of the building as well as the general character and liveability of the structure. While there have been a number of houses built in the southern part of the state, they are a little bit more difficult to identify because of lack of contact people.

There is a substantial activity in Northern Idaho in the mountainous area between Washington and Montana. The activity takes form in two different building types. In Northern Idaho there is a region around the community of Kellogg where there is extensive mineral deposits of lead, silver, zinc, and other precious metals. At the present time there are over 300 miles of tunnels that could be developed into some alternative use. The Bunker Hill Company is experimenting with using 500 linear feet of vacated tunnel to grow seeding Douglas fir trees fir trees for a reforestation project. Studies are being made at this time by a number of individuals in the Spokane, Washington area, as well as, at the University of Idaho for alternative uses for the mined space.

The University of Idaho is located in Moscow, Idaho very close to the Washington/ Idaho border. New interest was developed in the 1980/81 years and it appears as though there will be a number of houses built in the Moscow region in the next 2 to 5 years.

Around the Sand Point area, which is east and north of Spokane, Washington there has been a great deal of activity with low cost timber earth sheltered housing. The houses are reminiscent of the homestead type housing, but more sophisticated because of the use of contemporary materials, polyethylene film, foams and other products. Many people who have settled in the area with the idea of becoming self-sufficient.

Oregon

Interest in earth sheltering is gaining wide acceptance and popularity. Perhaps the most visible house in the Portland area is one designed by Norm Clark. Bill

156

Church, another Portland architect has recently designed 3 houses, one each in Portland, Salem and Medford. Dave Deppen, a former associate of Malcolm Wells has moved to Portland and is pursuing earth sheltering concepts where appropriate. The architectural firm of Wolfe, Zimmer, Gunsel, Frasca has recently completed a large garage complex for Multnomah County. At the University of Oregon there is a strong interest, both in the Department of Architecture and the Department of Landscape Architecture in Eugene. There is substantial interest on the part of the Landscape Architecture program at Oregon State University.

The eastern side of the State of Oregon is hot and dry and cold and dry. It is in that area of the state where we expect to see more development of earth sheltering as an energy conserving concept. The coastal areas and the Willamette Valley anticipate development of earth sheltered concepts because of the concern for ecologically related issues such as land use, erosion, and water retention.

Washington

The State of Washington is divided into 14 climatic zones. There are two primary zones, one the Western side near the ocean, and the Eastern side, east of the Cascade Mountains. The Western side contains the University of Washington near Seattle. Three quarters of the population of the state lives in this area. The climate is rather mild and the concept of earth sheltering is now receiving more vigorous attention than in years previous. An an example, the University of Washington, has created an interdisciplinary program entitled The Built Environment Study Teaching and Research group. There are three people who are working on research projects related to earth sheltering. They are Charles Kippenham, Professor of Mechanical Engineering, Ashley Emery, Professor of Civil Engineering, and Dan Heerwaden, Professor of Architecture.

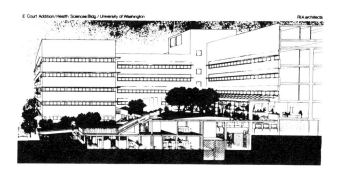

Fig. 3. Section - University of Washington,
Health Sciences Building, Seattle, Washington

A number of student projects, both at the undergraduate and master's level, have been accomplished but are not a focus of any design studio or educational program. It is simply one of the options that is generated and pursued by the students. Within the area there have been a number of buildings that have been generated in the last 10 years that use the earth sheltering concept as a way of achieving desired design goals. Perhaps the earliest project of note is a swimming pool for the City of Seattle that was designed by John Morse & Associates. The primary goal was to gain two uses for an urban site. The project has been extremely successful and well received by the profession and community.

In the San Juan Islands there is an earth sheltered house designed and built by the owners.

On the University of Washington campus an addition to the Health Sciences building (Fig. 3) has been designed and is in the process of construction by a young firm of architects, the Hull & Miller Associates. When they worked for the firm of RIA architects in Vancouver in the early 1970's, they assisted in the design of a major library facility at the University of British Columbia in Vancouver. Their more recent project is for the Robert Hansen family in the central part of the state at Moses Lake. (Fig. 4). They currently have a project under design for a house in the Cascade Mountains. The basic material will be logs.

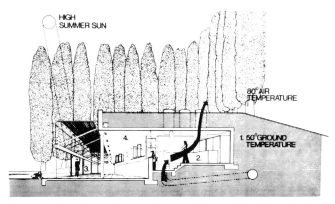

Fig. 4. Section - Hansen House, Moss Lake, Washington

Mr. Robert Kovalenko, architect, in association with Miles Yanek & Associates developed an earth sheltered pumping station for the Metro project in Seattle. The primary goal here was to subordinate the structure to the landscape. The building was located in a residential area.

There has been other designs executed by other architects such as on a fire station in Seattle and Chamber of Commerce building in Olympia. Again the primary goal was not energy conservation, but as a design solution to fit the facility into the landscape in a more graceful way. There are projects in the planning stage for as many as 200 to 300 houses in the Olympia area.

Dennis Niffert, a Seattle architect, has recently designed a home for a client in Boise, Idaho. Building permits in the Seattle area using earth sheltering are on the rise. Steve Denner, coordinator for the Energy extension service has been running a series of courses that have been well attended. They have been taught by Hull & Miller, Donald Nevins, and other Seattle architects. On one occasion, in the City of Bellevue in the fall of 1979, the audience was anticipated to be 75 to 100. Over 450 people sought attendance.

A new company to the Seattle area, called "Terradome" has built its first "earth sheltered demonstration home."

The center of activity, however, in the State of Washington occurs in Eastern Washington where the summers are extremely warm, up to 110° F and the winters can be cold, as low as -40° F. Primary activity centers in the Spokane area or along the Idaho-Washington border. In Spokane there are a number of architects who have developed commercial and institutional buildings. The most visible firm being Walker, McGough, Lyerla & Foltz. Perhaps their most well known building, and the building that has had the most profound impact on the region has been the home office of Central Pre-Mix Company, which is a 2 story office building in an industrial area, a section of which is shown in Figure 5. This project has received much favorable local response as well as a national prize for design excellance.

Fig. 5.   Section - Central Pre-Mix Office Building, Spokane, Washington

Recently they have just bid a project for a 2 story elementary school in Walla Walla, Washington (See Figures 6 and 7).

Fig. 6.   Aerial Perspective Elementary School, Walla Walla, Washington

Fig. 7.   Section - Elementary School, Walla Walla, Washington

Other firms in the area have had a long and sincere interest in earth sheltering as a design concept, and they include Warren Heylman, who designed a bermed house in the earth in the 1950's. Kenneth Brooks, FAIA, has consistently used earth as a design element and as a energy conserving element in buildings or proposals for buildings, including the redevelopment of an army post at Port Angeles, WA. The firm of Design Concepts Associates has designed 4 homes, has 5 homes on the board and has recently published a book called "Homes in the Earth" that has sold over 12,000 copies.

Don Stevens, a writer, editor, and designer lives in Spokane and has been a leader in education and in communicating with the lay public about the potential of earth sheltered housing. Seminars led by Don have reached over 5000 people through the Energy Extension Service. He is currently in his second year of teaching a class at the Spokane Falls Community College.

The Spokane Falls Community College is currently in the process of developing a project called "Earth Lodge," which was a competition and now is a project to be built and funded by the local community. Don has designed 65 houses, has consulted on over 300 others in California, Oregon, Washington, Idaho, Montana, and Colorado. Don is a University of Idaho graduate and has been interested in the subject of earth sheltered structures since his early undergraduate days.

There are a number of individuals who have built their own homes. Perhaps the most well known home is one built by Mr. and Mrs. Pendell. They formerly lived in an above grade solar structure designed by an architect in the early 1950's. The house has become a destination point for visitors. Actually thousands of people have seen the house and heard of the advantages of living in the soil. Mr. Pendell is probably the most colorful advocate of earth sheltering in our region.

Ken McCandless works for Bovay Engineers and has explored the mines and ideas of using the mines in the Northern Idaho area adjacent to the Spokane area. Perhaps the most significant fact about the mines is that there is 200 to 300 miles of tunnels available for some alternative use. At the present time 500 linear feet of these tunnels are being used on an experimental and demonstration basis to grow seedlings for reforestation projects taken place in the mining region. What is so significant about the use of these tunnels is earth sheltered greenhouses is that they are able to produce seedlings in 6 months that would normally take 2 years under conventional nursery conditions and do it at 60% of the cost.

South of Spokane is Washington State University where there has been an architecture program since 1917. The concept of building in the earth has been taught as an option in the studio course directed by David Scott since 1962. During the last 18 years approximately 700 graduates of the Department of Architecture have been introduced to the concept.

In 1968 the first earth sheltered house was constructed by a WSU faculty member, Donald R. Heil.

Early in 1977, a project to provide a park ranger house was initiated. Through the efforts of the County Park Dept. the project became a demonstration house. This project, located on the edge of the Snake River, in a county park, has become a community effort (See Figures 8, 9, 10, and 11). To this date the Corps of Engineers, Bonneville Power, Whitman County Park & Recreation Department, three banks, a power company, 10 professionals, 6 contractors, 50 suppliers, and approximately 200 volunteers have developed a modest, but attractive, 3 bedroom, earth sheltered house. Even though the project is not quite done and will not be dedicated until May 17, 1981, over 8000 people have journeyed a minimum of 20

160

miles and some for 400 miles to have an earth sheltered space experience. Perhaps the most gratifying comment that is heard time after time is, "This is so pleasnat, I could live here." The good news is that the banks and other people in the housing delivery system are beginning to be aware that earth sheltering is a concept that the public is demanding and are taking steps to provide for that possibility.

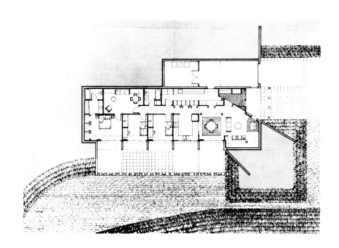

Fig. 8. Plan - Demonstration House, Whitman County, Washington

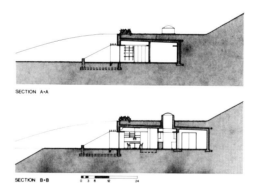

Fig. 9. Section - Demonstration House, Whitman County, Washington

Conclusion

Earth sheltering as a viable concept to save energy, to reduce maintenance, to improve land use, to provide secure environments from natural and manmade disasters as well as to provide a sensitive and visually satisfying result, is growing in popularity among the lay public, the professionals and the financial community.

The center of activity in the Western United States is in the States of Washington Northern Idaho and Western Montana. As each day and week goes by "the idea" is spreading south and inland.

Fig. 10. Elevation - Demonstration House, Whitman County, Washington

THE PRIVATE SIDE

Fig. 11. Elevation - Demonstration House, Whitman County, Washington

Each region, each community, and each person rationizes "the concept" to achieve different goals. The underlying theme however is energy. As one looks at the sections of the projects illustrated it would appear that the common denominator is orientation to the sun.

In the sixties and seventies going to the moon was a concept that appealed to us all and gave us a common sense of purpose.

Perhaps "the sun" will take the place of the moon and unit us as a community and give us a sense of purpose and a sense of order. (Fig. 12)

Fig. 12. Land, Water, Air, Space, Time,
Our Place---------Our Context

LIVING CONDITIONS IN UNDERGROUND HOUSES IN COOBER PEDY, AUSTRALIA

A. D. S. Gillies*, K. E. Mudd**, N. B. Aughenbaugh***
*Assistant Professor in Mining Engineering
**Student in Mining Engineering
***Professor of Geological Engineering
University of Missouri, Rolla, Missouri 65401

ABSTRACT

Over seventy per cent of the Australian continent is classified as arid or semi-arid and most of this land area is subjected to extremes of temperature during the summer months. Mining for opal has been undertaken at a number of locations throughout this region for the last hundred years, and to overcome the adverse climatic conditions of living, the residents have tunneled underground to establish homes. Many subsurface housing designs have evolved over the years.

Coober Pedy is the largest opal producing center in Australia with a population of between 3000 and 4000. Underground space has been developed within the town generally by excavating laterally into hillsides. By this method a hotel, a Church, and hundreds of homes have been built underground.

While some excavations are still mined using manual techniques, mechanized hammers and continuous mining machines, which are normally employed mining for opal, are in widespread use. Using these, the mining of an underground dwelling can be completed in a few days. Exposed wall and roof surfaces are left smooth so that the "squared" room geometries resemble the interior design of a conventional surface house.

A number of underground houses have been completed with features comparable to luxury homes and have been sold at prices exceeding $100 000.00. Important considerations in the design of a house include
(a) the insulating properties of rock cover and temperature extremes experienced inside,
(b) rock stability and the design of pillars, openings and roof spans,
(c) the use of auxiliary ventilation to reduce heat build-up and atmospheric condensation,
(d) the provision for waste water and sewage drainage,
(e) the treatment of interior walls, and
(f) the consideration of noise insulation.
Results from a survey of underground home owners in Coober Pedy will be presented. These illustrate the different approaches taken to overcome problems which can arise from living underground in this environment. While the geological and climatic conditions in this area create an advantageous environment for this form of housing, many of the design considerations present here have application to

164

subsurface construction in other parts of the world.

KEYWORDS

Subsurface housing, opal mining, mine ventilation.

INTRODUCTION

Over seventy per cent of the Australian continental land surface of 7 700 000 km^3 is classified as arid or semi-arid and most of this region is subjected to extremes in temperature during the summer months. Since the first significant discovery of opal bearing rock in 1890, Australia has been a major, if somewhat erratic, producer of precious opal. Important mining fields have been developed at Coober Pedy, Andamooka and Mintabie in South Australia, and Lightning Ridge and White Cliffs in New South Wales.

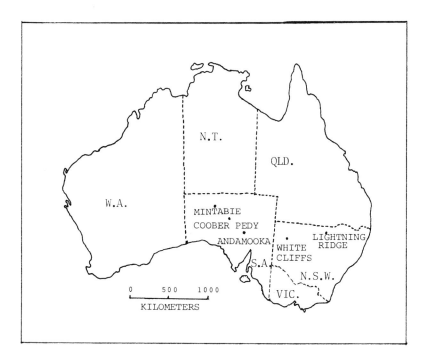

Fig. 1. Australian opal deposits

As the mining fields expanded, townships developed and some mined out underground areas were converted to domestic dwellings. This move to use of subsurface space resulted from both the difficulty and expense of obtaining conventional building materials in isolated areas, and the seeking of an environment protected from extremes of summer heat. Underground housing in arid regions can maintain interior environmental conditions at reasonably constant temperatures due to the insulating properties of the surrounding rock. This ability to hold dwelling temperatures at a comfortable level, and sometimes as much as 20°C below outside conditions, creates an attractive living environment in which energy consuming air conditioning is not generally required.

Coober Pedy is the largest opal producing center in Australia. It has a population which fluctuates throughout the year and reaches a maximum in the cooler months of 4000. Underground space has been developed within the town generally by excavating into hillsides. The outcropping rock resembles a fine-grained sandstone and is locally referred to as a "desert sandstone". Since it is characteristically soft and easily excavated, sub-surface openings have been developed where there is sufficient rock thickness for stable room spans to be mined out. Within the town area, hundreds of "dug-outs", the local term for underground houses, have been constructed as well as other structures such as a hotel and a Church.

The other opal mining towns in Australia are considerably smaller and do not have the outcropping sandstone ridges within their built-up area which facilitate the excavation of underground houses. Some subsurface dwellings have been constructed within these towns by forming a "basement house" normally on the side of a gently sloping hill. For these, a trench-like excavation forms the below-ground opening while a roof placed at ground level encloses the inhabitable space. While the insulating properties of the surrounding rock or soil cover protects these dwellings from the extremes of summer temperatures, the absence of rock cover above them means that they are not as effective in maintaining a tempered environment as the Coober Pedy underground homes.

Although some domestic excavations are still mined using hand tools, mechanized mining machinery, which is normally in employment mining for opal, is in widespread use. With these, the mining of an underground dwelling can be completed in a few days, exposed wall and roof surfaces are left smooth and "squared" room geometries which resemble the interior design of a conventional house can be achieved. A number of underground houses have been completed with features comparable to luxury homes and have been sold at prices exceeding $100 000.00.

THE PHYSICAL SETTING

Precious opal produced in Australia is won from rocks which were affected by deep weathering processes during the Tertiary period, 15 to 30 million years ago. In most areas of commercial mining, the host rocks for opal are bleached sandy clay-stones, originally deposited as marine sands and clays on the edge of the Great Artesian Basin during the Cretaceous period. The weathering process during the Tertiary period made three important changes to the sediments; it broke down the feldspar and complex clay minerals to produce kaolin thereby releasing excess amounts of soluble silica, it helped to create openings in the sediments by dissolving out carbonate lenses and fossil marine shells, and it facilitated movement along faults, joints, and bedding plane surfaces. These solution pores became passageways for groundwater carrying high concentrations of silica removed from the surrounding weathered rock. A general lowering of an otherwise fluctuating water table, which resulted from the onset of more arid conditions, carried silica-rich solutions downward to be trapped in openings above an impermeable claystone stratum. The growth of silica spheres probably occurred during this downward movement. Water eventually evaporated or filtered away, leaving the silica precipitated in the openings as precious opal.

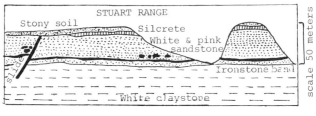

Opal level

Fig. 2. Diagramatic section depicting opal occurrence
(after South Australian Department of Mines)

Opal mining occurs at depths up to 30 m below the surface. A typical strata profile above the level consists of 2 m of red-brown stony clay at the surface overlying white sandy claystone 10 m to 15 m thick. This is followed by a pink sandy claystone with veins of alunite, gypsum and occasionally precious opal. It is in this sandy claystone geological setting that excavation for subsurface dwellings has taken place.

The significant opal mining fields presently being worked lie between latitudes 28°S and 31°S. Climatic conditions for the mining towns are typified by meteorological statistics for Coober Pedy.

TABLE 1 Coober Pedy Climatic Averages

	Jan	Feb	Mar	Apr	May	Jun	Jul	Aug	Sep	Oct	Nov	Dec
9 am Mean Reading												
Dry Bulb °C	27.3	27.0	23.8	19.8	14.4	11.4	10.3	12.0	16.1	21.5	23.1	25.8
Relative Humidity %	33	34	38	45	59	67	65	56	43	31	32	29
3 pm Mean Reading												
Dry Bulb °C	35.4	34.3	31.6	27.2	22.0	18.4	18.3	19.3	23.1	29.2	30.1	33.9
Relative Humidity %	17	21	21	27	34	40	37	32	24	17	19	15
Daily Maximum												
Mean Temperature °C	36.6	35.9	32.8	28.1	22.1	19.0	18.8	20.2	24.3	30.2	31.7	34.6
Daily Minimum												
Mean Temperature °C	20.2	20.4	17.8	14.1	9.6	7.3	6.0	6.7	9.3	13.7	15.6	18.4
Rainfall												
Mean, mm	15	25	11	5	13	15	7	8	7	12	10	11

Temperatures fluctuate from maximums which exceed 40°C during the summer months to mild frost free conditions in the winter. The air humidity levels are very low through most of the year. Vegetation is sparse, with few trees, and bushes, therefore dust storms are common during periods of high wind.

UNDERGROUND HOUSES IN COOPER PEDY

The town of Coober Pedy is encompassed within a delineated square of 1.6 km side length. While the first below ground houses within the town grew out of underground workings originally developed in the mining of opal, commercial mining within the town limits has not been allowed for a number of years. Nevertheless, construction techniques for domestic underground use follow closely the mining methods being practices within the area and it is not unusual for these houses to be excavated by commercial miners employing equipment normally in use for opal mining.

The underground houses, or "dug-outs" are generally located in the sides of naturally occurring hills where sandstone bluffs outcrop. Originally, mining techniques were simple hand excavation methods. In recent years, however, most excavation has been undertaken using mining machines.

Hand excavation of underground houses follows practices commonly in use in the opal mining. Rock is chipped out with chizel and hammer, loaded into a wheelbarrow by hand shovel and transported outside. Explosives are not needed because the rock is soft. Neither are they used on the mining field as percussion from the explosion would crack the precious opal. Use of light weight, hand held electric chippers improves the rate of mining. This unit is versatile and easily manouvered and allows the operator to develop any room shape. Rough wall surfaces left after use of these electric chippers give an attractive finish to developed rooms.

Fig. 3. Hand held electric chipper

Continuous mining machines which have been developed in the mining fields to cut the opal bearing strata have revolutionized the excavation of underground houses.

Fig. 4. Circular tunneling machine

Fig. 5. Rectangular tunneling machine

Circular head mining machines cut a tunnel of up to a maximum of 2 m in diameter. In the larger units, two cutting heads are attached to a rotating arm to excavate an outer and inner annular ring on each pass. While these machines can open out a smooth walled and structurally stress free passageway, for underground housing use, the corners have to be separately cut out to form serviceable rooms.

The rectangular tunneling machine can cut to 2.4 m in width and allows development of a "squared off" room in one pass. In this unit, cutting teeth are set into a rotating milling head drum which moves vertically, while forward motion is achieved by having the machine track mounted. Both the circular and rectangular mining machines are electrically powered. Broken material is moved from the front of the machines by either the use of loading arms and an internally fitted conveyor system which dumps material at the rear to a series of lightweight conveyor belt units or by a vacuum or suction system which picks up the rock spoil and passes it through lengths of large diameter piping to the surface. Suction pressure is achieved by use of centrifugal fan compressors operating on the surface; the spoil having been removed from the pressure system in passing through cyclone units set prior to the compressors. Baggs (1980) describes the operation of a cutting machine-compressor removal system pair which had the ability to excavate 67 m^3 per day and completed the mining of a 110 m^2 floor plan house in 3 days at a cost of $4500 in 1979.

Shaft sinking in the mining fields is by truck mounted churn drills which can excavate to a diameter of about 1.5 m. Ventilation and light shafts for underground houses are either developed using these drills, or are hand excavated to the surface from below.

UNDERGROUND HOUSING DESIGN

Excavation for an underground house in Coober Pedy commences with the bull-dozing of material at the edge of a sandstone bluff to create a vertical wall of sufficient height to leave a competent roof layer above rooms which will be mined at or below surface grade. In figure 6, the face of one of Coober Pedy's newer dug-outs owned by the Davidson Family can be seen. Entrance doors and windows are shaded by a verandah which runs the length of the house.

Fig. 6. Outside view of Davidson dug-out

From the front vertical face of this house, entrance ways are constructed and rooms opened out back into the bluff mass. Light and ventilation shafts are generally placed at the rear of the house. The floor-plan of the Davidson dugout shows how the various rooms of this house have been mined out, while pillars have been left which act as internal walls.

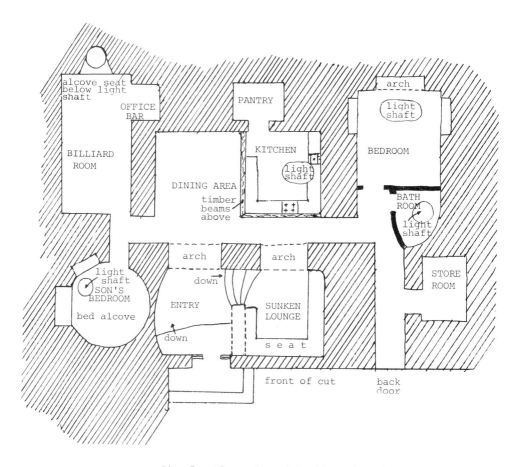

Fig. 7. Floor-plan of Davidson dugout

The Davidson dugout was mined by hand and wall surfaces remain at natural color and carry striated chisel marks. The floor surface is 75 mm thickness of concrete poured on plastic. Floor area is about 175 m². The kitchen-dining area has dimensions 5 m by 9 m while the billard room is 5 m x 8 m, with no interal roof supports. Some natural lighting to the dugout enters through the two exterior doors, while most is brought in through five interior light shafts. The internal surface of each of these has been painted white to increase light reflectivity. Each light shaft is covered and has set into it a louver regulator and small exhaust fan. Some air movement will occur through the house by natural convection, although the direction and rate of flow will vary from season to season. The small exhaust fans can be used to increase flow through the living area when a large number of people are present, although in summer the passing of high quantities of outside air into the house will reduce the ability of the surrounding rock mass to temper the interior environment.

In an effort to maximize rock strata thickness above the excavation, dugouts are normally developed at the level of the outside ground surface or below grade. In the Davidson dugout, the problem of drainage of sullage water is overcome by the sinking of covered storage wells beneath the bathroom and kitchen areas. These are provided with ventilating tubes to avoid gas buildup, and periodically need to be pumped out as surrounding rock is of low permeability.

The Davidsons have taken advantage of the underground environment to create an attractive living area with many unique architectural features. The living area is set a few steps below the rest of the house which gives it a high ceiling and expansive feeling. Lounge sofas have been constructed by leaving rock benches along walls and covering them with cushions. An arched opening has been cut between the living and dining areas. Alcoves and cupboards are readily recessed into rock walls and can be constructed later in the life of the house as the need arises.

Fig. 8. Sunken living room in Davidson dugout

172

Fig. 9. Billard room in Davidson dugout

Fig. 10. Kitchen-dining room in Davidson dugout

The interior wall surfaces within the house have been left at the naturally occurring rock color of beige with brown streaking. A clear sealant material has been applied to wall surfaces to reduce any feeling of dustiness to the touch, and reduce any absorption of moisture which may structurally weaken surface rock. In the kitchen-dining room, railroad ties form an interesting feature dividing the room and supporting a buffet. Although capable of such duty, these carry no structural roof load.

The Davidson dugout has been discussed in some detail to emphasize some of the attractive features that a well designed underground house can encompass. Important considerations in the design of an underground house in Coober Pedy will be examined.

The insulating properties of rock cover

People living underground in Coober Pedy have taken advantage of the tempering effect of soil and rock cover on interior temperatures.

I. The Davidson dugout lies beneath sloping ground under 3 m to 5 m of cover. They report that interior temperatures vary between a winter 21°C and a summer 28°C. They consider that the five vertical light and ventilation shafts serving their house do not lead to a significant breakdown of tempered environment characteristics as these are normally closed to airflow and the shaft air insulates the interior from outside conditions.

II. A second dugout visited had been lived in much longer than the Davidson home. Here, it was reported by the owner that interior temperatures increased slightly with the age of the dwelling. This effect, however is difficult to quantify as other influences such as the number of people living in the dugout and the heat output of lights and refrigerators can distort measurements. He reported that their interior temperatures were held below 27°C but this did entail the use, on occasions, of a small window mounted airconditioning unit in the hottest month, February.

III. A third dugout visited was situated under a sloping hill with the overlying cover varying from 0.2 m at the front to 8 m at the back of the house. The owner claimed that interior conditions remained relatively cool in summer when atmospheric relative humidity was low (as is the case most of the time). However, on occasions when outside conditions are humid in summer, the dugout will become stuffy and uncomfortable and use of air shaft fans under these conditions will only marginally improve conditions. Although humidity can be high, no moisture condensation has been observed on dugout rock surfaces despite their lower temperature. It is speculated that the unsealed rock surfaces may absorb any condensation before it becomes noticeable.

IV. The oldest dugout in Coober Pedy was originally the residence of a miner prior to changing hands and some remodelling to serve as a motel, "Radeka Dugout." Baggs (1980) refers to a temperature survey taken within three rooms sealed from the rest of the excavation. In figure 11, a section and plan view of this dugout area and a temperature profile through the rooms is illustrated. The conditions in the inside rooms closely reflect the surrounding rock temperature of 20°C while those nearest to the front wall, demonstrate the effect of the out-side summer heat. The exterior wall is a weatherboard faced, non insulated, un-lined timber frame construction with one single-glazed panel in a poorly fitted door. There is no external sunshading.

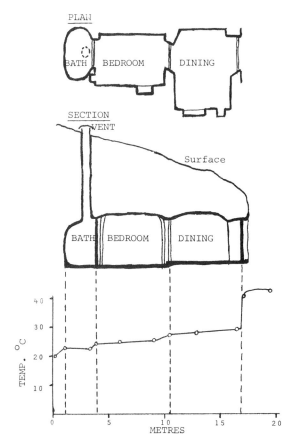

Fig. 11. Temperature profile through dugout (after Baggs)

Coober Pedy experiences summer conditions which are dry and very hot, and winters which are dry and mild. Underground dugouts designed with adequate overlying rock strata and sealed front walls and access ways can maintain comfortable interior conditions at up to 20°C below outside temperatures. Low atmospheric humidity levels through most of the year mean that relative humidity readings inside dwellings remain low and stuffiness is avoided. Condensation of moisture on "cold" rock surfaces is not a problem. In the prevailing hot desert climate, underground houses are insulated from outside temperature extremes and comfortable conditions are maintained with little or no mechanical airconditioning.

Rock Stability

Early underground dwellings which grew up as a by-product of opal mining were laid out to a mining pattern developed to maintain safe and stable rock conditions. With modern demands for large, open living spaces and ease of excavation with continuous mining machines, dugouts have been developed by men with little mining background which maximize the attractiveness of the excavation as a dwelling but fail to pay full attention to the long-term stability of the structure. Presently there are no Government or Municipal codes regulating underground construction for domestic use in this area although the South Australian

Department of Mines and Energy is considering this problem (Minogue, 1979).

Rock properties can vary over a very short distance within the township and a stable excavation configuration in one area is not necessarily repeatable in another.

Fig. 12. Roof rock fall within dwelling under construction

The result of a partial roof failure in a dwelling under construction can be seen in figure 12. Excavation for this dwelling was by rectangular cutting machines. The squared geometry of the opened rooms lead to stress concentration in corners and unconstrained sections of roof strata between pillar supports which can result in failures along planes of weakness.

To increase the amount of neutral light entering the dugout, many owners tend to maximize the number and size of windows and doors along exterior walls. Where rock wall and pillar sizes are such that stress loadings from the weight of over-lying strata on wide-span interior rooms exceeded rock strength, cracks from tensile failure can be seen above windows and doors, and across roof sections, and the effects of compressive failure can be seen in support pillars.

Although a general code written to regulate excavation practices cannot address all problems, specifications as to minimum pillar size, wall thickness, and roof strata cover, and maximum span width for rooms and passageways may improve the structural safety of new dwellings.

Auxilary ventilation

In well designed underground houses, a tempered and comfortable environment can be maintained inside the dwellings with use of auxiliary ventilation. Few houses use airconditioners, and with mild winters little supplementary heating is needed. In some houses no fans are in use to assist air movement, although most of these are either small in area, or have a number of outside windows and doors.

The heat level inside a dugout will be augmented from a number of sources.

I. Inflow of hot outside air through poorly insulated walls or doors and windows. Humid air of high enthalpy will be particularly uncomfortable.

II. Generation of heat from a large number of people confined in a small space.

III. Heat output from lights, motors, and hot water systems. To assist in reducing this build up, many dugouts make use of small fans or ventilators mounted on top of shafts.

Fig. 13. Air ventilator on surface

Use of these at night time to bring in cool outside air and overcome heat level increases during the day is advantageous although they may be used at all times to alleviate interior stuffiness.

Sullage drainage

With the floor level of the underground house often set below outside surface grade, provision of waste water and sewage drainage can be a problem. Various approaches are being tried.

I. Many dugout owners have placed a separate building attached to their outside wall to house a bathroom and laundry. These outer rooms are normally directly connected to the interior of the dugout, although being set at surface grade, conventional drainage can be used.

II. Boreholes can be sunk beneath the floor of the house to collect sewage. These need to be pumped out periodically as the strata is of low permeability.

III. The borehole for collection of waste water may be placed outside the dugout, with direct drainage, connection to the inside. This placement facilitates pumping out of the waste water, although it can lead to problems if set too low. At times of heavy rainfall, flooding and back up into the house can occur.

The treatment of interior walls and noise insulation

Internal walls in underground houses are either free-standing rock unmined during excavation of the dugout or conventional walls of brick or frame construction. The sandstone rock surfaces have a most attractive texture and are either left untreated or sealed with a clear binder to reduce any feeling of "dustiness." In some excavations walls have been painted in the lighter spectrum of colors to increase light reflectivity.

Underground houses are extremely well insulated from external noise. Further, interior free standing walls dampen noise to the extent that normal room to room conversation becomes difficult. Some owners have sought to overcome this difficulty by excavating additional openings and archways through rock walls, particularly in the case where communication with children is a problem.

CONCLUSION

In Coober Pedy it is somewhat of a status symbol to live in an underground dwelling. A sizeable percentage of the town population live underground, and many of the houses are most attractive with real estate market values in excess of $100 000.00. The success of this development in underground living is due to factors such as
(a)   the availability of mining equipment and expertise,
(b)   the low rock hardness which allows rapid excavation without the use of explosives,
(c)   the hot desert climate which has led people to seek a cool tempered environment, and
(d)   the low atmospheric humidity which allows inside conditions to remain comfortable without problems of condensation.

While the geological and climatic conditions in this area create an advantageous environment for this form of housing, many of the design considerations in use have application to subsurface construction in other parts of the world.

ACKNOWLEDGEMENT

The authors wish to express their appreciation to the staff of the South Australian Government, Department of Mines and Energy for their efforts and assistance in arranging visits to underground dwellings and opal mining operations in the Coober Pedy and Andamooka areas. Further, the residents of these towns are thanked for their help in the undertaking of this survey.

REFERENCES

Australian Government, Meteorology Bureau of the Department of Science and
    Consumer Affairs, Australian Climatic Averages, Aust. Gov. Pub. Service.
Baggs, D. W. (1980). The lithotecture of Australia: with Specific Reference to
    Thermal Factors. First Earth Sheltered Building Design Innovations Conference,
    Oklahoma State University, Tulsa, Oklahoma.
Minogue, J. P. (1979). Private communications.
South Australian Government, Department of Mines and Energy, (1977). Opal in
    South Australia. Government Printer, South Australia.

THE NEED FOR AN INTERDISCIPLINARY APPROACH TO THE DESIGN
OF EARTH-SHELTERED ENVIRONMENTS: THE ROLE
OF THE INTERIOR DESIGNER

Ron Raetzman, School of Architecture and Interior Design
Menelaos Triantafillou, School of Architecture and
Interior Design and School of Planning
University of Cincinnati
Cincinnati, OH 45221

ABSTRACT

Up to now, the design of earth-sheltered environments both, in practice and in education, has been centering primarily around the resolution of functional issues
with respect to energy conservation. Geotechnical, engineering, and architectural
concerns are the major areas of problem resolution. Yet the acceptance of earth-
sheltered environments by the broader public will be tested on the opportunity they
provide for increased livability in addition to energy conservation. This notion
includes the issues of the interior environment as it is interchangeable in design
with the site and architecture.

Three morphological zones need to be established and draw design attention from an
interdisciplinary perspective involving the user, the designer, and representatives
from the development and construction industries. These zones, representative of
the nature of earth-sheltered environments, are: 1) Site/Building Zones; 2) Transitional Zones; 3) Interior Zones. Key variables falling within these zones need
to be addressed equally. Interior designers face a professional challenge and,
with increasing sensitivity to the nature of the problem, need to participate in
all design decisions in order to increase the livability of earth-sheltered environments.

KEYWORDS

Morphological Zones; site/building zone; transitional zones; interior zones;
nature of earth-sheltered environments; livability; interdisciplinary; interior
design.

INTRODUCTION

With the rising attention and emphasis on energy efficiency, a great deal of
sensitive design and development activity has been rapidly enfolding into new forms
of energy-conscious building designs. During the last two to three years earth-
sheltered environments have been increasingly demonstrated as a major response to
the need for energy-conscious design. The momentum for earth-sheltered environments is growing and gradually gaining acceptance as new research, design and

developments display positive and promising performance levels with respect to energy-efficiency and cost-effectiveness.

The currently growing involvement of the Federal government is one such major demonstration in facilitating the phasing of this new technology to proceed from its initial experimental and "informal" stage to an institutionalized one.[1]

Currently, the design and construction of earth-sheltered environments is demonstrated in two distinct approaches: one which clearly maintains an ad hoc owner-designer-occupied profile, and which is responsible for the initiation of the movement and promotion of earth-sheltered environments; and one which is characteristic of the commissioned-designed environments.[2]

In the context of the efforts to promote the development of earth-sheltered environments, the present design approaches pose a critical question: How can the level of livability afforded in earth-sheltered environments be increased via design within the institutional context of land and housing development?

Up to now, the design of earth-sheltered environments both, in practice and in education, has been centering primarily around the resolution of functional issues with respect to energy conservation. Geotechnical, engineering and architectural concerns with respect to development and construction are the major areas of problem resolution. Yet the acceptance of earth-sheltered environments by the broader public will be tested on the opportunity they provide for increased livability in addition to energy conservation.

LIVABILITY OF EARTH-SHELTERED ENVIRONMENTS

The concept of livability is interrelated with the need for a theme indigenous to the essence of earth-sheltered environments which as of now has not yet been demonstrated. Any efforts to develop a theme and increase the livability of earth-sheltered environments will require designers to address simultaneously and interactively the energy, functional, environmental, aesthetic, and socioeconimic aspects of these environments within parameters established by the user's needs and the development and construction constraints. Design must be approached from an interdisciplinary perspective involving an exchange among designers, users, and representatives from the development and construction industries.

The utilization of earth as a barrier to temperature extremes is the most basic aspect of earth-sheltered environments. Unlike "conventional" environments, the nature of earth-sheltered environments requires that the landscape must make a stronger exchange with the architectural elements, space, and forms. Landform is sensitively blended with architectural materials in order to establish the interior environment. The context of the site poses strong constraints to the design of site elements, entry, transitional areas and interior orientation. The natural and cultural elements of the site and its context are brought "closer" to the living environment in a more direct, functional, aesthetic and symbolic exchange. In earth-sheltered environments the adjustment of the building to the

---

[1] Aiken, R., J. Carmody, and R. Sterling (1980). Earth Sheltered Housing: Code, Zoning, and Financing Issues. Published jointly by The Underground Space Center, University of Minnesota and U.S. Department of Housing and Urban Development.

[2] Dean O. Andrea, (April 1978). "Underground Architecture", Journal of the American Institute of Architects, 34-37.

site will develop a responsiveness which, in turn, will set forth the experience of the environment and the level of livability that it can afford. The interior environment becomes the focus of this responsiveness and interchange between site and building, with the exterior appearance and design being important in as much as they provide for the proper transition between the two and symbolize the use of earth as the prime design element.

In earth-sheltered environments the interior is both, the terminous of human activity and experience from the outside (above grade) to the inside (below grade), as well as the starting point for establishing the user's (consumer's) experience of living below grade (earth-sheltered environment). It is the interior environment which counts the most in addressing the need for livability, not as an after-thought to site and architectural design decisions but concurrent with such deci-sions. Architecture, landscape architecture, and interior design decisions need to be made simultaneously in order to establish the level of maximum livability which can be afforded given a set of user needs, development, financial, construc-tion and regulatory constraints. Most of the earth-sheltered environments that have been designed and built do not clearly demonstrate that such an approach was followed. On the contrary, most examples are simply the placement of a building into the ground in an effort to prevent heat loss.

In an interdisciplinary context, interior designers face a professional challenge with an exciting promise. Their involvement in all the phases of the design of earth-sheltered environments will require that they increase their sensitivity to the nature of the problem and address specific design issues. The transitional areas where the outside or aboveground meets the interior space; the organizing of space along small interior zones where natural light and solar radiation enter the earth-sheltered environment; the very important transitional area of the entry to the earth-sheltered environment where the outside (above ground) becomes inside (underground) requiring special design attention with respect to functional, aesthetic, symbolic and psychologic considerations are a few of the major issues that need to be addressed by interior designers through interaction with other disciplines during the design process.

EARTH-SHELTERED DESIGN STUDIO: AN INITIAL SEARCH FOR DESIGN ATTENTION

The subject of energy conscious design as related to earth-sheltered environments (passive solar) was presented to a group of interior design students as the educational objective of developing the level of cognitive skills necessary to: 1) communicate effectively, both graphically and verbally with clients and fellow professionals (team members); 2) design an earth-shelter of limited complexity (residential) with a focus on the interior-environmental issues of programming, and designing for underground habitability; 3) evaluate earth-sheltered environments on the basis of programming, psychology, environment, aesthetics, structural systems, building components, subsystems and building codes.

In order to make this assignment more feasible for interior design students the group was divided into five teams. In addition, five faculty members presented introductory technical information and served as interdisciplinary resource consultants to all groups. The following statement by one of the faculty seems to reflect the observations and experience encountered by the students during the earth-sheltered studio: "The class, as a whole, did an exceptional job and accessing both, the student and faculty observations, it would appear the interior design student had an initially somewhat painful experience in getting beyond the programming process and being able to think exterior design, in order to address the issues of site and transitional spaces that seem to lend themselves so well to livability. The problem was somewhat predictable considering the nature of interior design and its emphasis on livability, but the traditional linear design process which is the current state of the art did not seem to completely lend itself to the needed livability requirements of earth-sheltered environments.[3] This observation and others have led us to believe that it is important for interior design students as well as students from all other disciplines to begin to focus early on three distinct components each of which addresses the three zones of earth-sheltered environments: 1) Site/Building Zones; 2) Transitional Zones; 3) Interior Zones. The focus of these morphological zones should begin to occur during the pre-programming phase, while the interdisciplinary design team and client are starting to discuss and identify the user needs and should continue throughout the entire design process."

Following, are a few selective examples of students' work demonstrating design resolutions with respect to the three zones.

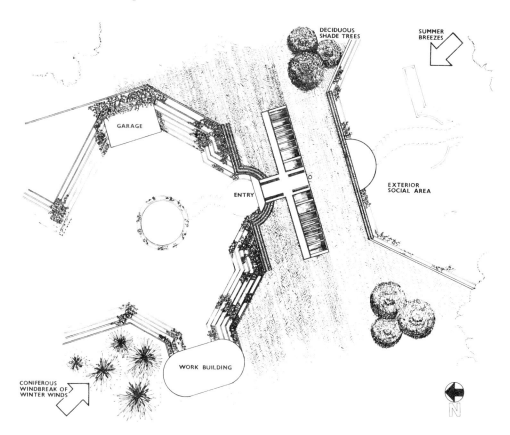

Building and site solution emphasizing relationships of automobile/entry points; grade changes; connection of building to the landscape via earth-berming, retaining walls and vegetation.

[3] Class, R.A. and Koehler, R.E. (1971). Current Techniques In Architectural Practice. Published jointly by the American Institute of Architects and Architectural Record (McGraw Hill). Chap. 3, pp. 19-25

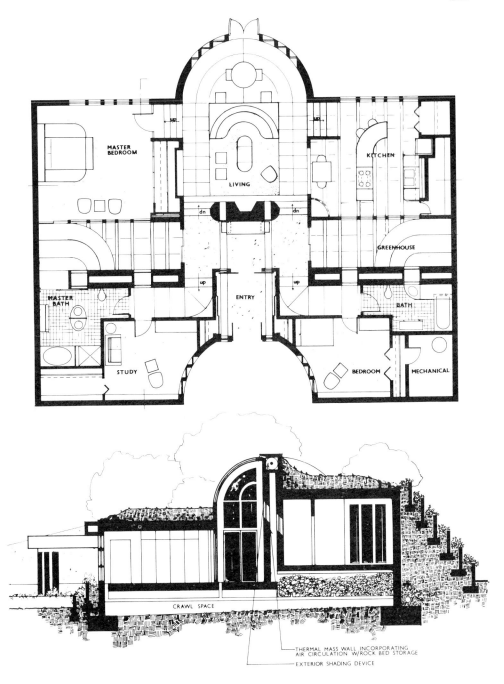

Plan and section of building shown in site plan emphasizing space organization and interior sequence in relation to building siting, grade changes, berming and use of greenhouses.

EARTH-SHELTERED ENVIRONMENT ZONES

Following the observations made during the interior design studios, an initial
effort is being made in this paper to establish three morphological zones
representative of the nature of earth-sheltered environments. These need to become the key areas of design concern inasmuch as they underline the theme of
these environments. In an interdisciplinary approach, the design attention needs
to center on problem resolution of physical elements and design issues with respect
to the three zones of 1) Site/Building responsiveness and the consideration of
site elements, landscape and environmental features, and the site context; 2)
transitional areas between landscape to building, architectural material to the
landscape, entry to the site and to the interior, architectural treatment of the
transition between the inside (below grade) and outside (above grade) with respect
to solar gain, views, etc.; 3) the interior environment with respect to its
sequence with the outside, color, light, organization, materials used, etc.

During the design process, the variables which reflect the needs of the user and
collectively define the nature of earth-sheltered environments will need to be
identified and addressed with respect to the three zones of attention. The
following lists key variables to serve only as initial considerations for design
attention:

Morphological Zones of Earth-Sheltered Environments for Increased Livability

Site/Building Zone	Transitional Zone	Interior Zone
●auto/entry points	●space grade changes	●space sequence and transition
●site grade changes	●solar penetration and solar gain	●human scale/psycho-logical comfort
●earth/berming, building connection to landscape	●entry points	●materials/texture/color
●construction materials	●airlock	●natural and artificial lighting
●insulation	●loading/unloading	●ventilation/heating/cooling/acoustics
●outdoor amenity	●indoor/outdoor transitions	●interior zones exposed to solar radiation/space utilization
●microclimate/modification	●construction materials	●floor material and design with respect to heat gain
●awareness of grade change to below ground	●ventilation	●awareness of below grade space
●building siting/solar exposure	●insulation	●relationship of adjacent spaces
●landform/site performance	●views	●furniture arrangement/relationship to views/solar exposure
	●awareness of the outside	
	●expectation of arriving at below grade space	
	●awareness of leaving the space below grade	

CONCLUSION

The main statement of the paper has been that the design of earth-sheltered environments needs to address the nature of these environments by isolating their key variables with respect to the three zones of design attention. Further, that this attention needs to be approached from an interdisciplinary viewpoint, by addressing equally the necessary design resolutions rather than treating them independently. Trade-offs and adjustments need to be evaluated against all three areas as they will affect their collective performances. Similarly, these three zones intrinsic to the nature of earth-sheltered environments need to be used for the formulation of criteria to be considered both, for design development as well as design evaluation and performance.

By examining current demonstrations and designs of earth-sheltered environments, the nature of the starting point in design is not always clear. Beyond the notion of utilizing earth for berming, design outputs do not clearly demonstrate how the nature and essence of these environments is incorporated into design decisions in order to produce a livable environment. The paper proposed that by focusing on the three zones, design decisions will result in 1) maximizing the level of livability needed in earth-sheltered environments; 2) increase their acceptability by the public; and 3) demonstrate the nature, theme and symbolism of these environments.

The expanding interest to earth-sheltered environments needs to foster a parallel set of activities with respect to their design, a design process, the development of a theme, and the definition of their nature. This paper was able to only make an initial statement and considerations. Many of the most important issues were beyond its scope and will need to be addressed in the years ahead. Three critical questions will require attention: How can programmatic design statements be developed which are able to address the concept of livability with the major emphasis been given at the three morphological zones? How can designers, users, developers and contractors interact to formulate a design process which addresses the theme and nature of earth-sheltered environments within an institutional context? What is the nature of earth-sheltered environments that requires a fresh and unique approach to their design?

REFERENCES

Aiken, R., J. Carmody, and R. Sterling (1980). Earth Sheltered Housing: Code, Zoning, and Financing Issues. Published jointly by the Underground Space Center, University of Minnesota and U.S. Dept. of Housing and Urban Development.
Baum, G.T., A. J. Boer and J. C. Macintosh, Jr. (1980). The Earth Sheltered Handbook. Tech/Data Publications, Milwaukee.
Campbell, S. (1979). The Underground House Book. Garden Way Publishing Co.
Carmody, J., and Colleagues. (1979). Earth Sheltered Housing Design. Prepared by the Underground Space Center, University of Minnesota. Van Nostrand Reinhold Company, New York.
Class, R. A. and R. E. Koehler (1971). Current Techniques In Architectural Practice. Published jointly by the American Institute of Architects and Architectural Record (McGraw-Hill).
Dean, O. A. (April 1978). "Underground Architecture", Journal of the American Institute of Architects.
Wells, M. (1977). Undergraound Designs. Wells, Cherry Hills, N.J.

EARTH SHELTERED HOUSING/COLD CLIMATE DESIGN

by L. J. Atkison

Earth Sheltered/Passive Solar Designer, Consultant, Builder

The proper designed earth sheltered residence incorporating passive solar and the envelope design should be a self sufficient residence in the terms of heating and cooling. My goal is to minimize the use of energy companies. The residence through proper design should heat itself during the winter months and cool itself during the summer months. It should be a living functional design for the climate in which it is placed.

KEYWORDS

Earth shelter; passive solar; envelope design; micro-climate design; passive solar preheat for domestic hot water; passive solar greenhouses.

INTRODUCTION

As we are all ware of, our life styles and our existance depends on energy supplies which are depleting and supplies in which we have little or no control of as far as price and continuing supply from other countries.

The ultimate design for an earth sheltered residence incorporates passive solar for heating as well as cooling. As we all are aware of, earth sheltered houses are virtually air tight. This therefore requires introduction into the house, some form of air change or fresh air for comfortable living atmosphere. The most efficient means of introducing air changes is through an earth tube. The earth tube entering my house will be constructed of 15" inside diameter concrete culverts. These culverts are laid at a depth of 8' below the surface. The run of the culverts will be approximately 200' long daylighting on the side of the hill and the other end comming into the greenhouse area. Depending on the climate and trane in which your tube is laid and depending on the soil conditions such as moisture and type of soil that the tube is buried in the air temperature introduced into the greenhouse will vary somewhat from the figures that I will give you. These following figures are based on the Kansas City area: the tube at 8' depth buried 200' in length and air quantity moving through it at 215 cubic feet per minute. Soil conditions in this area are sandy and damp. The tube slopes away from the house to drain off any ground moisture comming through the joints and also to help condensate air on the walls and drain it back out the exterior end. On a 2 degree air temperature outside, the air introduced into the greenhouse will be approximately 58.87 degrees. On a 110 degree outside air temperature day, the air introduced into the greenhouse will be approximately 64.50 degrees. The balance point where the air on the exterior and the air entering the house is the same as at a 62 degree outside temperature with the air introduced into the greenhouse being also at 62 degrees.

From this earthtube below the concrete floor slab, I have a 6" PVC pipe running from the tube to the back side of my wood burning stove. This is for combustion air. The earth tube is introduced into the greenhouse area because this area is the passive solar heating source of the house. This air will be heated during the winter time and circulated around the perimeter walls of the living spaces. Therefore, heating the perimeter and radiating through the gypsum board walls. Also this heat at the same time will be soaking into the concrete structure. The structure acts as a heat sink. During the summer time, this cool air and with the greenhouse shaded with the overhang, this cool air will be circulated again around the perimeter house cooling the concrete, therefore, cooling the interior of the house. With the envelope design, there is no dead air space throughout the entire structure. The walls are furred with 2 x 4's out from the concrete structure and the floor is raised on 2 x 4 floor joists shimmed off the concrete floor. This will give an even distribution of heat throughout the house in lieu of having a passive solar greenhouse in the front end with warm rooms in the front and cool rooms in the front and cool rooms in the back. This will still occur somewhat but not at the severity of what a typical earth sheltered house with a greenhouse heat source does.

The greenhouses let alone serves a cooling chamber and a heat collector, also serves as a pleasant green envirnoment year around. A moderate size greenhouse during the off growing season or the winter months can produce several hundred dollars worth of vegetables.

The structure of my house is poured in placed reinforced concrete. It is water-proofed with a liquid polyurthene and installed with 2 layers of 1" styrofoam in a staggered pattern. Visqueen will be layed on top of the insulation to act as a slip sheetfrom the earth movement and settlement. 4" of sand will be placed on top of the visqueen on the roof areas to retain moisture during the dry summer seasons for the above plants. The roof will be covered in the center with 4' of earth and then sloping off to the perimeter wall areas at a depth of no less than two feet. This creates a positive drainage from rains and snow melts.

My structure has a two exterior openings or two greenhouses. One facing southeast and one facing southwest. The purpose, one is to give a broader view of the ground in which we live on. The other purpose is that the southeast greenhouse will preheat the house early in the morning after a cold night. During the mid part of theday, both greenhouses will receive some heat but the interior will not require as much. Duringthe evening, we will pick up the sun again strongly in the southwest greenhouse to preheat the house for the cool night time. In the greenhouse area, there will be 55 gallon black polyethelene barrels to serve as a immediate heat storage or heat sink for the hreenhouses. These barrels are filled with water. Our back up heating source is a wood burning stove centrally located in the house. The stove will be surrounded three sides with a masonry wall to act as a immediate heat storage. Also the lowered ceiling which is throughout the house, will not be above the stove area. This allows for the excessive heat from the stove to rise and enter the air plentum or the envelope space of the house and will circulate somewhat by natural convection. As a greenhouse is heated, the air will tend to rise and flow above the ceiling at the same time replacing the extracted air with the cool air from beneath the floor. During the night time, the barrels will serve as the heat storage for the greenhouse in case of prolonged cloudy periods the wood stove heat rising above the ceiling and then falling into the greenhouse or around the perimeter walls will serve as their heat. There is no other mechanical systems involved.

In the garage area, there is a 5' x 6' skylight shaft penetrating the roof. The skylight above the roof has glass facing south sloping on a 45 degree angle. The shaft is our hot water preheat shaft. The shaft will house three of 40 gallon glass line steel tanks painted black. The cold water line entering the house will branch off and fill the lower tank then the line on the opposite end of the tank will fill the second tank. Again the second tank will tie into the third tank. From the third upper tank the line will come down and go over to a hot water heater tank. The hot water heater tank virtually acts as a storage tank and as a back up system. Between the hot water line leading from the upper tank to the hot water tank, a by-pass from the cold water line must be introduced with a cold water mixing valve. There will be periods of time which the water coming from the preheat tanks will be too hot for the hot water heater. On cloudy days, the preheat tanks will still act in their capacity by circulating the cold water coming into the house and bring it up somewhat near room temperature before dumping it into the hot water tank for its final heat. Also there will be a timer attached to the hot water tank to turn it off during the night time houts. 25% of our utility bills is the hot water heater. No one needs hot water in the middle of the night. With this passive solar hot water preheat system, there are no expensive collectors, there are no expensive pumps, no glycol needed and no exterior energy other than the water pressure entering the house to operate the entire system. The pay back period is much greater than the conventionally sold active solar systems. Remember the active solar system is the next generation of dinosaurs. Pour design requires active solar systems.

Above ground houses can be built equal in there thermol properties as earth sheltered houses, but earth sheltered houses properly designed require no maintenance, they are strom free, they serve as a fall out protection and also they have the advantage of the earth as a temperature modulator. Each above ground house that is built blots out that many square feet of nature, never to be recaptured. With the multiplying of the earth's population and the multiplying of our spraulding concrete jungle cities, nature is slowly losing the battle. The earth sheltered house minimizes the destruction of nature. In our society today we tend to manicure nature. We restrict plant type growth in our lawns and cut it to a height that is unnatural. We even mow the median strips of the highways. It has come to the point that out city parks have become a zoo for nature.

The following are items to remember during your own design and construction of an earth sheltered house:

1. Most important is to retain an architect who is knowledgeable of earth shelter and passive solar design. Too many of the architects of today are strictly visual creatures, hoping that their monument in which they have design to themselves will be published in a magazine.

2. The architect whom you retain should himself employ a knowledgeable structural engineer experienced in underground design. The fee in which you pay the architect will be saved due to proper design and through the years free from the common pit falls in which unexperienced people fall.

3. Remember that waterproffing should be liquid polyurathene the insulation should be styrofoam.

4. The drain tile should be rigid PVC and the backfilled earth should be compacted and not just losely filled and run over several times with a tractor wheel.

    The people living in earth sheltered houses today and particularly in the future, will survive very comfortable economically. Retired people particularly on a fixed type income will be able to maintain the life style in which they are acustom. One day our state and federal fovernments will wake up and construct earth sheltered low income housing which will do several things.

5. It will reduce the maintenance because these homes normally are not taken care of.

6. We the public will not be subsidizing their heating and cooling bills.

These structures therefore, will be good for hundreds of years with a minimal amount of maintenance or remodeling on the interior. Thus, this will give pride and dignity to the low income and elderly people who live in these structures.

L.J. ATKISON   DESIGNER
PASSIVE SOLAR - UNDERGROUND

Home to be toured during USCE-81
Designer: L. J. Atkison

AN EARTH BERMED - PASSIVE SOLAR HOUSE, CASE STUDY OF AN ACTUAL PROJECT

Timothy E. Montgomery* and C. Dale Elifrits**

*Architectural Designer (Project Architect), Montgomery and Associates
Bonne Terre, Missouri 63028

**Assistant Professor, Geological Engineering (Project Owner)
University of Missouri-Rolla
Rolla, Missouri 65401

ABSTRACT

This paper presents the reasoning and cronology associated with the decision to construct an earch bermed-passive solar house, the site considerations, the architectural considerations and a summary of the actual plans and specifications for the structure. Cost data are included, as are illustrations of plans, and discussion of the projects completion.

A design concept using conventional reinforced, poured in place concrete for the earth contact portion of the structure was used. The roof is a conventional, clear span, truss supported roof (trusses placed on twelve foot centers), with the south portion shorter than the north portion and sloped normal to the winter sun. A cathedral ceiling throughout most of the structure, developed around the twelve foot bays, provides ample passive solar gain through sky lights in the south facing roof.

Air handling is accomplished by a duct system moving air from the highest point in the ceiling to the eve level during winter and reversing this in summer. Supplementing this system is a wood burning stove and bathroom-located electric baseboard heaters. A one-and-one-half ton air conditioner is added for dehumidification. An earth tube for air tempering will be added to facilitate cooling whether using natural ventilation or the air handling system.

Costs are calculated at $27.50 dollars per square foot. These costs include site evaluation, all construction materials except floor coverings, all labor except that of painting and wood finishing, e.g. staining. Also included are a designed and steel cased well 235 feet deep, with pump and an aeriation waste water treatment plant with lateral field. Cost of land is not included, but excavation, backfilling, and grading costs are included.

INTRODUCTION

Perhaps the greatest concerns of individuals building homes today are the cost and availability of energy for heating and cooling purposes. Therefore, viable alternatives to conventionally-built houses must be investigated. One energy-efficient alternative is a home having an earth shelter/passive solar design. This utilizes the sun's "non-depletable" energy in a passively-heated structure, in combination with earth berming and super insulation, isolating the structure from the prevailing

191

external conditions.

DEVELOPMENT OF A DESIGN CONCEPT

The design and subsequent construction of this house were based on the following concepts:

a) The need to construct a dwelling which would be energy efficient, yet have a high degree of liveability,

b) The desire to demonstrate the state-of-the-art earth shelter and passive solar technology in a private dwelling,

c) The need to construct the dwelling economically, using standard and simplified construction methods.

Several specific parameters were considered in designing to meet these objectives. The initial consideration was the site plan. Drainage and geotechnical details were to be considered (through input from a geological engineer), as well as a-chievement of the desired passive solar concept. Ultimately, a south exposure was selected.

Several techniques with respect to the concrete work were researched. The "post-tensioning" technique, although quite viable, was considered to be too expensive and unfamiliar to local concrete contractors. Additionally, pre-cast concrete slabs were rejected, due to lack of availability, lack of local experience and problems of water proofing the structure. Consequently, the conventional poured in place reinforced concrete method was chosen.

The "earth-covered" vs. "earth-bermed" matter was studied. It was concluded that the additional cost of constructing an earth covered roof would not be offset in a reasonable period of time by the accompanying decrease in energy usage. Therefore an earth-bermed design was planned.

With the decision to have a conventional roof came the further decision that the roof was to be clear span truss construction. The major reasons for this were the quick closure and flexibility of interior structure allowed by such construction. (It should be noted that the fact of wintertime construction for the project helped to dictate this decision.)

The energy efficiency of an earth-sheltered home depends partially on the insulation throughout the structure. It was decided to insulate the roof to a value greater than R-33 and the exposed walls to at least R-24. Further, styrofoam was to be placed outside the concrete walls.

Finally there was the quite significant problem of achieving the passive solar effect. This was to be accomplished through the use of sky-lights and windows on the south side of the house.

Considering the above parameters, it was concluded that the project should also be cost-effective, with a goal being set of less than $30 per square foot.

ACTUAL DESIGN

With the objectives as guide lines, floor plans were researched, cross-section concepts were devised and as an end result the plans were developed. This took approximately two months including three evaluation/revision conferences between owners and architect. A structural engineer was retained by the architect to analyze the structural aspects of the various designs as the final concepts were developed and as shown in Figures 1a, 1b, 2 and 3.

At the same time site evaluation was being completed. The site was selected as the architect was retained. A two foot contour map of the building area of the

193

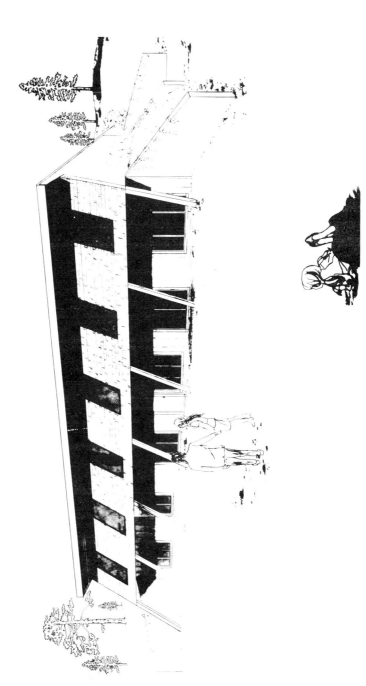

Figure 1a. Architect's Rendering of the Completed Project

194

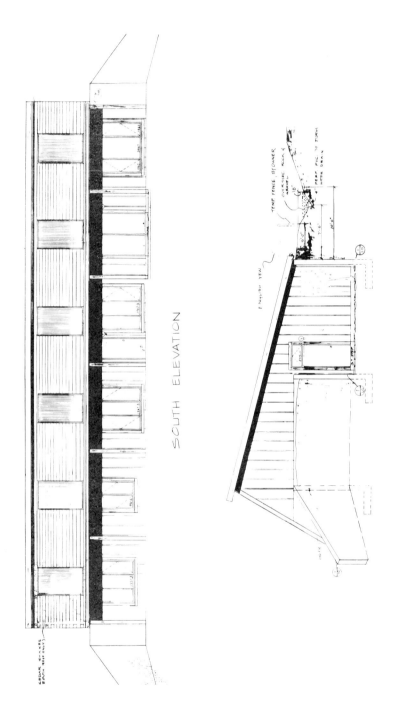

SOUTH ELEVATION

Figure 1b. South and East Elevations of the Structure

195

Figure 1c. View of the Completed Structure from the Southwest showing the South Face. Final landscaping and lot cleanup remained to be finished at the time of this photograph, 30 March 81.

196

Figure 1d. View of the Completed Structure from the East showing the Main entrance. Retaining wall and entrance area remained incomplete at the time of this photograph, 30 March 81. Landscaping and the incomplete items are to be finished by owner.

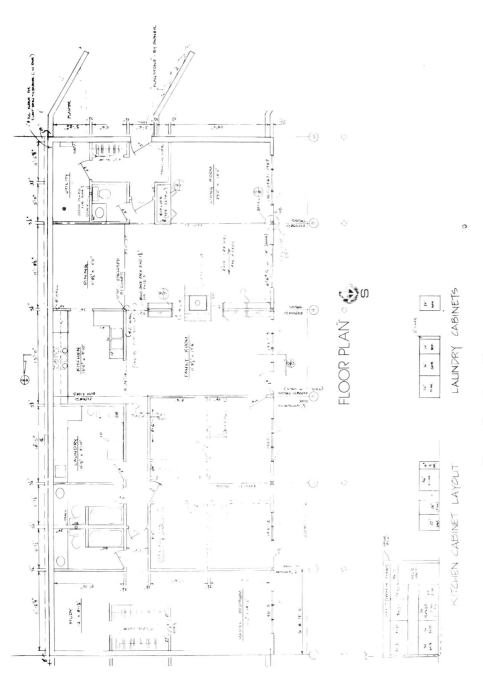

Figure 2.   Floor Plan

198

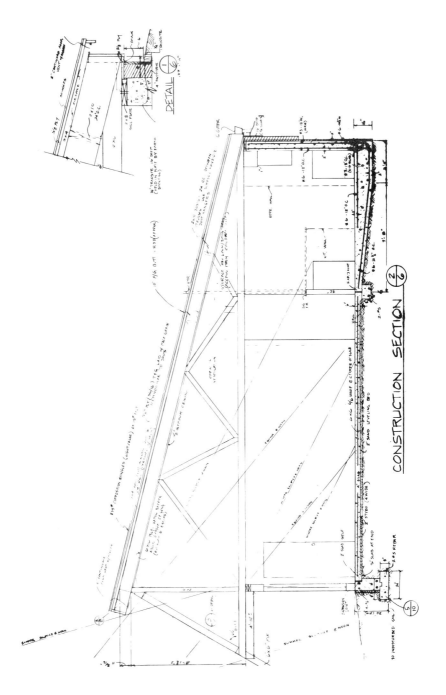

Figure 3.   Construction Section Through the Kitchen—Family Room Bay Showing Calculated Sun
Angles, Drainage, and Reinforcing.

4.34 acre lot was constructed from survey data. The well site and septic tank site were established.

The house was placed in a south facing break in slope to make maximum use of the south aspect without unneeded excavation. The excavation was designed and the volume of borrow was calculated for both cost estimates and backfill requirements. The site geology was evaluated to identify drainage design, foundation design, and excavation procedures. Results of this work are contained in Figures 4 and 5.

STRUCTURAL DETAILS

The concrete portions of the edifice were poured in place, conventionally, with a-bundant use of reinforcing bars. However, in an innovative move, the north wall, bearing the greatest loadings was designed to carry this load to the floor slab, thus saving steel and concrete, see Figure 3. Conventional walls would have re-quired a footing below and north of the wall, thus necessitating more excavation, as well as additional steel and concrete. It should also be noted that the con-crete walls have no footings protruding; this provides for better drainage, better application of water proofing and insulation and no necessity of pillars.

Drainage was of primary concern. The north wall will carry the largest volume of water, due to the roof and earth berming. Roof drainage is provided via a six inch gutter system, Figure 3, with care taken to slope backfill away from the house. Foundation drainage is provided through additional subgrade drainers be-neath the slab with protected, washed gravel filter as backfill adjacent to the wall, Figure 6. The entire concrete wall is waterproofed with coltar epoxy.

A simplified roof structure was desired, one which would maintain an "openness" and structural integrity. A wood truss system was designed to provide these qualities, being positioned on twelve-foot centers and supported by only the north concrete and south frame walls, Figure 3. Some of the trusses remain exposed (and stained with polyurethane stain varnish), contributing to better air circulation and reduced costs, i.e. reduced need for drywall and ceiling joists.

The insulation requirements were met in a variety of ways. A factor of at least R-33 was achieved in the ceiling through the use of 9½" bats. By using 6" bats, plus 3/4" thermax on all exposed walls, at least R-24 wall insulation was attained. Styrofoam, as shown in Figure 6, was used on all earth contact walls. Additional insulative measures include vapor barriers; dead air spaces; wood framed, thermo-pane casement windows and metal clad, insulated, weather proof doors.

The south exposure of the house allows for maximal passive solar gain, as well as admission of ample natural light. The north roof is approximately parallel to the winter solstice sun angle (for this latitude) to admit light to the rear (north) areas of the interior. The slope of the south roof is perpendicular to the same angle, to take full advantage of winter solar gain. Exolite panels were chosen for the sky-lights in the south roof, offering the same light-diffusion proper-ties as glass with better insulative and structural characteristics than glass.

The roof overhang on the south side is designed to minimize solar gain at the summer solstice. Thus maximum efficiency is reached. The roof itself is designed to be self-venting, with vent air passing through a series of screen vents in the north soffit (overhang), passing over the trusses and perlons, above the insu-lation, and out through a screen vent in the south soffit. This will contribute greatly to cooling the house in the summer.

A conventional heating, ventilation and air conditioning system was designed to augment the above-mentioned effect; it will be used primarily for ventilation throughout the year, Figure 7. In winter, a duct system will draw warm air from

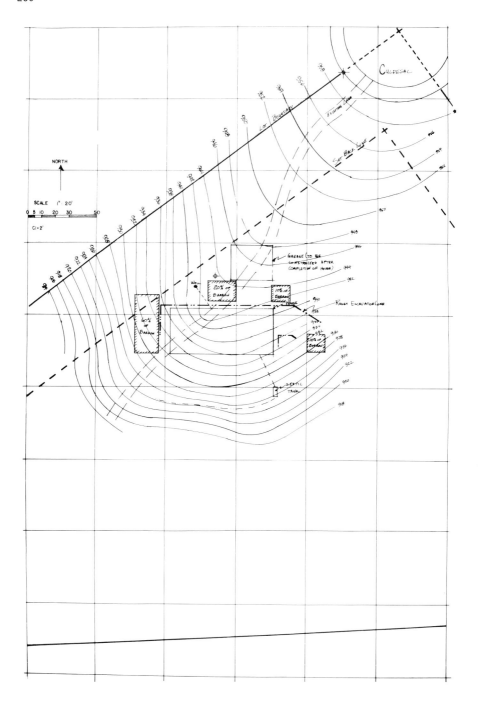

Figure 4.   Contour Map/Site Plot Showing Excavation Plan, Well and
Septic Tank Locations.

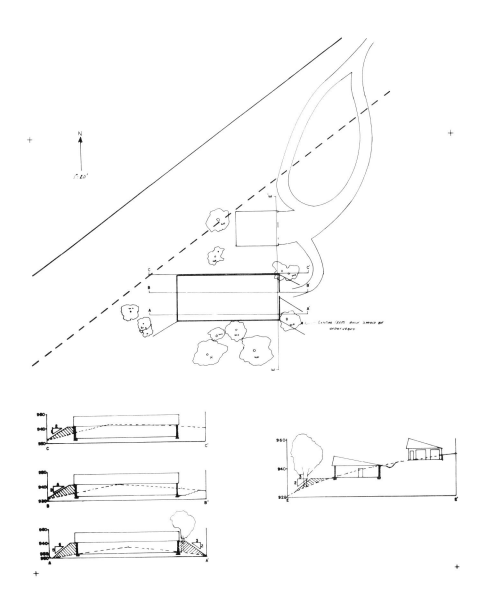

Figure 5. Site Plot Sections

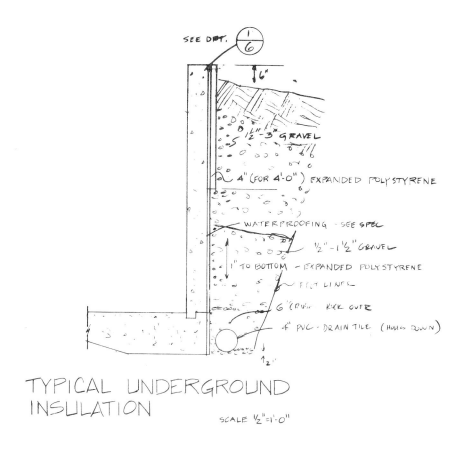

TYPICAL UNDERGROUND
INSULATION

SCALE ½"=1'-0"

Figure 6.   Earth Contact Wall Detail Showing Insulation, Drainer and Backfill

the south ceiling, transmit it through an air handling vent to ductwork distributing it along the north wall and thus, warm the back side of the house, in summertime, this air flow will be reversed, cooling the south side. A ceiling fan in the living room will provide additional air movement. Also, an opening was made in the north wall adjacent to the center of the north side duct. This was made to provide a means for connection of a yet-to-be-designed and constructed earth tempering tube. Vents were arranged beneath the south roof overhang in the south roof/wall to work in conjunction with this tube for natural ventilation during summer. These vents can be closed and insulated during winter.

The floor plan of the house had, as its basis, the Malcolm Wells "Terrasol" plan (Figure 2) with adaptations and implementation to suit the individual needs of the owner. Efficiency and simplicity were key factors considered. An airlock/entry is provided at the main (east) entrance to minimize infiltration of outside air. The living areas (kitchen, dining room, living room, family room) remain "open" to facilitate the use of the wood-stove and airhandling system and increase the liveability of the family portion of the dwelling. Built-in shelves and display cases in family and living rooms were designed to save space and to display books, collectibles, etc. The kitchen, although located towards the rear of the house, has been arranged to provide a good view of the outside, as well as other parts of the living area, while working at the sink. A unique venting system was devised, to eliminate the vent stack for the sink, again leaving this space open. The laundry room is located near the bedrooms, the sources of soiled clothes. It is notable that bathrooms, laundry and kitchen have been located to minimize plumbing materials. Space is provided above the laundry and bathrooms for heating/air conditioning equipment, as well as for storage. Children's bedrooms will be completed by the owner using modular wall sections; again, the clear span trusses provide flexibility for location of these walls. A play area is presently incorporated for the children. As they grow, rearrangement of the bedrooms will be accomplished easily. The master bedroom has its own bath and a large walk-in closet, which also serves as a partition to a study at the north wall.

The entire structure exceeds local, regional and state fire codes. Windows in the south wall are egress type. With the gypsum and concrete composition of most of the dwelling, any fire that might occur would burn very slowly.

ACTUAL COSTS

The construction costs discussed here do not include the purchase price of the 4.34 acre lot. It is located in a registered development southwest of Rolla, Missouri, and has no water supply or waste treatment systems available.

The cost of construction was $27.50 per square foot. This includes the following items:
1. well, with pump and pressure tank installed,
2. excavation and backfilling,
3. all construction materials,
4. all labor, except that of painting and wood staining,
5. aeriation waste water treatment system, installed with lateral field.

CONCLUSIONS

Construction of an earth sheltered dwelling with passive solar aspects need not be complex and unusual, but rather a novel application of conventional methods, with design and judgement. This project was built by a local housing contractor, Paul Lewis Construction, without problems or special management.

Earth sheltered homes take no longer to build than conventional dwellings. Due to its design, the structure was under roof within twenty-five working days (19 Nov.

204

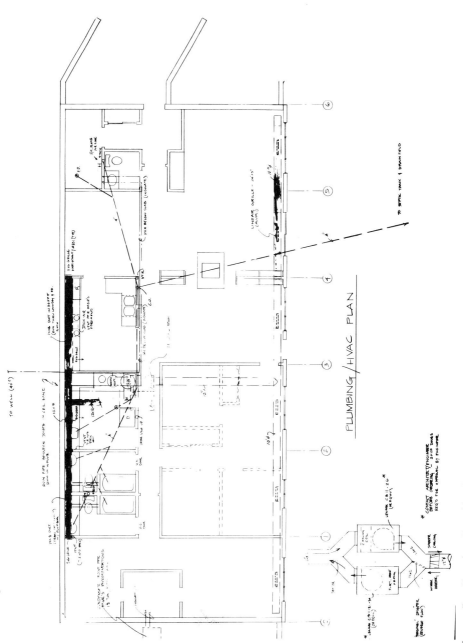

Figure 7. Heating Ventilation and Air Conditioning System with Plumbing Layout

31 Dec. 1980), from ground breaking. The total project was complete, three months thereafter as shown by the cronology in Table 1.

The cost of an earth sheltered structure is no greater, and possibly less, than a comparable conventional structure. Compare the $27.50 per square foot cost of this project to the $40.00+ per square foot average cost of homes in this locale at this time.

Although this dwelling is not the final nor all inclusive solution to energy efficient domestic housing, it is a valid step in that direction. Hopefully, many of its features can be adapted for other dwellings to suit different locations and personal desires.

TABLE 1. Abridged Cronology of the Project

January 1979 to June 1980	Building sites examined, evaluated using aerial photographs, existing topographic maps and site exploration/mapping. Basic concepts of structure's design established
3 July 1980	Purchased building lot, retained architect
25 July 1980	First plan evaluation/revision conference with architect, site plot constructed, excavation planned
8 Sept. 1980	Third and final evaluation/revision conference with architect (this was delayed three weeks due to the owner's vacation). Well drilled, excavation contractor selected from job estimates
20 Sept. 1980	Final plans ready for contractor bidding
16 Nov. 1980	Contractor selected, contract for labor and concrete materials signed, owner as general contractor
19 Nov. 1980	Excavation opened, completed 20 November.
28 Nov. 1980	South footings poured, remedial work completed where required on rock foundation
12 Dec. 1980	Slab/footing poured around east, north, west sides
19 Dec. 1980	Walls poured
24 Dec. 1980	Roof trusses set
31 Dec. 1980	Roof completed
13 Jan. 1981	Slab poured inside the structure to complete floor and concrete work
19 Jan. 1981	Majority of inside stud walls completed
20 Feb. 1981	Dry wall work completed, all outside finishing completed
23 Feb. 1981	Backfilling completed
13 Mar. 1981	Retaining walls, rough landscaping completed
20 Mar. 1981	Contractor finished; Painting, staining, final landscaping to complete.

APPENDIX - ABRIDGED SPECIFICATIONS

Excavation shall include all required material removal for structure, drainage at both footing level and surface as per site plot and plans. All trees and identified natural features are to be undisturbed. Borrow is to be placed to facilitate construction, drainage during construction, and backfilling per site plan. Placement and quality of backfilling are to be approved by geological engineer on site. Subgrade drainers in addition to those shown on plans, are to be placed on cost plus basis per geological engineer's on site recommendations.

Termite Protection shall include use of working solution of 0.5% aldrin as toxicant placed at recommended levels in subgrade below scabs, footing, etc. and in backfill by reputable exterminator who provides a five year or longer guarantee.

Concrete shall meet minimum 3,000 psi design strength at 28 days, six sacks cement per cubic yard mix, and maximum four inch slump on slabs, five inch slump on foundations and walls. Footings shall extend to undisturbed bedrock where excavation does not reach such material with piers required on southwest and southeast corners. Reinforcing bars shall conform to ASTM A615-40 grade. Granular fill shall be crushed limestone/dolomite or washed river gravel. Forms shall conform to shades, lines and dimensions shown. Concrete shall be placed in a manner which will prevent segregation of material and prevent occurrance of "honey-comb"; sections which have "honey-comb" will not be accepted. Protection of concrete from freezing shall be provided when required.

Carpentry: Lumber shall be number two or equivalent yellow pine or fir. Clear span, wood trusses, designed, built, and guaranteed by reputable truss manufacturer shall be used for roof support. Exterior plates, cants, etc. shall be pressure treated with pentachlorophenol. Siding shall be rough sawn cedar approved by owner. Boxing shall be 1/2 inch or better thermax, roof decking shall be three-quarter inch tongue and groove plywood or equivalent. All exterior stud walls are to be six inch walls with R-19 insulation, with partitions to be four inch walls. All work is to conform to plans in size and shape.

Waterproofing shall be coal tar epoxy equal to or better than Corotar (Cook Paint) and applied per manufacturer's specifications. Vapor barriers shall be six mil or heavier polyethylene. Drainers shall be PVC or approved equivalent. Drainers shall be placed per plans with changes as approved by geological engineer.

Roof shall have 15 pound asphalt felt membrane and 235 pound seal tab shingles. Insulation shall be R-33. All is to be installed to conform with material manufacturers' 20-year guarantee.

Windows shall be wood frame, thermopane, casement windows equipped with screens. Sky lights shall be Exolite (47 1/4" X 96") panels mounted in wood frames constructed on-site to manufacturer's specifications using 60 EPDM dry gasketing.

Exterior doors shall be metal clad, insulated, weather tight doors equal to or better than Perma-Door (American Standard).

Plumbing, electrical and heating/air conditioning shall be completed as per plans or codes where superseded by codes.

Site Improvements: The well shall be cased to a depth of at least 200 feet with six inch steel casing and grouted. The well bore shall be completed at a depth in excess of 200 feet to assure (a) 20 gpm minimum production and (b) a clean, open bore hole without cuttings or debris remaining in the well. The pump shall be minimum 3/4 horse power, 10 gpm capacity with lightning protector, sleeved electrical wire and minimum 1", sand packed, delivery line to an 80 gallon minimum, pressure tank.

Septic tank shall be aeration type installed per site plot with minimum of 100 feet of gravel packed lateral field. The waste line from foundation to tank shall be sand packed to protect it from rock damage.

Retaining walls shall be either concrete per plan or pentachlorophenal treated oak installed similar to concrete detail, with geological engineer's supervision. Backfilling and drainage shall be as for earth contact wall plus weep holes through the retaining walls. No eve or surface drains shall be discharged through sub-surface drainers.

Landscaping and final detail grading are to be completed by owner.

PASSIVE AIR CONDITIONING: AIR TEMPERING IN FLOW THROUGH ROCK
LINED TUNNELS

A. D. S. Gillies* and N. B. Aughenbaugh**
* Assistant Professor of Mining Engineering
** Professor of Geological Engineering
University of Missouri, Rolla, Missouri 65401

ABSTRACT

The insulating properties of rock surrounding underground openings have been used
to advantages in creating environments in which temperature is stable and characterized by the mean annual climatic condition at a particular geographic location.
However, significant flow rates of air at surface temperatures into subsurface
space will cause much of the insulated advantage of the tempered environment to be
lost unless the air is first passively conditioned by passing through a rock lined
tunnel.

Change in temperature and enthalpy as air passes along a length of rock lined
passageway will be affected by many parameters, the more important of which include
(a)   the airway length,
(b)   the amount of free water present on airway surfaces,
(c)   the thermal conductivity and thermal capacity of the rock,
(d)   the temperature of the rock, and
(e)   the air velocity through the airway.
Optimization of airway design to achieve efficient tempering of outside air ex-
hibiting extremes of climate involves examination of the influence of these conditions and the interrelationship of changes over time.

An in situ study to examine these conditions with time was undertaken by passing
air through a 185 m length of passageway in dolomite at the University of Missouri-
Rolla Experimental Mine. Continuous measurement of air rock temperature over a
2.5 day period gave data from which appropriate design conditions for a particular
application can be ascertained. These tests were conducted during the Summer, Fall
and Winter months to examine seasonal effects.

Thermodynamic conditions at the air/rock interface are important in ventilation
engineering of deep underground mines. Airflow heat transfer studies are of
particular significance in hot mines in which air refrigeration is necessary. From
various studies, theoretical relationships have been developed for predicting heat
flow across the interface. Results of this study are compared with the theoretical
relationships from which conclusions are drawn.

KEYWORDS

Subsurface space, mine ventilation, psychrometry.

INTRODUCTION

It is well known that the temperature of soil or rock at even a few meters depth
demonstrates a profile markedly different to the diurnal and seasonal readings
recorded on the surface. The earth tends to act as a thermal blanket because the
low heat conductivity of soil and rock slows changes in temperature so that at
depth of 10 meters an unfluctuating temperature is reached which characterizes the
annual mean conditions at a particular location. Figure 1, is a graph that illustrates the measured temperatures with depth and season at Rolla, Missouri.
Attention is drawn to the fact of the rapid damping effect the soil has on the
fluctuating air temperatures.

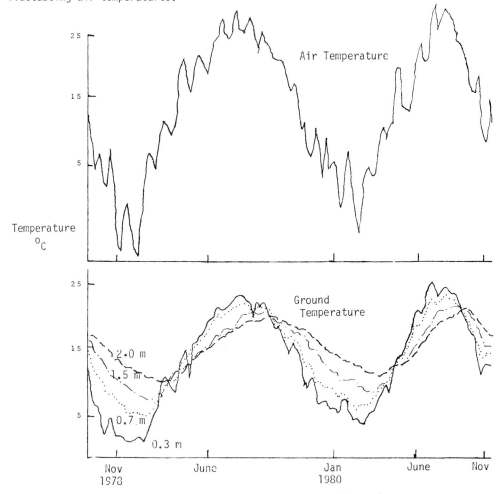

Fig. 1. Average daily temperatures, Rolla, Missouri.
Air temperature and ground temperatures at depth.

The placement of structures partially or fully underground is a method of incorporating this natural insulation to achieve tempered or controlled environmental conditions throughout the year.

In the static environment in which little or no air enters or leaves an underground space, thermodynamic theory can be applied to prediction of the closed system psychrometric conditions and a number of studies have been undertaken to empirical verify these relationships. Stauffer (1978), Lorentzen (1978), Boteau and Latta (1978) and Warnock (1978) have used these relationships in examining the effects of heating or cooling underground rooms to maintain conditions above or below the virgin rock temperature. In the design of underground housing, however, some account must be taken of air movement from outside. Positive ventilation is necessary to satisfy respiration requirements and in hot humid climates to reduce condensation of moisture on interior walls. The introduction of outside air will change the static environmental conditions within the underground opening. Significant flow rates of air directly into the dwelling at surface temperatures will cause much of the insulated advantage of the tempered environment to be lost. To overcome this dynamic change in conditions as air movement occurs, active air conditioning of surface air can be undertaken through the use of conventional mechanically operated cooling or heating equipment or alternatively passive systems can be incorporated in the underground opening design.

One of these passive systems for tempering air flow involves passing the intake air through tunnels or buried pipes before it enters the subsurface dwelling. Contact with rock or earth pipe surfaces transfers heat and moisture to or from the surrounding rock mass and adjusts air temperature to subsurface conditions. As part of an in situ investigation of roof stability in coal mines Bruzewski and Aughenbaugh (1977) found surface air, used for ventilating a mine, undergoes rapid tempering as it passes down the air shaft and along the passageways of a mine. In figures 2 and 3, some results from this study are set down.

Wells (1977) in discussing passive ventilation states
"...we think about earth pipes, about drawing fresh air into buildings-through long, buried pipes that would warm the icy winds of winter and cool the hot air of summer, making air conditioning and heating far less expensive...we know that a straight buried pipe, even a hundred-feet-long pipe, will not do the job very well...but if a whole maze of such pipes was laid in a buried bed of stones?"

In the ventilation of deep mines, consideration has to be given to the transference of heat from hot country rock as air passes through tunnels to ventilate working stopes. Practical engineering criteria used in the design of mine ventilation systems, or where necessary, underground refrigeration units has application to the tempering of intake air to subsurface dwellings.

To study the tempering effect of passing surface air along rock lined passages before it enters occupied subsurface space, a series of tests were undertaken using ventilation facilities at a small mine in dolomite rock at the University of Missouri-Rolla. Experimental results have been collated and compared with thermodynamic relationships and heat transfer equations used successfully in South African Gold Mines in an effort to determine their application to the design of inhabitable subsurface space at shallow depths.

EXPERIMENTAL PROCEDURE

The tempering effect of passing air along rock tunnels was investigated by the following in situ test procedure at the mine:

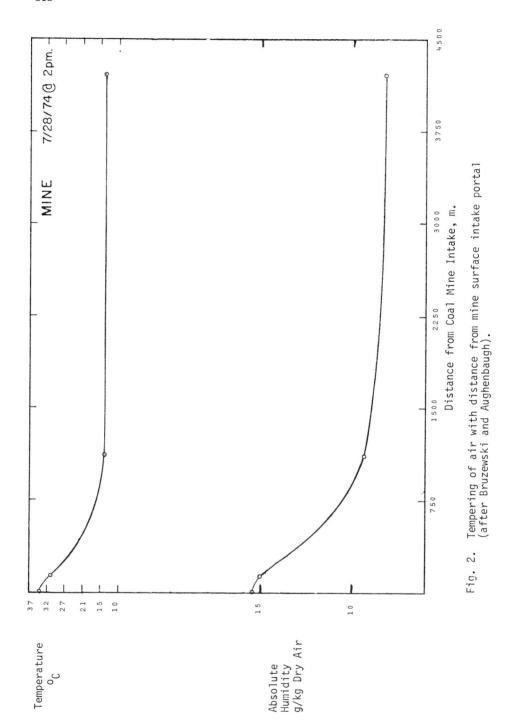

Fig. 2. Tempering of air with distance from mine surface intake portal (after Bruzewski and Aughenbaugh).

213

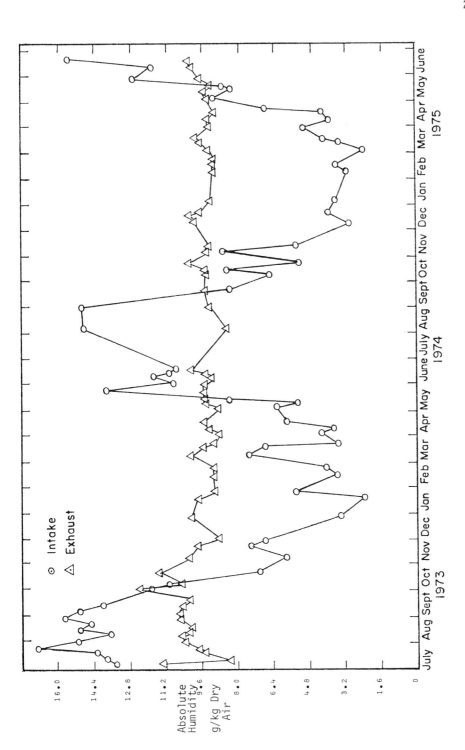

Fig. 3. Seasonal plots comparing absolute humidity at mine intake and exhaust stations
(after Bruzewski and Aughenbaugh).

1. A continuous airway 185 meters in length from the surface intake to the flow outlet in part of the mine was prepared. This continuous air flow pathway was composed of a 10 m length vertical shaft connected to a surface blowing fan and associated ductwork, and a horizontal section through the dolomite mine with average dimensions 2 m by 2 m. The airflow was forced to pass around a number of bends and the depth of the rock and soil cover was approximately 10 m throughout the airway length.

2. A variable speed vane-axial fan powered by a 15 kW motor forced air through the test airway. During the tests, an air velocity of approximately 1.0 m/s through the system was measured, giving a quantity flow rate of a little over 4 m^3/s.

3. To evaluate changes in air psychrometric conditions while considering the effects of diurnal and seasonal climatic influences, tests were run in the Summer, Fall and Winter with continuous measurements recorded over a 49 to 54 hour period on each occasion.

4. During the undertaking of each test, air temperature readings were taken at regular intervals (30 minutes initially and less frequently as differential changes became less) on the surface and at points along the underground airflow path. Further, airflow rate was checked regularly, and rock temperatures were taken at a number of test stations. At these stations rock temperatures were recorded at depths of 25 mm, 100 mm and 300 mm from the air/rock interface by use of previously installed thermister temperature probes. All tests were undertaken at an approximately uniform flow rate. Further details as to experimental test procedures are recorded in Smith, Orlandi and Gillies (1981).

EXPERIMENTAL RESULTS

The University of Missouri Experimental Mine is not an operating mine, ventilation fans are inoperative for most of the year and natural airflow passing through the workings is slight. Throughout the year, air temperatures in the underground passageways and rock mass temperatures (virgin rock temperature) are almost constant; varying from 10 to 15°C. The minimum value demonstrates a seasonal lag, occurring in early Spring, while the maximum takes place in early Fall.

The test sequences were undertaken at different times during the year,
(a) Summer: July 16, 17 and 18 1980,
(b) Fall: September 5, 6, 7 1980 and
(c) Winter: January 30, 31, February 1 1981.
Outside surface readings recorded during the first 24 hours of each test reflected diurnal and seasonal changes. Weather conditions were relatively stable during each 2.5 day test period with daily temperature patterns repeating themselves.

As the surface air passes along rock lined passageways in which the rock surface temperatures are colder in summer, and warmer in winter than the intake air, tempering of the air occurs. The rate of change can be seen on a plot which compares air temperature against flow distance along the tunnel where the measurements were made. The relationship plotted is that for readings taken at mid-afternoon after air had been flowing continuously for 49 to 54 hours. Statistical curve fitting demonstrates that the relationships follow an exponential function with a rapid rate of change in the first tens of meters of the air course and a slower rate of change towards the ends of the airway. For each season it can be seen that the flow air temperature at the end of the passageway closely approximates the virgin rock temperature.

215

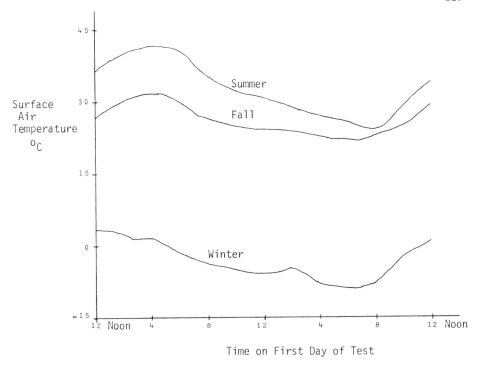

Fig. 4. Surface air temperature patterns on first day of Summer, Fall and Winter tests.

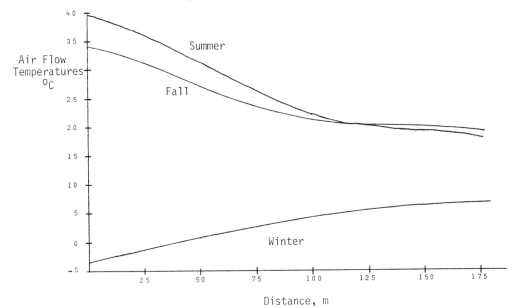

Fig. 5. Airflow temperature at distance from passage intake.

216

From statistical curve fitting, equations which describe these relationships are
(a) $t = 25.02e^{-0.014x}$, for Summer conditions,
(b) $t = 33.19e^{-0.0034x}$, for Fall conditions, and
(c) $t = 2.21 + 0.05x$ for Winter conditions, x in meters being the distance from
the air intake point.

The impact of the changing rate of airflow tempering at different points along the
passageway can be determined by plotting the change in air temperature which has
occurred at any point as a percentage of the total change which occurs between
intake and outlet of the passageway. Using data for the same time intervals as in
figure 5, it can be seen that 50 percent of flow temperature drop occurs within
the first 53 m in Summer and Fall and in the first 69 m in Winter. These results
emphasise the decreasing economic return that would be achieved by the lengthening
of an air tempering tunnel or tube.

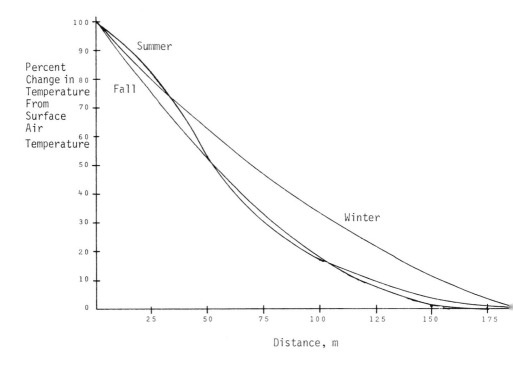

Fig. 6. Percentage change in air temperature between intake and
outlet airway points against distance along passageway.

As the airflow is tempered in moving along a rocklined passageway, two important
energy transfer processes occur. These involve that of exchange of heat energy
between the rock mass and the airflow and that of the latent heat change as water
vapor condenses in contact with cooler rock surfaces or free moisture on the rock
is evaporated by the warmed unsaturated air.

1. Energy conduction to or from the rock mass: The temperature sensing probes
set into the rock at points along the air passageway recorded temperature changes

in the rock mass during the tests. Using data recorded at the same times as in figures 5 and 6, a temperature profile was plotted at incremental distances into the rock mass. The readings were recorded in the rock mass at a point 56 m along the air passageway from the intake air shaft.

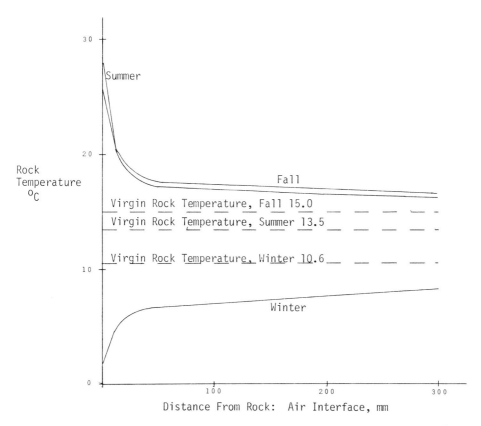

Fig. 7. Rock temperature at depth from rock/air interface.

These relationships demonstrate an exponential function with the outer rock surface layers adjusting at a much faster rate than the inner layers. Although measurements taken to construct the plot in figure 7, were recorded after the air had been continuously flowing for 49 to 54 hours, it can be seen that little change has occurred in the rock mass temperature at a short depth in from the air/rock interface.

2. Latent heat energy changes: Passageways in the underground mine in which measurements were taken had free water present; ground water inflow formed puddles and condensation was present in places on tunnel roof and walls. The water vapor content or absolute humidity level in the air at any point in the passageway was a function of the intake air level and the gain or loss which had occurred from evaporation or condensation along the airway. Humidity levels at all points in the system were determined by the use of wet and dry bulb thermometer readings and psychrometric constants. The amount of water vapor that air can hold is a function of its temperature (with an assumption of constant atmospheric pressure);

unsaturated air can hold an increased mass of moisture, while saturated air can only hold more if the air temperature is increased. Humidity levels can be calculated at points along the airflow passageway, and from this data energy transfer from or to the airflow system as latent heat of evaporation or condensation can be determined.   Humidity levels in the air along the passageway are plotted from measurements recorded at mid-afternoon on the last day of each seasons' test sequence.

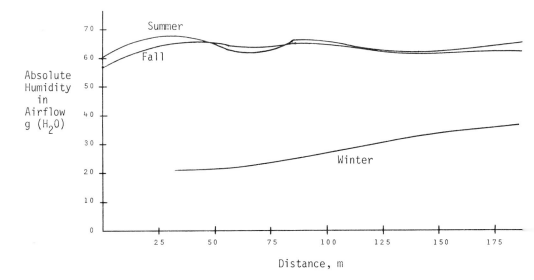

Fig. 8.   Humidity levels in air flowing along passageway, mid-afternoon readings.

During the night time, while surface air temperature is cooler and dropping to a minimum recording, (normally measured in the early hours of the morning), the air humidity level remains relatively constant. Humidity levels in the air flowing along the passageway are plotted from measurements recorded at 3 a. m. on the last day of each test sequence.

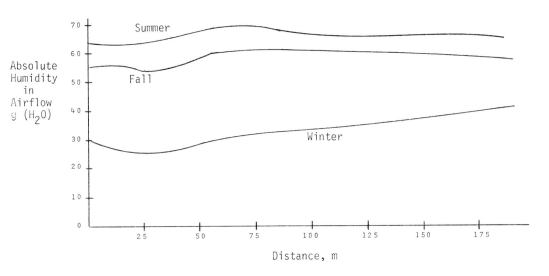

Fig. 9. Humidity level in air flowing along passageway,
3 a. m. readings.

Thermodynamic Balance

An examination of the energy changes occurring as air moves along the passageway demonstrates the existance of a number of heat transfer processes. To assist in understanding the characteristics of these processes, energy exchange pathways will be examined in an endeavour to quantify the magnitude of these changes.

Enthalpy of air

Enthalpy is a measure of Total Heat or Energy in a system. Enthalpy of air is readily calculated by measurement of wet bulb and dry bulb temperatures and reference to Psychrometric charts or tables. Energy change as air passes through the passageway system is found by calculating the difference in Enthalpy between intake surface air and outlet system air. Calculations have been undertaken to examine these changes. Data used is that from mid-afternoon and early morning readings on the last day of the Summer, Fall and Winter tests. These observed readings demonstrate day time and night time airflow characteristics after 36 hours of continuous passageway flow. System Enthalpy changes are given in table 1.

Energy changes in the air mixture

Energy changes in the air mixture can be described in terms of the heat changes in its components

1. Heat change in dry air or sensible heat change: This is quantified by multiplying air mass flow rate by air dry bulb temperature change through the system

TABLE 1   Energy Changes in Airflow Through Rock Lined Passageways

	SUMMER	FALL	WINTER
MID-AFTERNOON READINGS			
1. Enthalpy Change in Air	108.6	60.6	137.3
2. Air Energy - Sensible	111.7	69.7	60.0
- Latent	−12.3	−14.8	39.4
- Total	99.4	54.9	99.4
3. Rock Energy	57.9	50.3	57.4
EARLY MORNING READINGS			
1. Enthalpy Change in Air	36.1	9.4	50.2
2. Air Energy - Sensible	41.6	16.6	23.7
- Latent	−2.5	−7.4	22.1
- Total	39.1	9.2	45.8
3. Rock Energy	54.0	49.6	79.6

All calculated energy flow values are expressed in units of kJ/s.

and by the specific heat constant for air.

2. Heat change in water vapor: This is the energy to super heat water vapor from the wet bulb to the dry bulb temperature at a point. Quantified, this energy level is negligible and has been ignored in the study.

3. Heat to evaporate or condense water: Latent heat changes in an air mixture can be considerable and are quantified by multiplying moisture mass change in air passing through the system by the constant for water latent heat of condensation or evaporation.

Energy changes in the air mixture are quantified by summation of these components and are listed in table 1.

During the Winter test, cold dry air entered the air passageway, and in passing over relatively hot rock surfaces was warmed. The resulting warm air could hold more moisture, and this was evaporated from free water on the rock surfaces. Reference to figures 8 and 9, demonstrates that air moisture levels increase steadily along the passageway during winter. In the Summer and Fall tests, the warm outside air in passing over relatively colder rocks will lose its ability to hold moisture. The relative humidity of the ventilation air will increase until the dew point is reached with the result that condensation of moisture onto the cold surfaces commences. While this trend of decreasing moisture content can be seen in figures 8 and 9, in the later half of the passageway flow, another influence significantly affects air moisture level in the initial sections. The warm intake air to these sections is exposed to free water on the tunnel roof, walls and floor. Evaporation of moisture in these early sections causes an initial increase in moisture level; further, considerable energy in the air is required for this latent heat of evaporation. For the Summer and Fall test data examined, evaporation of moisture in the early sections of the passageway is more signifi- cant than the condensation stages later in the airway. Energy flow leading to this net increase in moisture level is given a negative sign in the data in table 1, as it is not in the direction of the assumed flow path from high temperature air to cold rock mass.

Energy changes in rock mass

As air flows along the passageway, heat is conducted to the surrounding rock in Summer and Fall, and from the rock mass in Winter. The effect of this flow on the temperature of the rock can be seen in figure 7. The total heat energy change within the rock mass can be calculated by multiplying the average observed tem- perature change by the mass of rock affected and by the specific heat constant for the rock. Results for each test are set down in table 1.

The temperature change in an incremental unit of rock is a function of the distance from the beginning of the passageway and the distance into the rock from the air/ rock interface. From figure 5, it can be seen that the maximum temperature change and the rate of heat energy flow occurs at the air intake point. For all test results, the 50 percent air temperature change point along the passageway took place between 50 m and 70 m. The closest rock temperature probe location, at 56 m has been taken as representative of average passageway air temperatures. The average temperature change experienced by the rock can be found from examina- tion of the temperature profiles at distance from the air/rock interface. For the test data recorded, profiles of these data were constructed and mathematical func- tions representative of "lines of best fit" were calculated. These equations are set down in table 2.

The average rock mass temperature was found by integrating the equation for the "line of best fit" with respect to "t". Examination of the profile data from the tests shows that at a distance into the rock of 1.0 m, the temperature change was negligible. Rock mass energy calculations were based on the assumption that heat flows in all directions from the passageway to that depth. Along a 185 m passageway length, 2061 m^3, or 5 709 000 kg of rock (with a density for dolomite of 2 770 kg/m^3, Clark, 1967) were affected. For the calculations, a Specific Heat constant of 0.825 kJ/kg-$^{\circ}$C was used, (Touloukian and Ho, 1981).

Comparison of thermodynamic balance results

Examination of the results in table 1, indicates that while the correlation be- tween air enthalpy and air energy is good, rock energy values do not show the same agreement

TABLE 2 Data on Rock Mass Temperature Profiles

	SUMMER	FALL	WINTER
MID-AFTERNOON READINGS			
Line of Best Fit	t=22.02-1.11 ln x	t=29.94-1.58 ln x	t=1.80+1.17 ln x
Rock Surface Temperature °C	26.3	26.0	2.0
Virgin Rock Temperature °C	13.5	15.0	10.6
Average Rock Temperature °C Change	2.4	2.0	2.15
EARLY MORNING READINGS			
Line of Best Fit	t=21.41-1.10 ln x	t=20.04-0.61 ln x	t=3.58+0.9 ln x
Rock Surface Temperature °C	22.0	21.0	5.5
Virgin Rock Temperature °C	13.5	15.0	10.6
Average Rock Temperature °C Change	1.66	1.44	2.19

t   denotes rock temperature, °C

x   denotes distance into rock from rock: air interface, mm

1.   Since enthalpy values are a function of the air wet bulb temperature, inaccuracies in taking this reading, or in calculating values from Psychrometric chart or table data can lead to significant error.

2.   Accuracy in determination of air energy Sensible Heat readings is dependent on reliability of dry bulb temperature readings in the air flow. The calculated latent heat readings rely on the accuracy in the measurement of both wet and dry bulb readings and the calcuation of air moisture levels. Complex heat exchange processes, with both moisture evaporation and condensation occurring simultaneously in the system, were observed in the passageway airflow. The calculated results are average readings for the system. Additional detailed studies and data for sections of the passageway would be of considerable value in evaluating the energy and heat transfer process.   Despite the limitations in experimentally determined readings for interpretation of the tempering phenomenon, the correlation between total air energies and Enthalpy results is good. For all tests, except one, results indicate agreement to within accuracy of 10 percent.

3.   Rock energy values are based upon accuracy of experimentally determined rock temperature data and assumptions with respect to rock properties of density and specific heat. Special effort was made to cement the temperature probes within rock boreholes with a procedure that would return experimental temperatures representative of the rock mass at depth. Interpretation of data is based on

averaging techniques to obtain system values representative of a complex heat exchange process. In interpretation of results, the assumption has been made that the heat flow rate of conduction through the rock mass is uniform and constant throughout the test time period. While the intention of this study has been to calculate average system results, this assumption, by not accounting for shunting and diurnal changes in rate of heat conduction, opens up a chance for potential error. Refinement of interpretation techniques to account for varying rates of heat flow conduction in the mine rock at different times in the day and night, and at different points along the air passageway, would considerably improve the model description of the heat exchange process.

Heat Flow in Underground Mine Airways

In the ventilation of deep underground mining operations, an understanding of the effects of heat flow from hot surrounding rock into airways carrying cool air to the working faces is of significance. Conditions in South African gold mines have led to a number of studies being undertaken to optimize design of mine refrigeration plants used to cool the air before it enters the mine stoping areas.

Whillier and Ramsden (1976) have proposed a formula for calculation of heat pick-up along mine airways which has been used with success by Hemp and Deglon (1980) and Steyn (1980) in the determination of heat load on mine refrigeration units. The relationship proposed is:

$$Q = 5.57(WF+0.255)(VRT-DB)CF \tag{1}$$

Q = Heat Pick-up, kW/100 m length of airway

WF = Wetness Factor, 0 for dry airway, 1.0 for very wet airway

VRT = Virgin Rock Temperature, ^{o}C

DB = Dry Bulb Temperature, ^{o}C

CF = Correction Factor for Airway Size, Airway Age and Type of Rock.

In application, the formula requires a number of assumptions to be made. A subjectively determined airway Wetness Factor is needed in the use of the equation and a correct estimation is necessary if heat flow values are to be accurately calculated. A correction factor can be applied to account
(a) for airway sizes varying from the normal South African dimensions of about 3 m by 3 m,
(b) for airway age where airway surrounding rock is cooler than normal due to heat liberation over time, and
(c) for rock type if thermal characteristics vary from those of quartzite.
The equation takes no account of air quantity or air velocity rate through the passageway.

The Ramsden equation has been applied to the study of test data from which calculated energy change values are obtained as shown in table 3.

In the application of the Ramsden equation, no allowance has been made for variation from the assumed airway size, airway age and type of rock and a Correction Factor of 1.0 has been applied. The Wetness Factor used for each determination is based on personal observation of airway surface characteristics as tests were in progress.

From the results, it can be seen that while there is reasonably correlation between energy flow levels determined by the Ramsden equation and air enthalpy calculation for winter test readings, discrepencies are apparent for other seasons' results.

TABLE 3 Energy Changes in Airflow Applying Ramsden Equation

	SUMMER	FALL	WINTER
MID-AFTERNOON READINGS			
Energy Change-Section 1*	62.2 (0.25)**	45.4 (0.25)	32.5
-Section 2	50.1 (0.75)	41.8 (0.75)	53.1 (1.0)
-Total	112.3	87.2	85.6
Air Enthalpy Change***	108.6	60.6	137.3
EARLY MORNING READINGS			
Energy Change-Section 1	27.5 (0.25)	16.7 (0.25)	15.7 (0.25)
-Section 2	36.2 (0.75)	25.1 (0.75)	32.2 (1.0)
-Total	63.7	41.8	47.9
Air Enthalpy Change	36.1	9.4	50.2

* Airway Section 1: Length 0 m to 85 m.

   Airway Section 2: Length 85 m to 185 m.

** Assumed Wetness Factor in Calculation: All Surface Dry-0

                                            All Surfaces Wet-1.0

***Enthalpy readings from table 1.

For the Summer and Winter data, energy change by application of the Ramsden equation exceeds that calculated from enthalpy considerations. It is appropriate in this case to look at the application for which the Ramsden equation was originally proposed.

1. The equation is put forward for usage in determination of heat flow from rock at high temperature to cooler air.

2. It is designed to be applied to situations of airflow through long passageways in which air is at a high relative humidity and airway surface moisture is from groundwater inflow.

The Summer and Fall studies undertaken examine a heat exchange situation in which flow is from warmer air to colder rock mass. Further, the passing air mass has not reached a high and stable relative humidity level. Considerable energy transfer is involved in latent heat of evaporation and condensation and the explanation of these changes is complex.

Further detailed interpretation of test results by examining the response of air-flow in different sections of the passageway may allow most discrepancies to be explained thereby improving the correlation.

CONCLUSIONS

As air moves through a rock lined passageway temperature changes take place as heat energy transfer occurs within the air and rock mass system. To obtain a better understanding of this process, an in situ study was undertaken to monitor and record the changing conditions at various locations along a 185 m passageway in dolomite rock with regulated airflow. The investigation consisted of three test periods of 2.5 days duration each that were conducted in July 1980, September 1980 and January 1981.

From the recorded data relationships have been determined that describe rate of change in air temperature and air moisture level as outside air was circulated through the passageway and identify temperature changes occurring in the surrounding rock due to interfacing of the exposed mine rock with the air. Also an attempt was made to calculate the thermodynamic balance of the system. The results of the study were compared with empirical and theoretical approaches used in deep mine air cooling investigations in South Africa.

Changes in air moisture level along the passageway, the latent heat of evaporation, and the condensation energy flow considerations are important to a full understanding of the physical principles involved in the energy balance of a ventilation system. This involves detailed examination of changes in air behaviour along the passageway on a section by section basis. Experimental determination of conditions must be undertaken with care and accuracy, and interpretation made on a basis which examines unit changes within the system. Good correlation has been obtained between experimentally determined and theoretically based results. Areas of study where further effort in interpretation is needed have been identified.

ACKNOWLEDGEMENT

The authors express their deep appreciation to the students in Mining Engineering, Mine Atmosphere Control, for undertaking as a class project the round the clock monitoring and data collection required during the tests. The authors also wish to thank Mr. M. A. Schimmelpfennig, a graduate student in Mining Engineering, for his assistance throughout the project.

REFERENCES

Boileau, G. G., Latta, J. K. (1978). Calculation of Basement Heat Losses, Underground Utilization, University of Missouri-Kansas City.

Bruzewski, R. G., and Aughenbaugh, N. B., (1977), Effects of Weather on Mine Air, Mining Congress Journal, Vol 63, No. 9, Sept. 1977.

Clark, S. P. (Ed.) (1967). National Research Council Physical Constants of Rocks.

Hemp, R. and Deglon, P. (1980). A Heat Balance in a Section of a Mine, Proceedings Second International Mine Ventilation Congress, University of Nevada-Reno.

Lorentzen, G. (1978). The Design and Performance of an Uninsulated Freezer Room in the Rock. Underground Utilization, University of Missouri-Kansas City.

Smith, D. J., Orlandi, W. T. and Gillies, A. D. S. (1981). Airflow Cooling in Rock Lined Tunnels. Missouri Academy of Science, Annual Meeting.

Stauffer, T. (1978). Efficiency in the Use of Energy has been Effected Through Industrial Use of Subsurface Space. Underground Utilization, University of Missouri-Kansas City.

Steyn, J. H. (1980). The Use of Mid-Shaft Cooling in the Strathmore Shaft System, Buffelsfontein Gold Mining Company Limited, Journal of the Mine Ventilation Society of South Africa, March, 1980.

Touloukian, Y. S. and Ho, C. Y. (1981). Physical Properties of Rocks and Minerals. McGraw-Hill, New York.

226

Warnock, J. G. (1978). New Industrial Space-Underground: Thoughts on Cost/
   Benefit Factors. Underground Utilization, University of Missouri-Kansas City.
Wells, M. (1977). Underground Designs Box 183, Cherry Hill, NJ.
Whillier, A. and Ramsden, R. (1976). Sources of Heat in Deep Mines and the Use
   of Mine Service Water for Cooling, Proceedings, International Mine Ventilation
   Congress, The Mine Ventilation Society of South Africa, Johannesburg.

ENERGY USE OF NON-RESIDENTIAL EARTH-SHELTERED BUILDINGS
IN FIVE DIFFERENT CLIMATES*

George D. Meixel, Jr.
Underground Space Center
University of Minnesota
Minneapolis, MN 55455, USA

ABSTRACT

Computer predicted HVAC energy consumption of earth-sheltered single story office
buildings and dry-storage warehouses indicate that increased earth-sheltering may
significantly reduce energy requirements and peak demand. Buildings in Boston,
Massachusets; Washington, D.C.; Jacksonville, Florida, and San Diego, California in
the United States and Manila in the Philippines were modeled. For the office
building in Boston where the largest heating energy reductions were calculated, the
heating energy for the bermed and covered building was reduced to 45% of that
calculated for the slab-on-grade configuration. The corresponding peak load was
reduced to 35% of the slab-on-grade value. Cooling energy reductions were also
predicted for the office buildings. The Washington, D.C. bermed and covered office
building had predicted cooling energy requirements that were 61% of those for the
slab-on-grade building.

Warehouse results are dominated by uncontrolled infiltration at the loading dock
doors. Consequently, the impact of earth-sheltering is less on a percentage basis
than for the office building. For Boston weather conditions, the bermed and
covered warehouse has a heating energy requirement that is 70% of that for the
slab-on-grade building. However, wall, roof and floor thermal performance improved
dramatically with increased earth-sheltering because the low 55°F heating
thermostat setting, which is suitable for the dry-storage warehouse, is close to
the annual mean ground temperature.

Computer analysis is used to associate energy loads with the wall, floor, roof and
windows, to determine the energy load due to infiltration and ventillation, and to
estimate the internal load. Systematic changes in the energy requirements
associated with these components of the HVAC load that correspond to changes in the
degree of earth-sheltering are analyzed to quantify the improvements in thermal
performance due to the major features of earth-sheltering.

---

*An initial version of part of this paper was presented at the 3rd Miami
 International Conference on Alternative Energy Sources, December 15-17, 1980 in
 Miami Beach, Florida.

228

KEYWORDS

Earth-sheltered buildings; HVAC energy analysis; computer simulation; office building; warehouse.

INTRODUCTION

Increased levels of earth-sheltering can be expected to result in reductions in the heating, ventilation and air-conditioning (HVAC) loads associated with heat gains and losses at the building perimeter. Features of earth-sheltering that contribute to this improvement in building thermal performance are the moderation of the buildings external thermal environment by the soil mass together with the surface vegetation, reduced infiltration and windchill, compatability with passive solar heating and the use of the ground as a thermal sink [1, 2, 3]. The objectives of the paper are (1) to present preliminary data from a systematic computer study of the predicted reductions in HVAC loads that are obtained with increased earth-sheltering and (2) to use the computer to estimate the portions of the loads that are due to the different architectural elements and to the building operating schedule. Five different locations--Boston, MA; Washington, D.C.; Jacksonville, FL; San Diego, CA in the United States and Manila in the Philippines--are analyzed to develop a broad perspective on the quantitative influences of earth-sheltering and the relative potential for energy savings.

Synthesis of the results of these objectives will provide an assessment of the relative contributions of the various features of earth-sheltering to reduce the building HVAC energy consumption. Two different building types, a one story office building and a dry storage warehouse, are analyzed for four degrees of earth-sheltering in each of the five different locations--to develop a broad perspective on the quantitative energy impact of earth-sheltering.

Because the thermal mass of the earth moderates the temperature fluctuations that occur at the air-earth interface, the peak conduction losses (or gains) through earth contact surfaces may be reduced. On the ground surface winter snow cover and summer evapotranspirative losses through vegetative ground covers further moderate the thermal environment in the ground in comparison with above grade conditions. For example, computer predicted isotherms around a below grade portion of a building are plotted for a cold winter period in Minneapolis, MN on Fig. 1. On an hourly or daily time scale the isotherms are stationary, daily peak heat loss through the below grade walls is the same as the daily time average loses. The large drop in outside air temperature characteristic of winter nights in Minnesota is averaged out in the ground thermal mass. Consequently, below grade earth-contact walls do not experience the large diurnal fluctuations that occur outside above grade walls, and peak loads are concomitantly reduced. Note, however, that this qualitatively different peak load performance characteristic associated with earth-sheltered walls can not be directly generalized to heating energy performance. To evaluate overall heating energy performance the load must be integrated over the entire heating season. Earth-sheltered walls do not experience the coldest part of the above grade winter night. Similarly, earth-sheltered walls do not experience the warmest part of the winter day. Generalizations about the seasonal heating performance require a full heating season analysis such as those reported in this paper.

Figure 2 plots the summertime isotherms around the same building shown during the winter in Fig. 1. In Fig. 2, heat flows from the below-grade building walls perpendicular to the isotherms into the relatively cooler soil mass. Winter coolth has been stored in the thermal mass of the soil and is responsible for this unique behavior. While above-grade walls would be experiencing heat gains, earth-contact walls can provide free cooling or, atleast, no heat gain.

$T_{AIR} = 11.2°C$

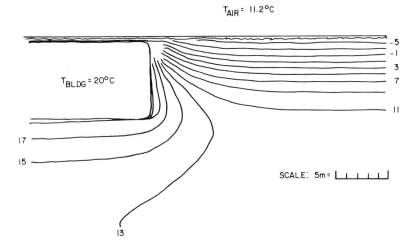

$T_{BLDG} = 20°C$

-5
-1
3
7

11

17

15

SCALE: 5m = └┴┴┴┴┘

13

Fig. 1. February soil isotherms around an underground building, air temperature is -11.2°C.

$T_{AIR} = 23.8°C$

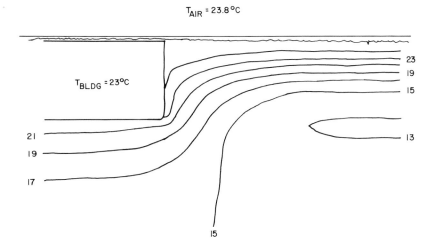

$T_{BLDG} = 23°C$

23
19

15

21

13

19

17

15

Fig. 2. August soil isotherms around an underground building, air temperature is 23.8°C.

During the summer period, plant cover on the surface of the ground further improves cooling performance because of evapotranspiration heat loss. Under conditions where there is sufficient surface moisture, the ground surface temperature is often considerably below the air dry-bulb temperature. This may result in additional cooling benefits as discussed by Speltz and Meixel [4].

Implicit in earth-sheltered designs are window distributions that differ markedly from more the traditional relatively even spacing of above grade buildings. Frequently, in cold climates much of the glazing in an earth-sheltered building is located on the southern exposure to maximize passive solar gain, and, as the amount of earth-sheltering increases, window area generally decreases. Reduction of the window perimeter as well as grouping the windows on one or two sides of the building can greatly reduce infiltration. Therefore, the one story office building model that was studied in the work described in this paper, and documented subsequently, has a constant floor plan, but both the window placement and the amount of window area change with varying levels of earth cover.

OFFICE BUILDING MODEL

The one story office building modeled in this study is shown for four degrees of earth-sheltering in Fig. 3. The building designated "No Berm" is conventional slab-on-grade construction. The remaining three are half bermed, fully bermed to the roof and, in the last configuration, bermed and covered with twelve inches of soil and four inches of gravel on the roof. In each case, the south facing facade is left fully exposed.

The thermal resistance R, expressed in conventional units of $hr*ft^2*°F/BTU$, for the above grade walls was R=13 for Boston, R=10 for Washington, D.C., and R=7 for Jacksonville, San Diego and Manila. Rigid wall insulation is placed along the outside of the below-grade walls down to the base of the footings to the following insulation levels: R=11 for Boston, R=8 for Washington, D.C. and R=4 for Jacksonville, San Diego and Manila. All roofs are R=20.

Window area is based on typical areas for offices, and, in all but the constant exposed south facade, decreased by half in the half-bermed model. The windows are eliminated on three elevations in the fully-bermed and in the covered models. One-inch insulating glass is used for all windows in all climates except for San Diego and Manila. These two have one-fourth-inch single glazing. All windows are assumed to be operable for purposes of figuring infiltration. However, the windows are modeled in the closed position to facilitate comparative analysis and to simplify the modeling. Entrance doors at the south wall are modeled to account for infiltration due to entrance traffic.

Partition locations significantly influence the heat transfer through the various components of the building envelope. Consequently, a regular system of interior partitions and corridors is defined. The layout consists of four concentric areas of constant depth. These are, starting from the outermost ring: small rooms, larger open-plan spaces; corridor; and an interior zone that contains medium sized rooms. Miscellaneous features such as toilets and janitor's closets are not modeled. The partition layout results in a number of mechanical zones which can be Lighting and equipment loads are based on typical loads for office buildings. The outside air ventilation rate is 5 cfm per person. When the building is unoccupied, the ventilation system is off. Refer to Tables 1 and 2 for a summary of physical data and for the thermostat settings.

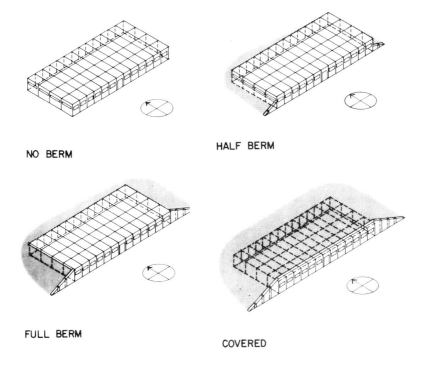

NO BERM

HALF BERM

FULL BERM

COVERED

Fig. 3. Office building isometrics showing the four levels of earth sheltering.

TABLE 1   OFFICE BUILDING SUMMARY

Area	31,200 sq ft
Volume	249,600 cu ft (below ceiling)
Length	260 ft
Width	120 ft
Clear Height	8 ft from floor to ceiling
Height to bottom of roof deck	13 ft
Bay size	20 ft x 30 ft

ABOVE-GRADE CONSTRUCTION

Roof	Built-up roof with rigid insulation on metal deck, open-web steel joist roof structure
Wall	Face Brick, rigid insulation, 12 inch concrete block, metal furring, $1/2$ inch gypsum board
Openings	a. No Berm 7'4" x 4'0" windows at 10'0" O.C. (22% of above grade wall)
	b. Half Berm 7'4" x 2'0" windows at 10'0" O.C. (11% of above-grade wall)
Glazing	One inch insulating glass in all climates except San Diego and Manila which use $1/4$" single glazing

BELOW GRADE CONSTRUCTION

Roof	Twelve inch soil and four inch gravel cover, rigid insulation, membrane waterproofing on 6" concrete slab, concrete beam and girder framing
Walls	Twelve inch concrete with furred gypsum board, rigid insulation located on outside of wall, 14" concrete where earth is up to roof or higher

ENVIRONMENTAL CONTROLS

Internal load	Three watts/sq ft average, including equipment and lighting, excluding occupants
Ventilation rate	Five cfm outside air per person
Temperature, humidity	
	a. Cooling 78°F DB, 60% maximum RH
	b. Heating 68°F DB, 25% minimum RH (No thermostat setback)

TABLE 2 OFFICE BUILDING USE SCHEDULE

This Table applies to all zones. Figures for all one-hour periods not shown are the same as for the period between 7:00 a.m. and 8:00 a.m. Thermostat settings for heating and cooling are constant at 68°F and 78°F, respectively. Lines designated M-F are for days Monday thru Friday. S-S denotes Saturdays, Sundays and holidays. To obtain the total load for any hour, multiply the total peak load by the factor in the table.

		A.M.					P.M.				
		7	8	9	10	11	12	1	2	3	4
Occupancy Schedule	M-F	0.	1.	1.	1.	1.	1.	1.	1.	1.	1.
Maximum Number of Persons=266	S-S	0.	0.	0.	0.	0.	0.	0.	0.	0.	0.
Lighting Schedule											
Total Peak Lighting Load=78.0 kw	M-F	.05	1.	1.	1.	1.	1.	1.	1.	1.	1.
(based on 2.5 watts/sq ft)	S-S	.05	.05	.05	.05	.05	.05	.05	.05	.05	.05
Equipment Schedule											
Total Peak Equipment Load=15.6 kw	M-F	0.	1.	1.	1.	1.	.05	1.	1.	1.	1.
(based on 0.5 watts/sq ft)	S-S	0.	0.	0.	0.	0.	0.	0.	0.	0.	0.
Ventilation Schedule											
Maximum Total Outside Air=1330 cfm	M-F	0.	1.	1.	1.	1.	1.	1.	1.	1.	1.
(based on 5 cfm/person)	S-S	0.	0.	0.	0.	0.	0.	0.	C.	0.	0.

WAREHOUSE MODEL

The dry-storage warehouses modeled in this study are shown for the different degrees of earth-sheltering in Fig. 5. In each case the south facade is left fully exposed to accommodate the loading docks. Above grade walls are R=11 for Boston, R=7 for Washington, D.C., and R=4 for Jacksonville, San Diego and Manila. Below grade walls are R=11 for Boston, R=8 for Washington, D.C. and R=4 for Jacksonville, San Diego and Manila. All roofs are R=20.

All warehouse models are windowless. Interior mass is estimated at 20 million pounds total and is based on the number of pallets likely to be stored in the warehouse at one time. Typically, in warehouses, uncontrolled infiltration occurs chiefly at the overhead doors; an infiltration factor is used to account for this. This factor presupposes dock seals that fill the perimeter gap between the truck door and the warehouse door. It is assumed that the doors would be closed immediately after the truck departs. No consideration is given to the temperature of incoming pallets. Also not included in the model is the specific location of the storage racks. Refer to Tables 3 and 4 for a summary of physical data, temperature requirements and operating schedule. Figure 6 is a floor plan of the warehouse.

TABLE 3 WAREHOUSE SUMMARY

Area	102,400 sq ft
Volume	2,329,600 cu ft
Length	400 ft
Width	256 ft
Clear height	20 ft
Height to botom of roof deck	23 ft
Bay size	32' x 50' for steel framing
	32' x 25' for below-grade concrete framing
Interior mass	Total of 20 million pounds of storage based on 80% capacity
Pallet stacking	4 pallets high

ABOVE-GRADE CONSTRUCTION
Roof	Built-up roof on metal deck, open-web steel joist roof structure
Wall	Field-assembled insulated metal panels
Openings	No windows, sixteen 7'-6" x 8'-0" (high) insulated overhead doors at one end of building (256' side)

BELOW-GRADE CONSTRUCTION
Roof	12" soil and 4" gravel cover, rigid insulation, membrane waterproofing on 5" concrete slab, concrete beam and girder framing
Walls	Poured-in-place concrete

ENVIRONMENTAL CONTROLS
Internal heat load	1 watt/sq ft average for equipment and lighting
Mechanical system	Fossil-fuel-fired unit heaters at spot locations
Temperature	55°F maximum for heating, no cooling required
Ventilation	No mechanical ventilation required in heating season

TABLE 4 WAREHOUSE USE SCHEDULE

These tables apply to all areas of the warehouse. Figures for all one-hour periods not shown are the same as for the period between 7:00 AM and 8:00 AM. The thermostat setting for all periods is 55°F for heating. No cooling is provided. Lines designated M-F are for days Monday through Friday. S-S denotes Saturdays, Sundays, and holidays. To obtain the total load for any hour, multiply the total peak load by the factor in the table.

		A.M.					P.M.				
		7	8	9	10	11	12	1	2	3	4
Occupancy Schedule	M-F	0.	1.	1.	1.	1.	.5	1.	1.	1.	1.
Maximum Number of Persons = 20	S-S	0.	0.	0.	0.	0.	0.	0.	0.	0.	0.
Lighting Schedule											
Total Peak Lighting Load = 102.4 kw	M-F	.05	1.	1.	1.	1.	1.	1.	1.	1.	1.
(based on 1.0 watts/sq ft)	S-S	.05	.05	.05	.05	.05	.05	.05	.05	.05	.05
Equipment Schedule											
Total Peak Equipment Load = 0. kw	M-F	0.	1.	1.	1.	1.	.5	1.	1.	1.	1.
(equipment load is negligible)	S-S	0.	0.	0.	0.	0.	0.	0.	0.	0.	0.

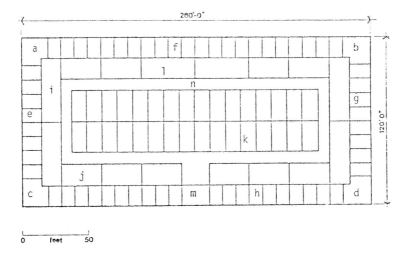

Number of occupants by room type

perimeter zones:
    a. 5
    b. 5
    c. 5
    d. 5
    e. 1
    f. 1
    g. 1
    h. 1

interior zones:
    i. 10
    j. 7
    k. 2
    l. 7
    m. 0
    n. 0

Fig. 4. Diagrammatic Building Plan.

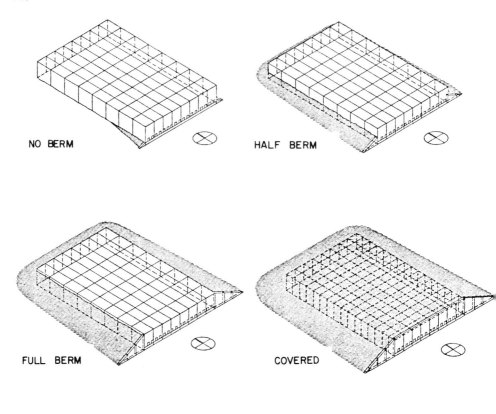

NO BERM   HALF BERM

FULL BERM   COVERED

Fig. 5. Warehouse isometrics showing four degrees of earth sheltering.

COMPUTER ANALYSIS

The principle objectives of the computer studies are (1) to predict the annual HVAC energy requirements of each configuration, building type and geographic location and (2) to predict the peak loads that must be met by the HVAC system. Computer analysis is further utilized to quantify the approximate contribution of the different elements of the building envelope--e.g., roof, walls, floor--to the total HVAC energy requirements. This ability to investigate the systematic variation in the performance of the principal facets of the building envelope facilitates an overall understanding of the potential for energy conservation through earth-sheltered construction.

BLAST [5], NBSLD [6] and DOE-2 [7] are the public domain computer programs that have been developed to carry out the comprehensive calculations necessary to determine the HVAC loads for conventional above-grade buildings. Each of these programs performs an hour-by-hour load calculation for a specified building configuration based on actual weather data for the location under study and subject

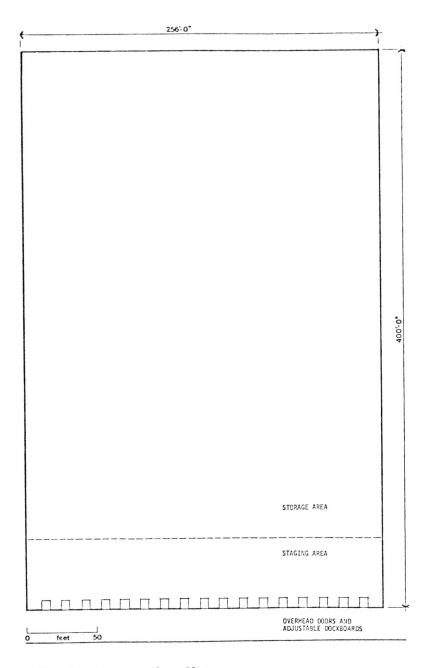

Fig. 6. Warehouse Floor Plan

to the designated building operating conditions. Each of the load calculation programs has been experimentally validated for above-grade buildings. In early 1980, when the initial analyses described in this paper were carried out, BLAST and NBSLD appeared to be more accurate than DOE-2 [8].

Although BLAST and NBSLD perform accurate calculations of the HVAC loads associated with above-grade building exposures, both of these computer programs have crude and inaccurate methods for calculating the heat flow through floor slabs and earth-sheltered walls. The Underground Space Center and the Department of Mechanical Engineering at the University of Minnesota have jointly developed a computer program to calculate the heat transfer through underground walls and floors [9, 10]. This program, designated TWOD, performs a transient, finite difference calculation of the heat conduction throughout a two dimensional spatial domain surrounding the earth-sheltered portion of the building. The success of TWOD in describing the heat loss from a large underground building has been documented by Shipp [9]. However, its validity for all climates has not been experimentally determined. Nevertheless, TWOD was the best calculation scheme available for these studies.

To determine the HVAC energy requirements for the earth-sheltered buildings investigated in this project, the computer program TWOD was used in conjunction with the computer program BLAST. TWOD computed the heat flux through the underground walls, floor and earth-sheltered roofs. BLAST computed the HVAC energy requirements of the traditional above grade facets of the building as well as the internal and ventillation loads. The results of these two programs were manually integrated into a composite description of building performance.

The building geometries, materials, and operating schedules have been set forth in the preceding section on the building models. This information defines the data base for the computer calculations of HVAC energy requirements. In the computer program BLAST the computations are organized so that heat balance equations are solved for each thermodynamic zone of the building. Appropriate wall and window dimensions, material specifications, thermostat settings, equipment operating schedules, etc. are ascribed to each zone during input to the computer.

Because of the large thermal mass of the ground, heat transfer through earth-sheltered walls and floors is relatively constant for extended periods of time such as one day, one week, or one month, depending on the climate and the time of the year. Thus a sufficiently detailed description of the weather for a particular location can be supplied to the program TWOD by specifying the average temperature for each day, the daily radiation balance and latent exchange, and, if applicable, the depth of snow cover. This information can be obtained from an analysis of the published Climatological Data [11] for each location and, in addition to the thermal properties of the soil, are typed directly into the computer. However, because the above-ground environment changes rapidly, the program BLAST calculates heat fluxes from hourly weather information read off weather tapes. These tapes store recorded weather information for complete years for particular geographic locations and include data on the dry-bulb temperature, wet-bulb temperature, wind speed, wind direction, relative humidity, barometric pressure and cloud cover. For each location in the United States the Test Reference Year (TRY) weather tape for that location was used, i.e., Boston, MA = 1969; Washington, D.C. = 1957; Jacksonville, FL = 1965; San Diego, CA = 1974. For Manila the computer simulations were run for 1958.

Following a real world building construction and HVAC system start-up, the earth-sheltered building would undergo a period of initial non-equilibrium interaction with the ground, lasting from a few months to more than one year. Modeling that was based on this initial ground warm-up period may result in greater than normal cooling benefits to the building or slightly greater required heating energy depending on the season in which the HVAC system was turned on. Consequently, the time frame used in the evaluation of building energy performance is the third year after system start-up. Further details of the TWOD program operation and documentation can be found in references 9 and 10.

OFFICE BUILDING RESULTS

The results of the computer analysis of the one story office building will be presented by location. Boston results are discussed in detail because the impact of earth-sheltering is greatest for this location in comparison with the other four that were analyzed, and, consequently, the performance of several specific facets of the building associated with earth-sheltering can be most clearly delineated. Subsequent discussions of other locations serve to highlight new features of the HVAC energy performance.

Boston, Massachusetts

Figure 7 is a plot containing two sets of four bar graphs illustrating the total annual heating energy which must be supplied to the conditioned spaces of each of the four configurations to maintain the specified thermostat settings (68°F minimum, 78°F maximum). The four bars in the right half of Fig. 7 plot the total heating load (energy units of 10^6BTU) for each level of earth cover. In the left half of Fig. 7, the heating energy requirement associated with the major factors in the energy balance are plotted. The energy that must be supplied to balance heat losses through the roof, walls, floor, and windows, as well as losses to infiltration and ventillation air, are plotted above the zero line. Internal gains that may be subtracted from the total gross energy demand to obtain the net heating energy requirement are shown as negative values below the zero line. By following the changes in the roof, wall, floor, window, outside air and internal gain contributions to the annual heating season energy requirements that occur with changes in the level of earth-sheltering, an understanding of the potential for energy conservation with each level of earth-sheltered construction is facilitated.

Before examining the specific facets of the building envelope in detail, it is important to note that only the perimeter zones, zone types (a,b,c,d,e,f,g,h) and (m) in Fig. 4, require heating. Interior zones have internal gains larger than their energy losses resulting in a net cooling requirement during the heating season. To simplify the analysis, it is assumed that this cooling requirement can be met by an economy cycle through cooling with outside air during these heating season months. Energy required to warm-up the internal zones at the beginning of each working day, since it is the same for all levels of earth cover and because it is difficult to predict, is left out of the energy comparisons.

On the graphs of annual energy performance by component, in the left half of the succeeding figures, some major components may not appear, for example, the window component (C) for the full-berm and the covered configurations in Fig. 7. If a component does not appear, the net seasonal energy transfer through that component is negligible. Note also that, when the available internal gains exceed the heat losses for a given time increment and building zone, the internal gain can only be partially utilized. The total of these partial internal gains is the portion that can be subtracted from the total HVAC energy requirement. Consequently, the internal gains shown in the graphs of heating load components vary. Conversely, since internal sensible gains are never an energy credit in the cooling season, they remain constant for a given climate. Similiar accounting applies to other component loads, and, therefore, these quantities are described as cumulative net energy losses or gains.

240

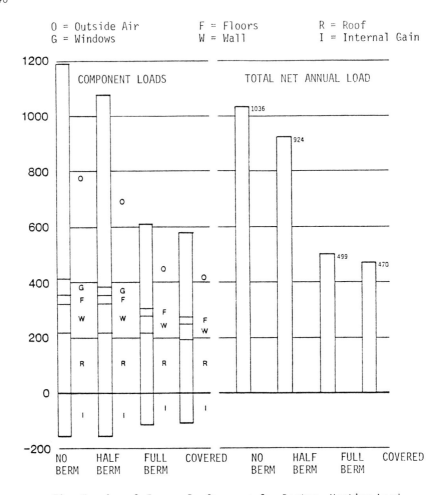

O = Outside Air   F = Floors   R = Roof
G = Windows   W = Wall   I = Internal Gain

Fig. 7. Annual Energy Performance for Boston, Heating Load
(Units of 10⁶BTU). On the left, component loads are shown
for cumulative net energy losses or gains that will affect
the overall load for a given zone. Heat loss is shown as
the overall load for a given zone. On the right, component
loads are summed resulting in the total annual heating load.

Note that only sensible loads are plotted on the graphs of HVAC energy performance.
Only the sensible loads associated with the major building components may be
algebraically summed to obtain the net sensible energy requirements. Latent
loads must be considered separately in the HVAC energy analysis. To limit the
complexity of the graphs and thus facilitate their interpretation, the influence
of the latent load component on the energy budget is not included.

Returning to the left half of Fig. 7, note that since the roof component for the
no-berm, half-berm and full-berm buildings are of the same construction, the

heating energy associated with heat lost through the roof is the same in each case. However, the covered configuration has a more massive earth-covered roof that performs better due to its mass and the capacity of the sod to hold snow. A 12% reduction in heating energy results.

Tracing changes in wall performance from left to right on Fig. 7 corresponding to increases in the level of earth-sheltering, an apparently unexpected result is observed. The walls for the half-bermed building represent a larger energy requirement than the unbermed walls as observed both on Fig. 7 and on an expanded scale on Fig. 8.a. In fact, on a per square foot basis as shown in Fig. 8.b., the half-berm walls perform better than the above-grade walls. Because berming the walls has necessitated reducing the window area there is proportionately more total opaque wall area (above and below grade combined) and therefore slightly more energy loss associated with the wall component. As the building walls are moved deeper into the ground their performance progressively improves as shown with the fully-bermed and the bermed-and-covered walls in Fig. 7 and Fig. 8.

The energy associated with heat loss through the floor decreases slightly as the floor is moved further from cold ambient temperatures at the ground surface. Because of the large floor area and the relatively constant building temperature of 68°F, the floor heat loss, except near the perimeter, is not strongly coupled to the ground surface, but rather establishes a quasi-steady-state heat transfer rate with the deep-ground environment. This results in a small heat loss with a very small dependence on depth below the ground surface.

Windows influence the HVAC energy requirements in three ways: by heat conduction, through radiation exchange and by providing additional air infiltration pathways. In this study, heating energy required to balance infiltration losses is lumped with the ventilation air heating energy and plotted as outside air, labeled (O), in the component load graphs. Net heating energy associated with conduction and radiation through the window glass is plotted as the window component--labeled (G) for glazing. As described in the building model section, window area changes from a maximum in the above-grade building to a minimum for the fully-bermed and the covered buildings with south-facing windows only. Decreasing the north, east and west window area (south window area remains constant) decreases the window component load. For the fully-bermed and the bermed-and-covered buildings in Boston, the constant glazing on the south facade represents no significant net contribution to the required heating energy. Gains from incident solar radiation balance conduction losses due to the lower outside air temperatures.

Because the presence of windows in an otherwise uninterrupted wall is linked with infiltration, more and larger windows correlate with increased levels of infiltration. Based on ASHRAE guidelines [12], nominal infiltration values for the various window configurations were input into BLAST. The resulting different contributions to the heating energy, shown as the outside air component in the no-berm and the half-berm configurations, represent a decrease in the crack length around the windows corresponding to a decrease in window area. Elimination of the windows on all but the south facade greatly reduces infiltration for the fully-bermed and the bermed-and-covered building types.

Internal gains reduce the heating load. Building zones that have heat losses greater than or equal to the internal gains for all time increments in the heating season utilize all of the available internal gain to reduce the heating load. This is the situation for the perimeter zones of the above grade building in Boston (As noted earlier, the internal zones have a cooling energy requirement all year).

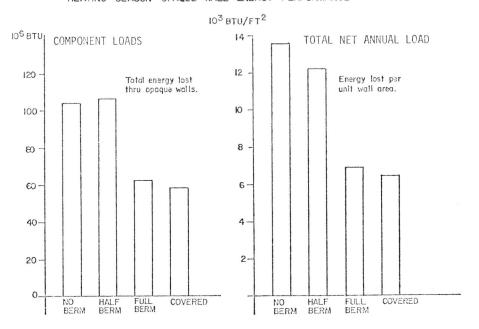

Fig. 8a. Energy lost through open walls for each degree of earth sheltering.

Fig. 8b. Energy lost per sq. ft. of opaque wall area for each degree of earth sheltering.

However, as the level of earth-sheltering is increased to the fully-bermed and the bermed-and-covered configurations, the available internal gains in some of the zones--zones (a, b, e, f, g,) in Fig. 4--exceed the heating energy requirement to maintain the 68°F inside air temperature for certain periods during the heating season. Only internal gains that can be utilized to maintain the 68°F thermostat setting are included in the net HVAC energy balance to determine the heating season energy. This is the reason for the smaller internal gain component for the fully-bermed and the bermed-and-covered building types as shown on the annual performance graphs such as Fig. 7.

Adding the net component loads to obtain the total heating season energy requirements, the results displayed on the right half of Fig. 7 are obtained. The dramatic reductions in heating energy requirements demonstrated by the fully-bermed and the bermed-and-covered buildings are mainly due to the decreased infiltration associated with increased earth-sheltering and to the improved performance of the walls because of the thermal buffering provided by the soil mass.

Figure 9 presents cooling season energy requirements for the Boston building in the same format and to the same scale as the heating season data on Fig. 7 (energy units of 10^6BTU). HVAC cooling energy associated with heat lost from the building components is plotted as positive on the left half of Fig. 9 (in the summer heat loss is a benefit where previously, Fig. 7, during the heating season it was a liability). Cooling season gains to the building result in an energy component plotted as a negative value. Total sensible cooling energy requirements are

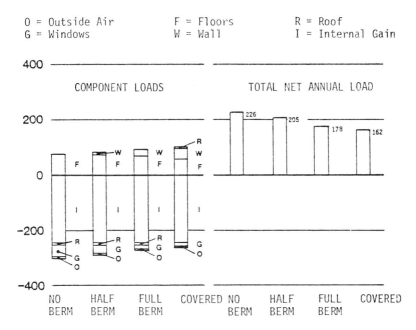

O = Outside Air   F = Floors   R = Roof
G = Windows   W = Wall   I = Internal Gain

Fig. 9. Annual Energy Performance for Boston, Cooling Load
(Units of 10^6BTU). On the left, component loads are shown for
cumulative net energy losses or gains that will affect the
overall load for a given zone. Heat lost is shown as positive;
heat gained as negative. On the right, component loads are
summed resulting in the total annual cooling load. Note that
the plots are for sensible heat only.

plotted as positive energy on the right half of Fig. 9. To maintain the proper
perspective for these energy graphs, the difference between peak or instantaneous
load (BTU/HR), which determines system size, and these time-integrated energy
values (BTU), which reflect overall energy consumption, must be kept constantly in
mind. In Boston the cooling season is much shorter that the heating season,
therefore the total cooling energy is much less that the heating energy. The peak
cooling load, however, is comparable in magnitude to the peak heating load as will
be pointed out later.

The left half of Fig. 9 shows that the cooling component energy loads are dominated
by the cooling requirements due to the internal gains for people, lights, and
equipment, which is constant for all levels of earth-sheltering. The cooling
season liabilities of windows due both to the conduction and radiation gains and to
infiltration, though much smaller that the internal gains, decrease as the window
area is reduced. Note that the floor represents a cooling season benefit as do the
earth-sheltered walls and, for the bermed-and-covered building, the earth-covered
roof. The right-hand side of Fig. 9 summarizes the cooling benefits accrued by
increased earth sheltering. These benefits are due to the capacity of the earth to
act as a cooling season heat sink and to the elimination of the above-grade wall
and window heat gains.

An examination of the peak heating and cooling loads and HVAC energy, as summarized in Table 5, will conclude the description of the thermal performance of the buildings in Boston. In part A of Table 5, both the heating energy and the peak heating loads for the different levels of earth cover are tabulated both in energy units, and with a normalization relative to the slab-on-grade (no berm) building.

TABLE 5   HVAC ENERGY AND PEAK DESIGN LOADS FOR BOSTON

A. Heating Season Summary

Building Configuration	Heating Energy		Peak Heating Load	
	10^6BTU	Normalized %	10^6BTU/HR	Normalized %
No Berm	1036	100	0.677	100
Half Berm	924	89	0.578	85
Full Berm	499	48	0.274	40
Covered	470	45	0.240	35

B. Cooling Season Summary

Building Configuration	Cooling Energy		Peak Cooling Load	
	10^6BTU	Normalized %	10^6BTU/HR	Normalized %
No Berm	226	100	0.598	100
Half Berm	205	91	0.548	92
Full Berm	175	77	0.502	84
Covered	162	72	0.493	82

Comparing the pecent change in peak heating load to the change in heating energy for the bermed-and-covered configuration, the peak heating load has been reduced to 35% of the above grade value. This is an even greater reduction than the 45% normalized figure for heating energy for the bermed-and-covered building. This reduction in peak load is due to the moderating effect of the soil mass. As the amount of earth-sheltering is increased the components of the heating load that have diurnal peaks, (i.e., windows, infiltration, above-grade walls) are replaced by earth-sheltered conponents that do not experience the diurnal peaks, infiltration is controlled and, since the peak design heating load is smaller, a less costly heating system may be employed.

For Boston, the HVAC energy and peak load do not correlate in the same manner for the cooling season as they did for the heating season. From the cooling season summary in part B of Table 5, it can be seen that the peak load does not decrease as rapidly with increased earth-sheltering as the cooling energy. This is because the largest cooling benefits from the earth-sheltered facets of the building occur during the first part of the summer, while for Boston in 1969 the peak cooling load occurs at the end of the summer in September.

The preceding discussion of the Boston results is intended to give an introduction to the relationships that exist between the various component loads for an earth-sheltered building and to suggest their relationship to the building thermal

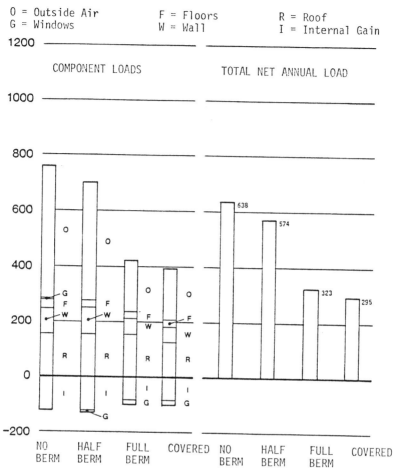

Fig. 10. Annual Energy Performance for Washington, Heating Load (Units of 10^6BTU). On the left, component loads are shown for cumulative net energy losses or gains that will affect the overall load for a given zone. Heat lost is shown as positive; heat gained as negative. On the right, component loads are summed resulting in the total annual heating load.

performance. The following comments highlight unique facets of the results for the four other locations.

Washington, D.C.

Two features distinguish the heating season performance of the buildings in Washington from the same building types in Boston which they otherwise qualitatively resemble. In Washington, D.C., glazing on the south facade is a substantial energy benefit. This is clearly seen on the left side of Fig. 10 for both the fully-bermed and the bermed and covered buildings. Secondly, note that the wall performance in Washington, D.C. does not improve as much with increased earth-sheltering as it did for Boston.

Snow covers the surface of the ground for much of the Boston winter. The presence of snow provides an additional insulating blanket that helps the earth retain its warmth thereby improving the heating season peformance of the earth-sheltered walls. Since there is little snow to provide this insulation in Washington, D.C., the air more rapidly cools the ground and, consequently, earth-sheltered walls do not have as large a thermal advantage over above-grade walls.

Except for the aspects of building energy performance described in the preceding paragraph, the Washington, D.C. buildings resemble the Boston buildings during the heating season. The contributions of each component to the total HVAC heating energy are simply reduced due to Washington's milder winter. Washington's cooling season, however, is slightly warmer and of longer duration than for Boston. Consequently, during the cooling season, Fig. 11, windows are more of a liability. Also, earth-sheltered walls are less of a cooling season benefit than in Boston. Because the Washington, D.C. winter is not as severe as in Boston, the ground does not get as cold. The potential for the ground to act as a summertime heat sink is thereby reduced.

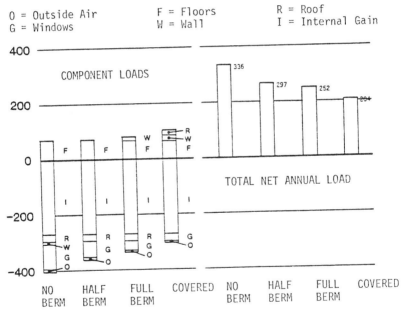

Fig. 11. Annual Energy Performance for Washington, Cooling Load (Units of 10^6BTU). On the left, component loads are shown for cumulative net energy losses or gains that will affect the overall load for a given zone. Heat lost is shown as positive; heat gained as negative. On the right, component loads are summed resulting in the total annual cooling load. Note that the plots are for sensible heat only.

Jacksonville, Florida

Figure 12 shows that the HVAC heating energy component loads in Jacksonville due to outside air, walls, and roof follow patterns similar to Boston and Washington, D.C. Because the heating season is short and the weather is mild, the magnitude of the associated energy consumption is low. Unique features of the thermal performance as shown on the left side of Fig. 12 concern window and floor performance. Both of these facets supply net energy benefits during the heating season.

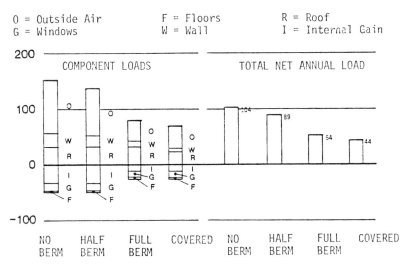

Fig. 12. Annual Energy Performance for Jacksonville, Heating Load (Units of 10^6 BTU). On the left, component loads are shown for cumulative net energy losses or gains that will affect the heat gained as negative. On the right, component loads are summed resulting in the total annual cooling load. Note that the plots are for sensible heat only.

Figure 13 presents the HVAC cooling energy results for the lengthy cooling season experienced in Jacksonville. The internal load due to people, lights and equipment is the dominant factor in the energy balance and, of course, is not modified by earth-sheltering. The thermal performance of all other major factors in the building energy balance are improved by greater degrees of earth-sheltering. Increased earth-sheltering reduces the window area thereby reducing the cooling load due to the glazing. Walls shift from a cooling season energy liability to a benefit as they are moved deeper beneath the ground surface. The floor is an energy benefit for all building types. The earth-covered roof is six times better then the traditional roof. Improvements in cooling season performance obtained with earth-sheltering are plotted on the right side of Fig. 13.

San Diego, California

San Diego results are similar to the Jacksonville results. San Diego has a less extreme climate than Jacksonville, cooler during the summer months and warmer during the winter. The HVAC energy requirements are modified accordingly as

248

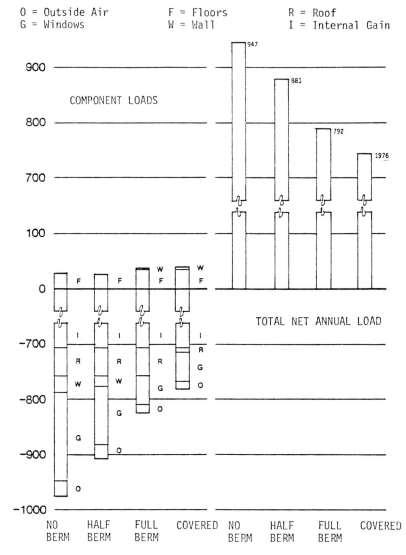

Fig. 13. Annual Energy Performance for Jacksonville. Cooling Load (Units of 10⁶BTU). On the left, component loads are shown for cumulative net energy losses or gains that will affect the overall load for a given zone. Heat loss is shown as positive; heat gain as negative. On the right, component loads are summed resulting in the total annual cooling load.

indicated on Fig. 14 and Fig. 15. Note the changes in scale and the breaks in the scale on these and other figures.

Manila, Republic of the Philippines

In Manila the mean air temperature and the mean ground temperature are both above 78°F--the cooling thermostat setting. All facets of the building contribute to the year long cooling load as shown in Fig. 16. Reducing the window area is the most significant energy saver associated with earth-sheltering for this climate. Reduction in window area not only reduces the radiation and conduction heat gains but also reduces infiltration. Earth-sheltered walls are a somewhat smaller contributor to the required cooling energy than above-grade walls. An earth-covered roof again results in a large improvement in thermal performance during the cooling season.

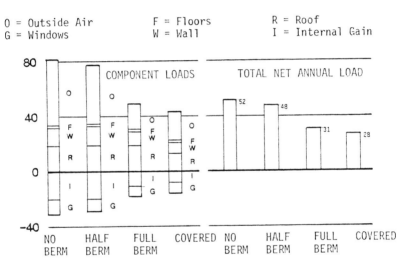

Fig. 14. Annual Energy Performance for San Diego, Heating Load (Units of 10^6BTU). On the left, component loads are shown for cumulative net energy losses or gains that will affect the overall load for a given zone. Heat lost is shown as positive; heat gained as negative. On the right, component loads are summed resulting in the total annual heating load.

250

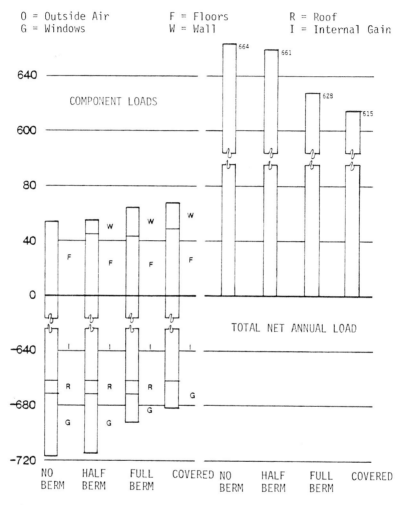

O = Outside Air   F = Floors   R = Roof
G = Windows   W = Wall   I = Internal Gain

COMPONENT LOADS

TOTAL NET ANNUAL LOAD

NO   HALF   FULL   COVERED   NO   HALF   FULL   COVERED
BERM   BERM   BERM   BERM   BERM   BERM

Fig. 15. Annual Energy Performance for San Diego. Cooling Load
(Units of 10^6BTU). On the left, component loads are shown for
cumulative net energy losses or gains that will affect the
overall load for a given zone. Heat lost is shown as positive;
heat gain as negative. On the right, component loads are summed
resulting in the total annual cooling load. Note that the plots
are for sensible heat only.

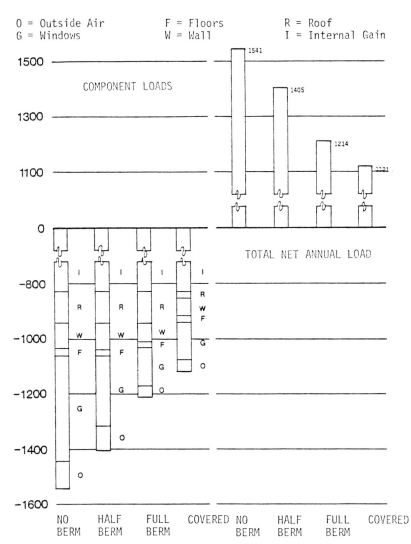

O = Outside Air   F = Floors   R = Roof
G = Windows   W = Wall   I = Internal Gain

COMPONENT LOADS

TOTAL NET ANNUAL LOAD

Fig. 16. Annual Energy Performance for Manila. Cooling Load
(Units of 10^{6}BTU). On the left, component loads are shown for
cumulative net energy losses or gains that will affect the
overall load for a given zone. Heat lost is shown as positive;
heat gained as negative. On the right, component loads are
summed resulting in the total annual cooling load. Note that
the plots are for sensible heat only.

Table 6 presents summaries of the total cooling energy requirements and the peak cooling load for Manila. An important potential impact of earth-sheltering in the tropical climate of Manila is the reduction of the peak cooling load. As seen from Table 6, these reductions in HVAC mechanical system sizes are greater than the energy reduction.

TABLE 6   COOLING ENERGY AND PEAK COOLING LOADS

MANILA, PHILLIPINES

Building Configuration	Cooling Energy		Peak Cooling Load	
	10^6BTU	Normalized %	10^6BTU/HR	Normalized %
No Berm	1541	100	0.571	100
Half Berm	1405	91	0.494	87
Full Berm	1214	79	0.394	69
Covered	1121	73	0.360	63

WAREHOUSE RESULTS

The dry-storage warehouse is a simpler building configuration than the office building. There is only one building zone; the heating thermostat setting is 55°F; there are no cooling requirements, and there are no windows. Because the mean ground temperature in Jacksonville and San Diego is greater than 55°F (the warehouse heating thermostat setting), and because of the large thermal mass in the warehouse to moderate temperature swings, all heating requirements for the modeled warehouses in these locations can be met by heat gains through the floor or by heat stored in the building. In Manila, both the average air temperature and the ground temperature are considerably above 55°F. Consequently there is no heating energy required for the Manila warehouses. This leaves only the results for warehouses in Boston and Washington, D.C. for presentation.

Figure 17 shows the component breakdown and the total annual HVAC energy required to heat the four warehouse configurations to a minimum temperature of 55°F in Boston. As was the case for the Boston office building, the dominant component load is due to outside air that infiltrates through walls and at the loading docks. Increased earth-sheltering reduces the infiltration through the walls and also decreases the conduction heat loss through the walls. The decrease in wall heat loss for the bermed and covered building to 17% of the slab-on-grade value is again due to the thermal buffering supplied by the soil mass. Because the warehouse must only be heated to 55°F, which is near the mean ground temperature for Boston, the earth-sheltering is especially effective in reducing conduction heat loss. Note also how the earth covering on the roof reduces heat loss through that component to 30% of the above-grade roof.

Table 7 compares the reduction in peak load to the reduction in required heating energy. The close similarity between the reductions for both HVAC energy and peak load, when taken in the context of the peak load results for the office building configurations, makes it clear that even though clearly significant reductions in the infiltration load and improvement in the wall thermal performance exist on an integrated energy basis, one must be very careful about guessing the peak load reductions.

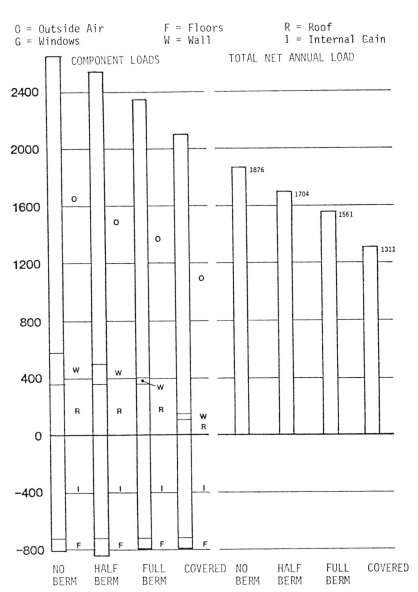

Fig. 17. Annual Energy Performance for Boston Warehouse, Heating Load (Units of 10^6BTU). On the left, component loads are shown for cumulative net energy losses or gains that will affect the overall load for a given zone. Heat lost is shown as positive; heat gained as negative. On the right, component loads are summed resulting in the total annual heating load.

TABLE 7 Heating Energy and Peak Design Loads for the Boston Warehouses

Building Configuration	Heating Energy		Peak Heating Load	
	10^6BTU	Normalized %	10^6BTU/HR	Normalized %
No Berm	1876	100	4.784	100
Half Berm	1704	91	4.356	91
Full Berm	1561	83	3.995	84
Covered	1311	70	3.131	65

The Washington, D.C. warehouses performed similarly to the Boston warehouses with the milder Washington, D.C. climate reducing the heating associated with each facet of the building energy balance as shown in Fig. 18.

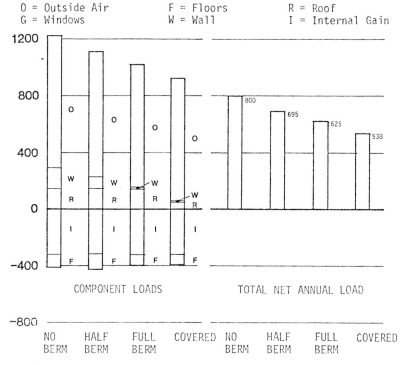

Fig. 18. Annual Energy Performance for Washington, D.C. Warehouse, Heating Load (Units of 10^6BTU). On the left, component loads are shown for cumulative net energy losses or gains that will affect the overall load for a given zone. Heat lost is shown as positive; heat gained as negative. On the right, component loads are summed resulting in the total annual heating load.

CONCLUSIONS

Computer analysis of HVAC energy requirements for a one story office building in five different locations--Boston, MA; Washington, D.C.; Jacksonville, FL and San Diego, CA in the United States and Manila in the Philippines--for four different levels of earth-sheltering has predicted significant energy savings associated with increased earth-sheltering. Table 8 summarizes the annual HVAC energy consumption for each building.

TABLE 8 Office Building Energy Summary for Five Locations

A. Annual Heating Energy Summary (Energy units are 10^6BTU)

Location	No Berm		Half-Berm		Full-Berm		Covered	
	Energy	%	Energy	%	Energy	%	Energy	%
Boston	1036	100	924	89	499	48	470	45
Washington	638	100	574	90	323	51	295	46
Jacksonville	104	100	89	86	54	52	44	42
San Diego	52	100	48	92	31	60	28	54
Manila	0		0		0		0	

B. Annual Cooling Energy Summary (Energy units are 10^6BTU)

Location	No Berm		Half-Berm		Full-Berm		Covered	
	Energy	%	Energy	%	Energy	%	Energy	%
Boston	226	100	205	91	178	79	162	72
Washington	336	100	297	88	252	75	204	61
Jacksonville	947	100	881	93	792	84	746	79
San Diego	664	100	661	99	628	95	615	93
Manila	1541	100	1405	91	1214	79	1121	73

Examination of the percentages of the slab-on-grade ("no-berm") heating energy for both the full-berm and the bermed-and-covered ("covered") configurations in each climate as presented on Table 8.A., indicates that earth-sheltering a one story office building can save approximately 50% of the annual heating energy in each of the climates studied. For this type of building, the cooling season reductions in energy consumption, shown on Table 8.B., are not so large.

It is important to distinguish the reductions in HVAC energy consumption from the changes in the HVAC peak load characteristics which are summarized in Table 9.

From this table of peak loads it is readily apparent that earth-sheltering to the fully bermed or bermed and covered levels has a very positive benefit of reducing the peak load and thereby reducing the size of the HVAC mechanical system.

From the computer predictions in Tables 8 and 9, the efficacy of earth-sheltering in reducing both HVAC energy consumption and peak load requirements has been estimated. The energy advantages of increased earth-sheltering in these simulations are clear; it is important to review why they have occurred. Improvements in heating season performance associated with earth-sheltering as it has been modeled in this investigation are due to the major reduction in uncontrolled infiltration, improvement in wall thermal performance and the elimination of windows on all sides except the southern exposure. For these

TABLE 9   HVAC Peak Loads for Office Buildings

A. Peak Heating Load Summary (Power units are 10^3BTU/HR)

Location	No Berm		Half-Berm		Full-Berm		Covered	
	Energy	%	Energy	%	Energy	%	Energy	%
Boston	677	100	578	85	274	40	239	35
Washington	572	100	486	85	275	48	242	42
Jacksonville	367	100	306	83	206	56	155	42
San Diego	163	100	142	87	62	38	32	20
Manila	0		0		0		0	

B. Peak Cooling Load Summary (Power units are 10^3BTU/HR)

Location	No Berm		Half-Berm		Full-Berm		Covered	
	Energy	%	Energy	%	Energy	%	Energy	%
Boston	598	100	548	92	502	84	493	82
Washington	612	100	551	90	364	59	331	54
Jacksonville	616	100	487	79	376	61	320	52
San Diego	573	100	498	87	419	73	407	71
Manila	571	100	494	87	394	69	360	63

simulations the biggest positive factor in reducing the heating load was the reduced infiltration. Typical reductions in infiltration for the fully bermed and the bermed-and-covered configuration were nominally 50%. There is great uncertainty in predicting the absolute level of infiltration, but its potential impact on the energy budget is large, especially when it is coupled with ventilation requirements as was done in this study. Increased levels of earth-sheltering will definitely facilitate control of infiltration/ventilation.

The thermal performance of earth-contact walls improved with increased levels of earth-sheltering. Thermal buffering due to the soil mass typically reduced the energy lost through the walls by at least 50%. Although the quantity of energy saved in this study because of earth contact walls is considerably less than the savings shown that were due to reduced infiltration, these savings are important. For buildings with more exterior wall surface area for the same volume, e.g. a two story building, or for a building with less infiltration, the impact of this improvement in wall thermal performance would be enhanced.

The improvements in cooling season performance associated with earth-sheltering are due to reduction of uncontrolled infiltration, reduction in heat gain through the walls, the opportunity to use the ground as a heat sink, and, for the buildings with earth-covered roofs, reduced heat gain through that component. The estimates of cooling season benefits in this study are conservative because of limitations inherent in the computer models, both for heat loss to the ground around the building and for heat loss through the earth-covered roof during seasons when significant evapotranspiration is occurring at the ground surface. Computer storage restrictions have thus far restricted analysis to two dimensional simulations of ground heat transfer. Consideration of the third spatial dimension will increase cooling season heat loss to the ground. In addition, as understanding of earth-covered roofs increase, integration of more accurate models of the evapotranspirative cooling potential at the ground surface will enable the irrigation of earth-covered roofs to be managed efficiently to further improve cooling season performance.

ACKNOWLEDGEMENTS

Major contributions to the analyses discussed in this paper were made by the author's colleagues Raymond Sterling and Jeanne Hladky at the Underground Space Center and by Ed Frenette and Dan Kallenback of the architectural and engineering firm Setter, Leach and Lindstrom, Inc., Minneapolis, MN. Valuable guidance in the development of the building models and overall project direction was given by Richard Vasatka of Setter, Leach and Lindstrom, Inc. Discussions with Thomas Bligh of the Department of Mechanical Engineering at the Massachusetts Institute of Technology were instrumental in interpreting the earth-sheltered thermal performance. Work on this project was supported by the United States Department of the Navy, Northern Division.

REFERENCES

[1] Bligh, T.P., A Comparison of Energy Consumption in Earth Covered vs. Non-Earth Covered Buildings, Alternatives in Energy Conservation: The Use of Earth Covered Buildings, pp. 85-105. National Science Foundation, ed. F. L. Moreland, Arlington, Texas (1975).

[2] Sterling, Raymond L., Current Research Into the Effectiveness of Earth Sheltered Buildings as a Passive Energy Conservation Technique, Proceedings of the 4th National Passive Solar Conference, ed. Gregory Franta, pp. 425-428. Kansas City, Missouri (1979).

[3] Underground Space Center, Earth Sheltered Housing Design, Van Nostrand Reinhold Company, New York (1978).

[4] Speltz, J. and C. D. Meixel, "A Computer Simulation of the Thermal Performance of Earth Covered Roofs," presented at the Underground Space Conference, Kansas City, MO, June (1981).

[5] U.S. Army Construction Engineering Research Laboratory (CERL), BLAST, The Building Loads Analysis and System Thermodynamic Program Users Manual-Volume One, Technical Report E-153, June (1979).

[6] Kusuda, T., NBSLD, The Computer Program for Heating and Cooling Loads in Buildings, NBS Building Science Series No. 69, Washington, D.C.

[7] Building Energy Analysis Group, DOE-2 Reference Manual, Energy and Environment Division, Laurence Berkeley Laboratory, Berkeley, California (1978).

[8] Carroll, W.L., Annual Heating and Cooling Requirements and Design Day Performance for a Residential Model in Six Climates: A Comparison of NBSLD, BLAST 2 and DOE-2, Proceedings of the DOE/ASHRAE Conference: Thermal Performance of the Exterior Envelopes of Buildings, December (1979).

[9] Shipp, P.H., Thermal Characteristics of Large Earth-Sheltered Structures, Ph.D. Thesis, University of Minnesota (1979).

[10] Meixel, G.D., Shipp, P.H., and Bligh, T.P., The Impact of Insulation Placement on the Seasonal Heat Loss Through Basement and Earth-Sheltered Walls, Underground Space, Vol. 5, pp. 41-47, 1980.

[11] U.S. Department of Commerce, Climatological Data, Asheville, North Carolina.

[12] American Society of Heating, Refrigeration and Air-Conditioning Engineers, ASHRAE Handbook of Fundamentals, p. 21.5, 1977.

ON-SITE BUILDING MATERIALS USING UMR WATER JET TECHNOLOGY

B.H. Green* and D.A. Summers**

*Environ. Plan. Eng., Civil Eng. Dept.,
Univ. of MO-Rolla, Rolla, Missouri   65401
**Rock Mech. & Explos. Res. Ctr.,
Univ. of MO-Rolla, Rolla, Missouri   65401

ABSTRACT

The creation of underground space, concurrently generates a volume of "waste" rock. This paper discusses the energy savings that can be achieved if this rock mass is cut up into blocks for use on On-Site Building Material. The use of high pressure water jets as a means of cutting out these blocks, is evaluated and shown to be energy economical. The energy thus conserved is considerable and can be reallocated for other more productive uses. It also generates new local employment and energy efficient buildings through its use.

KEYWORDS

Water jets; underground space; stone; energy; building; construction; excavation; housing; masonry.

INTRODUCTION

The hole for an earth-sheltered or underground structure is commonly excavated either by a machine, in soil, or by blasting in rock. This paper is directed to those instances where the site will contain rock or where the entire structure is to be sited in rock. Common practice is to remove this rock by drill and blast techniques. Such practice has a number of disadvantages which offset the primary benefits of blasting; i.e., its relatively low cost; the relatively small amount of equipment and the low skill levels required to do the job, Fig. 1.

In this paper we would like to suggest an alternate method of approach. Our thinking has been galvanized by the recent commercial use of high pressure water jets as a means of cutting rock. It is proposed to use this technology for two purposes. The first is to excavate the profile of the spaces in the rock, and, the second is to use the water jets to cut up the central rock mass into discrete volumes so that it can be used as on-site building material.

It is perhaps pertinent to point out a number of recent advances in the state-of-the-art, which pertain to this topic. Firstly, the advantage of pre-profiling an excavation, as a means of improving rock stability, has been recognized for some considerable time. The recent program in France (Pera, 1978), where the profile of

259

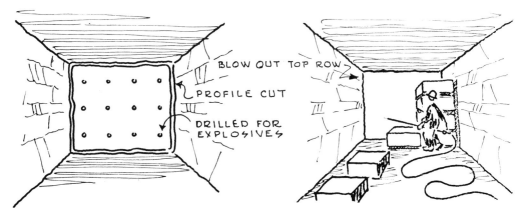

A. Profile and Blast.   B. Profile and Block Removal.

Fig. 1.   Chamber type excavations.

a tunnel was excavated with a mechanical cutter, indicates a continued interest in
this concept. However, mechanical cutters are limited in application and much
more sensitive to rock type than is a water jet cutting system.

The increased range of application of a high pressure water jet system can be
demonstrated by the results of a recent test. A team at the University of Missouri-Rolla has shown the viability of water jets for cutting granite, a rock not
normally considered practical to cut with mechanical cutters (other than diamond
saws, Summers, 1979). In the experiment, in a granite quarry, it was shown that
water jets at a pressure of approximately 100 MPa (15,000 psi)[1] were capable of
slotting granite at rates between 2 and 3 times that conventionally achieved, at
half the energy input, and with much lower noise levels. The rock compressive
strength was approximately 200 MPa (30,000 psi). It should be emphasized that the
equipment used is now generally commercially available. The advent of high pressure water as a cleaning mechanism has led to the growth of a number of different
manufacturers of the high pressure pumping system and ancilliary equipment required for this purpose.

In the granite cutting trial, the quality of surface left by the water jet was
considered sufficiently smooth to provide a satisfactory initial surface finish
for many uses, and so while the exact quality of the finish is a function of the
rock being cut, it is not anticipated that this would pose too great a problem.

Based on the design of a house, located in granite, currently available production
rates (which have not been optimized) suggest that three 3 m x 4.5 m rooms, 2.4 m
high (10 ft x 15 ft x 8 ft high) could be excavated by a single unit in one week.

The use of water jets for excavation of the central rock core is not entirely
novel. Many of the tunnels, for example, under Minneapolis, in the St. Peters
sandstone, have been excavated using low pressure water (Nelson, 1975). Unfor-

---

[1] Only approximate conversion values are given for the SI units, since only "range"
values are being given.

ᴛunately, however, the majority of the rock one is likely to see in surface and close to surface excavation is apt to be considerably harder than that of the St. Peters sandstone, requiring that a larger pressure be developed. Moreover, it is our intent that the water jets would also be used to kerf out blocks of rock rather than slurrying the entire rock volume as is the case in Minneapolis. Thus one can anticipate a different approach than that of current tunnel technology.

DEVELOPMENT OF THE CONCEPT

If water jets are to be accepted as a rock slotting and excavating system, then they must first satisfy the same criteria as with explosive technology, if not improve upon it. Water jetting must be relatively simple and easy to operate, and also not be very expensive. Until recently, those claims could not be made. However, as has been mentioned the granite industry has demonstrated the mobility, relative simplicity, and cost effectiveness of this technology. It is, perhaps, pertinent that jets operating at pressures up to 100 MPa are now used by divers to clean oil rigs underwater in the North Sea (Thomson, 1980). The stage is therefore set for the application of water jets in the housing and building industry, for those spaces which must be excavated in rock.

The major purpose of this paper, however, is not to discuss the primary benefits of water jet technology for the wall excavation (that point has been made earlier, Green, 1980(a), Green, 1980(b)), but rather to examine the potential for using water jets to carve out the central rock mass which occupies the projected building volume, such that it may be used for on-site building material. Again, such a suggestion is not novel. In most instances historically, local rock has been used as construction material. However, the more recent tendency to remove rock by explosives, in the most common types of excavation has reduced the availability of this supply. Modern architects have also shied away from this valuable resource, to turn instead, to energy intensive materials, such as steel and glass.

ON-SITE BUILDING MATERIAL

The increased expense of building material, the high energy cost for its acquisition and processing, and the transport costs that go with it are factors which must be involved in the decision on the type of building material to use at a site. The move towards treating underground space as a useful void has had to overcome several shibboliths; the re-evaluation of old judgements should, however, not stop with the use of the space.

While creating the underground space is important, it also, in the process generates a secondary resource, the rock removed. Normally where the rock is blasted out, its commercial use is limited, and the value is correspondingly small. For many home or small business sites, the volume of rock is too small to sell, and the owner must pay to have it removed. The use of rock blocks, as a building material, has recently been limited to a thin decorative layer, and is a highly expensive cover because of its processing. However, where the rock is cut out, to building block size, it could be used for that purpose (Fig. 2). This would, however, be a judgement that must be made on both aesthetic and cost grounds. The cost viewpoint, for many, is however a sine quae non, and so this is addressed here. A calculation is made of the energy cost since, now and much more in the future, this will be the major portion of the cost.

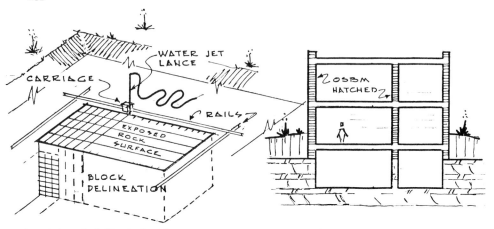

A. Automated Block Cutting.　　　　B. Use of On-Site Building Material.

Fig. 2.　Open cut type excavation.

If one looks at the excavation efficiency of a high pressure water jet, in slotting rock (Appendix 1), one can calculate, once the original excavated volume is defined by a profiling lance, the energy required to cut the remaining volume into building blocks. In the energy calculation we have not included this primary profiling as a charge to the block manufacture, since it is done to delineate the rooms. (It would increase the energy level about 15 percent.) The energy requirements calculated are those to carve out individual blocks, from the solid, of conventional building block size (Table 1).

TABLE 1　Energy Requirements to Excavate a Rock Block

It is assumed that a specific energy of 2000 j/cc is required to excavate the rock*. The rock is assumed to have a density of 2.5 gm/cm^3 (156 lb/cu ft).

Type	Dimensions (cm)	Block Weight (kg)	Specific Removal Energy	
			(j/gm)	(Btu/lb)
Brick	5 x 10 x 20	2.5	800	344
Block	20 x 20 x 40	40	300	129
Large Block	20 x 40 x 60	120	100	43

It is presumed that a slot 2.5 cm wide is required to establish a slot around each block (Fig. 3). This requirement is to allow the jet lance access into the block. Based on the actual energy requirements to cut slots in granite and coal, materials near each end of the likely rock spectrum, then an approximately value of 200 Btu's per pound of block can be reached. In estimating this value, it is recognized that the block will weigh more than conventional concrete, however this increase in building mass, is beneficial in stabilizing building temperatures.

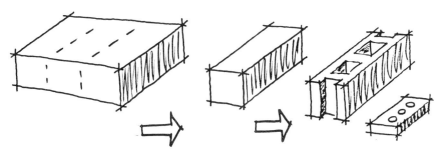

A. Large Block   B. Small Block   C. Hollowed Block   D. Brick
8" x 16" x 24"   8" x 8" x 16"

Fig. 3. Cutting and milling to dimensional sizes.

The energy required for such a block is relatively low in contrast to other forms of construction material (Table 2). Further, the material is available on site, and requires no transportation. The resource is domestic, in contrast to the current trend, because of cost, to import material from aborad, such as Belgian cement, Japanese plywood, or steel.

TABLE 2   Unit Production Energy of Common Building Materials (Catani, 1980)

Material	Joules/gm	BTU per Pound
aluminum	95,350	41,000
steel	30,230	13,000
brick	30,230	13,000
glass	29,300	12,600
drywall	5,020	2,160
ceiling materials	3,490	1,500
concrete	960	413
concrete blocks	960	413
"On-Site Building Material" (rock) (the value varies with the rock)	465	200

It is accepted that the aesthetics of the surface may not be quite as congenial; however, the very low costs of the excavated material in terms of energy and site availability, allow that the structure can be manufactured from the on-site building material and that once it is completed, an aesthetic cover can be placed over it. A thin coat of plaster or plywood will render the surface visually acceptable (to those who do not like the look of the "Living Rock") and still bring in the total project, at a competitive price.

CONCLUSIONS

In summary, this paper examined a change in the method of excavating underground space in rock. Historically, this material has been excavated by drill and blast and the spoil removed (rubble) has been utilized for road fill and other inexpensive resource usage. Otherwise, it must be disposed of at a cost to the owner. Because of the ability of high pressure water jets to generate a relatively smooth surface in rock, at a relatively low energy cost, two major advantages to their use in underground excavation can be foreseen. These, specifically, are the creation of a competent wall in the natural rock, which will not require as much re-inforcement, or other treatment. Of equal importance, is the ability of the water jet system to excavate the remaining volume of rock from the space so that it can be used as a viable building material. This material can then be utilized on site for construction of those portions of the building which lie above the rock horizon, at an energy cost much below that of competing artificially manufactured materials.

Several other advantages can also be conceived for this technique, and the use of natural stone for the surface construction. These include: the rock strength, fire resistance and wear resistance, and the lack of noise during the excavation process. Because of the very limited equipment required for this technical approach, its simplicity and use of naturally available resources, it may be that this could prove a very effective way, also, of creating useable space in Third World Countries with little capital or normal construction materials (eg. wood).

CLOSING

We realize that the idea of using water jets to cut rock is hard for many people to accept. We do have some films and videotapes of some of our work, and we also have an active research program. We would invite you to visit us, or seek to borrow our visual records, in order that you may better appreciate the simplicity of our approach.

REFERENCES

Pera, J.W. and M. Bougard (1978). Some Recent Progress in Tunnel Construction in France. Proc. Int. Tunnel Symp., Pergamon Press, Tokyo.
Summers, D.A. and C.R. Barker (1979). The Slotting of Granite. A Report to Georgia Granite Association.
Nelson, C. (1975). Present Tunnelling Techniques, St. Anthony Park Stone Sewer Project. Proc. Workshop on the Application of High Pressure Water Jet Cutting Technology, University of Missouri-Rolla, Rolla, MO.
Thomson, J. (1980). Rig Cleaning in the North Sea. Water Jetting in the '80's, British Hydromechanics Research Association, Sheffield, UK.
Catani, M.J. and G.B. Barney (1980). The Concrete Advantage. Concrete Construction Magazine.

APPENDIX

Most likely rock encountered at a site will lie between coal and granite in strength. Based on field trials it is possible to establish excavation rates for both materials.

Coal

A 33.5 KW jet cutting lance will cut a slot 5 cm wide, 60 cm deep and 6 m long in 1 min. The specific energy of cutting is thus:

$$\frac{33.5 \times 0.06}{5 \times 60 \times 600} \ MJ/cm^3 \ = \ \underline{11.2 \ j/cc}$$

Granite

A 112 KW cutting lance will cut a slot 5 cm wide, 30 cm deep and 10.5 m long in 1 hr. The specific energy of cutting is thus:

$$\frac{112 \times 3.6}{5 \times 30 \times 1050} \ MJ/cm^3 \ = \ \underline{2560 \ j/cc}$$

If the rock is presumed to require a specific cutting energy of 2000 j/cc, we can compute the energy required to carve out the block as follows:

Typical block dimensions are 20 x 20 x 40 cm.

4 faces will be 2.5 cm wide to allow the lance into the cut.

Excavated volume of material is thus:

$$\{[2 \times (20 \times 20)] + [2 \times (20 \times 40)]\} \times 2.5 = 6000 \ cm^3$$

<u>Total energy required</u> = 6000 x 2000 = <u>12 MJ</u>

If the rock has a mean density of 2.6 g/cm^3 then the block will weigh 41,600 gms.

The energy required for block removal is thus:

$$\underline{288 \ joules/gm}$$

(For those who feel more comfortable in other units, this converts to:

$$\frac{288}{1055} \ = \ \frac{453}{1} \ = \ 124 \ Btu/lb).$$

NOTE: If the material is closer in specific energy values to coal, then the energy required for removal will be on the order of <u>0.5% of that calculated herein.</u>

GEOTECHNICAL CONSIDERATIONS OF AN EARTH SHELTERED
MANUFACTURING FACILITY IN NORTHERN ILLINOIS

Daniel J. Aucutt, P.E. - Senior Engineer
Soil Testing Services, Northbrook, Illinois

ABSTRACT

A case history is presented which involves a partially buried commercial manufacturing facility. The northern Illinois site was selected primarily for its advantageous location with respect to existing business centers. A partial geotechnical site selection study was conducted. The physical characteristics of the selected land parcel rapidly complicated building design. Ground water control and stabilization of loose glacial silt soils created the need for a specially designed gravity drainage system and low bearing pressure foundations. Close cooperation between the owner, architect, geotechnical consultant, and contractor was required to minimize cost overruns and bring the manufacturing facility on line. A discussion concerning the testing and selection of geotextiles is presented as well as the importance of providing a strict on-site construction monitoring program.

KEY WORDS

Geotechnical exploration; earth sheltered structures; ground water control; glacial silt; geotextiles.

INTRODUCTION

The office and manufacturing complex designed by O'Donnell Wicklund Pigozzi -Architects for the MacLean-Fogg Company, was constructed on a 22-acre land parcel located just north of Richmond, Illinois near the Wisconsin State line (Fig. I). The complex consisted of a 4000 square foot, one story slab-on-grade office area with an attached 20,000 square foot, high bay, manufacturing area. The manufacturing area was approximately 100 ft wide and 200 ft long and was to be inset into a natural hillside such that the north and west walls would be buried almost to the roof line while the south and east walls would consist of thermal pane windows with an exterior, adjustable, sunscreen or awning.

The facility would be used for production of various fasteners for use primarily in the automobile industry. The manufacturing area included an enclosed, double bay, recessed truck dock which would allow raw materials and final products to be received and shipped entirely indoors.

The primary reason for constructing an earth sheltered facility was that operational costs would be reduced. However, another factor influenced the decision. The company believed that overhead cost might be reduced if an improved working environment would reduce employee turnover rates.

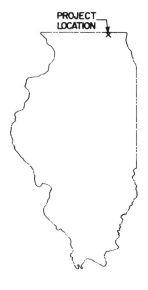

Fig. I. Location diagram -
State of Illinois.

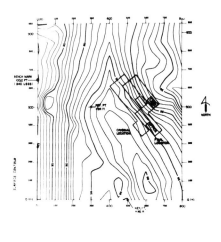

Fig. 2. Computer interpretation
of site topography.

MacLean-Fogg employee turnover has been a problem at an existing conventional, one story, office and manufacturing facility located approximately 30 miles south, in Mundelein, Illinois. Each time an employee is replaced, the company faces an increase in overhead due to new employee training costs and initial low productivity. This problem is especially chronic among the production employees. It was believed that the high employee turnover rate was partially due to working conditions in that the existing manufacturing facility was over-crowded and the heat generated by the manufacturing process produced uncomfortable working conditions during the summer months.

Since some form of plant expansion was necessary, the possibility of constructing a new earth sheltered facility was explored. An earth sheltered structure could minimize energy costs and proper air exchange could maintain more uniform, cooler temperatures which could, in turn, reduce overhead by minimizing employee turnover.

SITE EXPLORATION

The exploration area encompassed approximately 175 acres of agricultural land. The initial site study was conducted by another consulting firm in 1974 and consisted of seven soil borings. Only one of these borings was placed within the 22-acre parcel which eventually became the subject site. That boring indicated that footings positioned near the ground surface could be proportioned for a design bearing pressure of about 2500-3500 psf. The boring also indicated that the depth to ground water after completion of the boring was only 1.5 ft. The high water table was attributed to temporary perched water table conditions. Comments were presented regarding the probable necessity to raise the grade 2 or 3 ft if septic fields were going to be installed.

The final site was selected for two main reasons:

      1. The overall relief of the site (20 ft) appeared to be sufficient for a partially buried structure.

      2. The site exhibited good southern exposure (i.e. the north side of the structure would be buried while the south side would be exposed to the sun as well as to U.S. Route 12, a major highway).

Early in 1979, Soil Testing Services, Inc. (STS) prepared a preliminary site evaluation for the subject site. This study included four additional borings. The soils at the site were classified as part of a gravelly glacial outwash plain. These soils were deposited by glacial melt water which spread loose silty sand, gravel and cobble-laden debris westward from the ice front and produced outwash fans that eventually coalesced to produce the outwash plain. The deposits were found to be highly variable in texture and thickness but generally graded finer from east to west. In most areas the loose granular deposits were capped with a thin silty and clayey layer.

Since the location of the structure was not known at the time the exploration was conducted, recommendations were still quite general. The design bearing pressure for foundations was given as 2000-3000 psf. The ground water table was reported at a depth of approximately 5 ft in the higher elevation areas and above the ground surface in the lower elevation areas.

By August of 1979, most of the design details had been prepared and a preferred location for the structure was selected. STS then performed an additional nine soil borings. These borings indicated that a design bearing pressure of 2500 psf was appropriate for the selected location. However, ground water conditions would make construction very difficult.

Figure 2 presents a topographic map of the site. The elevations shown are referenced to an assumed benchmark at +100 ft located near the northwest corner of the site and corresponds with approximate U.S.G.S. elevation 840. The contour lines indicate a gentle drainage swale extending southward along the approximate north/south center line of the site. Also shown on the map are the proposed and as-built building locations. The proposed floor slab elevation was +82, requiring a recess of at least 1 ft at the far west corner and approximately 10 ft at the far east corner. The soil borings revealed that the water table in this area occurred at about elevation +85 and that soft soil conditions existed near the drainage swale.

The use of perspective computer mapping enables the architect and engineer to better visualize the geotechnical problems. Figure 3 depicts the location of the proposed excavation with respect to the surrounding topography and shows the relationship of the ground water table at elevation +85 and the proposed floor slab at elevation +82. The excavation shown is representative of the factory area. The office area and the recessed truck dock in the manufacturing area were omitted for reasons of clarity. The approximate orientation of the drainage swale has been highlighted and several reference points are given to aid in visualization.

In light of the high water table and the soft soils, it was proposed that the entire structure be relocated to higher ground. The architect concurred and the design was changed to relocate the structure to its present position (Fig. 2). At this location only two of the original borings remained within the limits of the structure.

A new boring program was set up and another five borings were performed by STS. These borings indicated a slight improvement in soil conditions and confirmed the high water table conditions. Altogether, nineteen (19) borings were performed for the 20,000 square foot manufacturing facility.

As part of the design revisions, the floor slab elevation was changed from +82 to +87. Figure 4 depicts the new location of the building excavation and shows the improved relationship between the floor slab and the ground water table. It also reveals that a considerable amount of new fill would be required around the structure in order to retain the sheltered design.

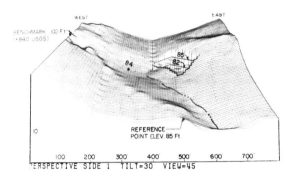

Fig. 3. Computer perspective of site topography indicating proposed
excavation and relationship with ground water table.

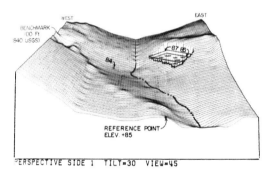

Fig. 4. Computer perspective indicating relocated excavation and
ground water conditions prior to dewatering.

TRUCK DOCK DESIGN

The recessed truck dock extended 4 ft below floor level, to approximate elevation +83. This design placed the floor of the dock below the water table. Excavation into the loose, saturated glacial silt soil would have been very difficult and some form of dewatering undoubtedly was necessary.

As part of the plant design, a retention pond was to be located along the natural drainage swale. In an effort to pull down the water table, it was decided to deepen the drainage swale and retention pond. For more direct dewatering, an underdrainage system was incorporated into the design. Normally, the underdrainage system is installed after the foundations and walls are poured. However, in this particular case it was believed that construction difficulties could be reduced by installing at least a portion of the underdrainage system prior to construction.

The design for the underdrainage system will be discussed in a subsequent section. For now, it is enough to say that the underdrainage system successfully lowered and has maintained the ground water table at approximate elevation +82 such that a conventional recessed truck dock design was possible. Figure 5 dipicts the final configuration of the manufacturing facility and retention pond. Note that a considerable amount of fill was required around the north and west walls of the structure while the south and east walls remain exposed. The original ground surface from Fig. 4 can be seen along the rear face of the excavation. The floor slab is shown at elevation +87 and the lowered water table is shown at about elevation +82. The figure also indicates that the drainage swale has been regraded and the retention pond has been placed with a design water level elevation of +80.

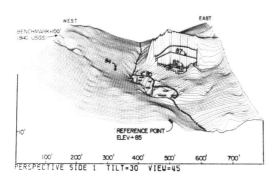

Fig. 5. Computer perspective dipicting as-built topography and improved ground water conditions after placement of retention pond and gravity drain lines.

DRAIN LINE CONSIDERATIONS

The original drain line design called for a standard drain tile and gravel pack configuration leading to a sump pump and pit. This system was not desirable for two reasons. First, the electrically operated sump pump would probably run on a more or less continuous basis which would be quite expensive. Since the building was designed to be an energy efficient structure, continuous operation of the pumps would be counterproductive. Second, a temporary power failure during a heavy rainstorm could permit a rapid rise in the water table with subsequent flooding or rupture of the truck dock. The ideal drainage system, therefore, would be a gravity system which would not require an external energy source.

The soil borings at the site indicated that the predominant soil type was a loose to medium dense fine sandy silt. A gradation analysis of this material is indicated on the right side of Fig. 6. This material will generally remain stable under confined conditions. However, because of the high percentage of silt, this material can rapidly become unstable in the presence of free water. In other words, water moving through this soil toward a drain line could pass through the natural pore spaces without dislodging soil particles. However, at the interface between the soil and the gravel pack around the drain line, free water flowing from the silt could carry soil particles with it. The mobile silt particles would move into the gravel pack and drain lines, and would eventually clog the entire system. Furthermore, the loss of silt particles from the natural soil would most likely lead to subsidence of the ground surface near the drain lines.

In an earlier effort to improve construction conditions, the contractor elected to install a temporary drainage system. This system consisted of placing open joint drain tiles in gravel-filled trenches. Approximately 400 lineal feet of drain line was installed in this fashion. Initially, a considerable quantity of water could be observed exiting from the drain lines. However, within a few weeks the water flow had reduced to a trickle, yet moist ground conditions continued to be observed. Upon excavation of the drain lines it was found that the lines were almost entirely filled with silt and sand particles.

272

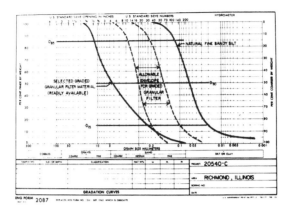

Fig. 6. Particle size analyses for natural soil and selected
granular filter material.

In order to minimize clogging and loss of natural soil it is theoretically possible to place a graded granular filter around the drain lines which would permit passage of water but would restrict the movement of soil particles. The U.S. Army Corps of Engineers (Ref. I) has developed a design criteria for graded granular filters (see Fig. 7). While their design criteria was formulated for earth dams it can be used for other drainage systems as well.

Using the Corps method, a gradation envelope was prepared as shown on Fig. 6. A material having a gradation within this envelope should function as a suitable filter for the natural sandy silt. However, three important questions must be answered before design of the drainage system is complete:

        I. Is the selected gradation compatible with the size of the openings used in the drain lines? If individual particles from the filter material are permitted to move into the drain lines, flow could eventually be restricted and clogging problems could develop.

        2. Is the selected graded granular filter material readily available near the construction site?

        3. Does the filter material have a permeability sufficiently great to permit effective dewatering of the site? If the filter does not have a permeability that is appreciably greater than the surrounding soil, the flow may not be great enough to produce a significant drawdown.

The failure of the open joint drain tiles at the site has already been discussed. We believed that superior performance could be achieved by using closed joint drain line with a controlled perforation size. Our drain line recommendation called for 6 inch diameter closed joint PVC plastic pipe with 3 staggered rows of 1/4 inch diameter holes placed on 1 inch centers along the invert of the pipe. This design was selected in order to provide sufficient perforation area while maintaining pipe strength. The use of commercially available, flexible, slotted drain line did not appear to be suitable because it could collapse under the weight of the additional fill to be placed around the building. Furthermore, the flexible tubing would be difficult to rod out if the drain lines ever did become clogged.

According to accepted design practices (Ref. 2) if 1/4 inch diameter perforations are used, then the D_{85} particle size of the filter should be at least 1/4 inch diameter. D_{85} refers to the particle size from the gradation curve at 85% finer by weight. As shown on Fig. 6 the entire allowable filter envelope lies within a particle size range less than 1/4 inch diameter. This means that while the design envelope is compatible with the natural soil at the site, it is not compatible with the openings in the drain line. Using a graded granular filter which falls within the design envelope would probably result in washing of the filter material into the drain lines.

One possible solution was to construct a second (coarser) filter envelope using the gradation of the first envelope as a basis. With this alternative, seepage water would pass from the natural soil into the first filter and continue through the second filter before entering the drain line. The second filter would have a gradation which would prevent migration of soil particles from the first filter and would also prevent migration of particles moving from the second filter into the drain line. In theory, this system should be quite effective. In practice, however, it would be very difficult to install.

GEOTEXTILES

As a second alternative, it was proposed that a man-made filter fabric (geotextile) be used to provide additional filtration. In this case a granular filter material is selected which is compatible with the drain line openings. This filter fabric membrane is placed between the natural soil and the filter gravel such that the filter gravel is completely encapsulated by the fabric. However, there are hundreds of geotextiles presently available. The majority of the analyses and data that are generally available have been generated by the manufacturers for promotion of their products, so extreme caution must be used when evaluating product performance. We have found that selection of a geotextile must be carefully matched with the specific project. The best method to determine which fabric(s) will provide optimum performance for a specific project is to perform comparative studies which simulate actual field conditions.

On this particular project the granular material was selected by the contractor. Gradation testing (Fig. 6) indicated that the D_{85} particle size was greater than 1/4 inch perforation size, which satisfied Question 1 above, and that the material contained a sufficient percentage of medium and fine sand particles to just enter the original design envelope. This material was also determined to be readily available (see Question 2).

Having satisfied two of the three important considerations, we had only to assure that the permeability of the sand and gravel filter material would be sufficient to dewater the site. Falling head permeability tests were conducted on both the natural soil and the granular filter material. The permeability of the natural soil ranged from 10^{-4} to 10^{-5} centimeters per second (cm/sec) depending upon the method of testing and the density of the material at the time of testing. The permeability of the sand and gravel filter material was typically more than 10^{-3} cm/sec, or approximately 10 to 100 times the permeability of the silt to be filtered.

As part of our laboratory analysis, we also evaluated the filtration capacity of the selected granular filter material relative to the site silt. The material was tested in a falling head permeameter in which the hydraulic gradient could be controlled. When a very thin (1/4 inch) layer of silt was placed over the sand and gravel and subjected to a hydraulic gradient of over 100, piping or washing of the silt through the sand and gravel filter occurred. However, when a second test was set up in the same apparatus with a 3 inch layer of silt over the sand and gravel filter, and with a hydraulic gradient limited to approximately 10, no penetration of the silt into or through the sand and gravel filter was observed. In addition, grain size tests performed on the sand and gravel before and after the filtration test indicated almost no change in the grain size distribution. This observation can serve as a further indication that the natural silt was not washed into or through the sand and gravel filter material.

It was determined that under low hydraulic gradients, the proposed filter sand and gravel would perform adequately. This material, however, would not perform well under severe hydraulic gradients. Therefore, in those areas where severe hydraulic gradients were possible, the use of a suitable geotextile was required. However, cost could perhaps be minimized by eliminating the fabric in those areas which would likely experience only low hydraulic gradients.

In order to determine which geotextile should be used to retain the natural silt, the permeability tests were performed with different geotextiles to evaluate their comparative performance. The geotextile we ultimately recommended for this project was a thick, non-woven, needle punch fabric manufactured by the Monsanto Textiles Company.

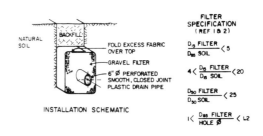

Fig. 7. Gravity drain line installation and filter gravel specification.

DRAIN LINE INSTALLATION

The fabric was obtained in a roll approximately 166 inches wide. This enabled the fabric to be installed in one piece in a 30 inch wide trench. The fabric was then wrapped up the sides of the trench for about 3 ft and overlapped on top of the interior gravel filter material (see Fig. 7). In this way any seepage water entering the drain line would have to pass through at least one layer of filter fabric and at least 6 inches of granular filter material. One wrapped drain line was installed which ran from beneath the truck dock directly to the retention pond. Another wrapped line was installed as a footing drain and ran along the north and west walls of the structure before being directed to the retention pond.

The slope of the drain line was maintained between 1/2% and 1%. This is equivalent to a drop of 1 inch every 13 ft. In order to maintain this slope, the invert of the footing drain was permitted to drop below the bottom of the exterior footings. This is not normally permitted because movement of soil into the drain line could seriously undermine the footings. In this case however, the system was designed such that as the elevation of the invert dropped 1 inch, the center line of the drain pipe was moved laterally 3 inches away from the edge of the footing. Therefore, along the 200 ft rear wall, the drain line invert dropped approximately 15 inches and was angled away from the wall approximately 45 inches. At the north corner the drain line then turned 90 degrees and ran along the west wall, toward the retention pond, maintaining the same slope and angular offset.

Within the interior portion of the structure, two underslab drain lines were installed which eventually connected with the footing drain line before being directed to the retention pond. The underslab drain lines were installed at a higher invert elevation than either the truck dock drain line or the exterior footing drain line. Therefore, it is expected that the underslab drain lines would only function during extreme high water conditions. For this reason, the underslab drain lines were not installed with the filter fabric membrane separating the natural soil from the granular filter material. Care was taken, however, to   locate and remove the original, clogged, drain lines which had been installed by the contractor. Even though these drain lines had essentially failed, they could potentially permit some additional loss of natural soil below the floor slabs.

RESULTS

The drain lines were installed during February, 1980. The lines effectively dewatered the site and have been flowing continuously for over a year. The truck dock itself has never indicated any seepage, even during the spring thaw of 1981. In addition, water samples collected from the drain line exit points have maintained clarity and do not indicate pass-through of sand or silt particles.

The operational success of the drain line speaks well for the engineering which went into the design. However, accurate implementation of the design during construction is essential. The contractor on this project was extremely conscientious and sensitive to the needs of the owner and the requirements of the architect and engineer.

It should be stated, however, that as construction costs go up and competitive bidding becomes more and more intense, the need for close on-site observation during construction by a qualified, independent, geotechnical and materials engineering firm becomes more and more essential. Even a conscientious contractor can misinterpret the design drawings and specifications. Since earth sheltered buildings sometimes cost more to construct, the impact of minor inadvertent errors and omissions tends to be greater than for a conventional building. By increasing the level of construction monitoring, many of these problems can be avoided.

ACKNOWLEDGEMENT

The author would like to express his gratitude to the MacLean-Fogg Company, to O'Donnell Wicklund Pigozzi-Architects, and A. May Construction Company for support for the preparation of this paper. Special thanks also are due to Joni Lewis who assisted in typing and proofreading of the manuscript.

REFERENCES

1. Corps of Engineers, Department of the Army (1952). Seepage Control, Soil Mechanics. Washington, D.C.
2.   Naval Facilities Engineering Command, Department of the Navy (1974).   Design Manual-Soil Mechanics, Foundations, and Earth Structures. Washington, D.C.

APPLICATION OF THE BERNOLD SYSTEM TO
BARREL SHELL EARTH SHELTERED ARCHITECTURE

Dr. John B. Langley, A.I.A.

Sun Belt Earth Sheltered Research
P.O. Drawer 729, Winter Park, FL 32790

ABSTRACT

Cast in place concrete barrel shells have been built by placing reinforcing over earth mounds and reusable wood or metal forms. The first method is cumbersome, the second time consuming and expensive. BERNOLD Plate was designed for mining, tunneling and culverts. The goal of this study is to devise an application of the BERNOLD Plate System for concrete barrel shell earth sheltered architecture.

The method of this paper is to identify nomenclature, product, process, and design detail parameters. This has been done by showing the application of the BERNOLD Plate to a specific prototype project. The detail application is shown and the conclusion drawn that the BERNOLD System, as manufactured by The Metal Products Division of U.S. Gypsum Company can reduce the requirement for time and skilled labor without economic penalty for earth sheltered construction.

KEYWORDS

BERNOLD System; concrete barrel shells; earth sheltering shotcrete application; thermal breaks in concrete shells.

Fig. 1. BERNOLD TUNNEL APPLICATION

HISTORY

In this very small world of ours the idea of a Swiss tunnel engineer who was getting behind schedule with a Swiss vehicular tunnel project and the idea of a research minded Florida architect who was trying to increase roof loading from 50 pounds to 500 pounds so he could design Sun Belt Earth Sheltered Buildings were brought together by a major American company who had seen the value of the Swiss idea and its application to the earth shelter problem.

My concrete barrel shell solution to the greatly increased roof loading was first described in a paper presented at the Oklahoma State University Earth Sheltered Conference in the spring of 1980. It was later outlined in an article in the Underground Space, Vol. 5, 1980, and fully developed in a small book, Sun Belt Earth Sheltered Architecture, published in December of 1980.

APPLICATION

Our first concrete barrel shell was constructed with a bar reinforced shotcrete operation over reusable wood forms. The wood forms were rather expensive and required a great deal of skilled labor to place and relocate, as was the placement of the steel reinforcement. Although the forms were finally used to cover the above grade garage, this solution was obviously time consuming and required three trips by reinforcing crew and the shotcrete applicator to the job site.

You can imagine my delight when Brad Burnside, of the Metal Products Division of United States Gypsum Company who had read the Underground Space article, sent me the illustrations shown in Fig. 1. These are the BERNOLD Plates in their tunnel applications. These curved plates provide form and reinforcement all in one. They can be placed for the total shell area and covered in one continuous operation.

Their biggest advantage is that they are rolled to the specific curvature required for the individual project.

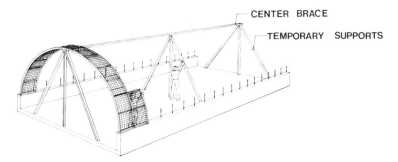

CENTER BRACE

TEMPORARY SUPPORTS

Fig. 2. STARTING ASSEMBLY

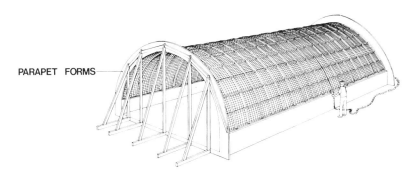

PARAPET FORMS

Fig. 3. READY FOR SHOTCRETE

ERECTION (Fig. 2)

Their woven form allows simple (unskilled) erection procedures whereby one plate overlaps with the next one with only an angle pin to secure the joint. The plates are approximately four foot square and weigh 40 to 50 pounds each. This means that no crane is needed to set them in place. A two person crew, even an owner-builder couple, could place them ready for coverage with concrete.

COVERAGE (Fig. 3)

When the plates are in place, end forming and inserts may be positioned. The plates may then be given the shotcrete application. When the removable wood forms are used, care must be taken that other forming and inserts such as anchor bolts, electrical conduit and boxes do not interfere with the removal of the forms. With BERNOLD Plates this is no longer a problem.

The concrete coverage is obviously different than in the tunnel application, where the shotcrete is applied from the inside and behind each plate as they are erected. The earth sheltered shotcrete application is applied to the outside first, then to the inside which is worked to a smooth curved finish.

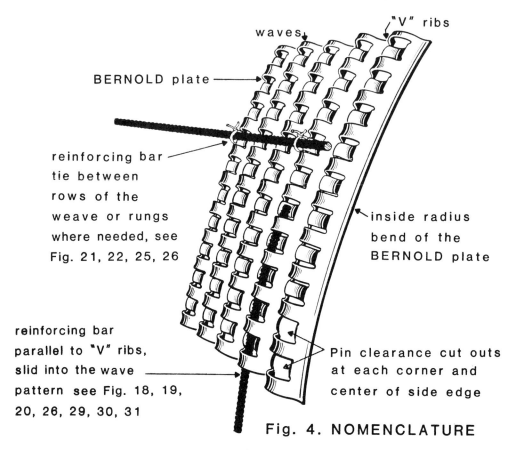

"V" ribs

waves

BERNOLD plate

reinforcing bar
tie between
rows of the
weave or rungs
where needed, see
Fig. 21, 22, 25, 26

inside radius
bend of the
BERNOLD plate

reinforcing bar
parallel to "V" ribs,
slid into the wave
pattern see Fig. 18, 19,
20, 26, 29, 30, 31

Pin clearance cut outs
at each corner and
center of side edge

Fig. 4. NOMENCLATURE

NOMENCLATURE

If you understand weaving, BERNOLD Plate is the woof of the fabric. The reinforcing bars introduced into it might be considered the warp. If that is not helpful, then we will use the term the manufacturers use, "wave", as the form seen in the end view of the plate (Fig. 4). The waves end in a "V" rib (Fig. 5).

The plate to plate connection is shown in Fig. 8. At footing, if required (Fig. 18), reinforcing bars running parallel to the "V" ribs are inserted into the space between the high and low waves. At parapet walls (Fig. 21) and thermal breaks (Fig. 22), reinforcing bars required to run perpendicular to the "V" ribs may be easily wired into the weave of the plate or between the rungs if you picture the plate as a ladder.

Notice also that the BERNOLD Plate is rolled so that the "V" ribs are on the inside when it is placed in the construction position. The connection detail (Fig. 8) is a view of the inside of the plate and the tie pins are installed on the inside face. In the drawing above, lower right hand corner, you can see that two of the interior weaves or rung pieces have been cut out to receive the weaves or rungs of the next plate that is to be lapped. This should help you understand some of the reinforcing details which are to be shown later.

<u>DIMENSIONS</u>

U.S. Gypsum's BERNOLD Plates are stamped out of 14 gauge steel plate. The stamping provides a "V" rib between a hi-low basket weave profile section (Fig. 5). The plates come in standard nine wave (42.2") and 11 wave (51.6") widths. Fig. 6 shows only 5 waves in plan view. The ribs are one and one-half inches deep. Parallel to the ribs the sheets are 48.2 inches before rolling to the required radius. After rolling the section appears as shown in Fig. 7.

5 waves : 23.45"

Fig. 5. END VIEW OF PLATE

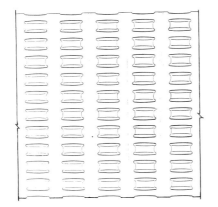

Fig. 6. PLAN VIEW OF PLATE

Fig. 7. SECTIONAL VIEW

These dimensions are not those needed to provide economical (low waste) design. First the top and side lap of the plates must be deducted, then the affect of the radius bend must be included. For design development the net width of the nine wave plate can be figured at 3.13 feet and the 11 wave plate at 3.91 feet.

The circumferal dimension modulus is about two and one-quarter inches so there is a great deal of freedom in radius selection. This can be "fine tuned" further with adjustment in the footing detail to be discussed later. Do not let all these dimensions scare you. The extra lapping flexibility to fit a special design program will not be all that high in small and medium size installations. However, if you are going to put up an acre of BERNOLD Plate, this extra lapping might become a factor. Details will be shown later about angular meeting of plates where waste considerations may be a factor as in Fig. 30. Plates may be field cut with a torch.

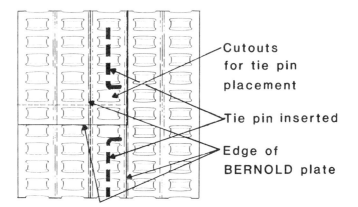

Fig. 8. PLATE CONNECTION WITH PINS

One way that BERNOLD Plate beats wood forming is the simplicity of the erection technique. The plates are lapped top and side, three rungs top and one wave to the side. The plate is punched to receive a metal tie pin which is dropped into place. The connection is made. The plates nest firmly enough so that in many cases the tie pin is only a safety factor against knocking the connection loose.

Should the structural engineer decide, for span considerations or additional overburden, that he requires a slightly higher section modulus in the reinforcing, he may choose to extend the side laps for more than one wave. This will develop a stiffener within the plates. At very large spans a rolled steel channel or "H" section may be used to stiffen the plate installation.

Engineering techniques will be discussed later.

Complete product information is available from the manufacturer, United States Gypsum Company/Metal Products Division, Underground Support Systems, 101 South Wacker Drive, Chicago, IL 60606. For further design information consult the author.

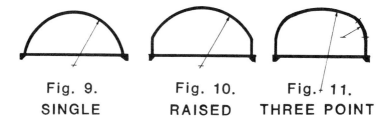

Fig. 9.	Fig. 10.	Fig. 11.
SINGLE	RAISED	THREE POINT

From the architect's standpoint the freedom to select a design radius is an obvious advantage. There are a few restrictions, however. Present tooling requires a minimum plate bending radius of about four feet. The maximum radius is limited only by the structural elements of concrete and additional rib bracing. Up to a 15 foot radius seems to be possible without the addition of internal steel or external concrete arched ribbing (Fig. 9).

It is also possible to start the radius with a straight vertical section which will provide additional close-to-the-wall headroom or furniture location (Fig. 10).

THREE POINT RADIUS

An elliptical arch may be developed with the three point method so that the different plates are rolled to different radii. The designer must remember the minimum radius restriction when selecting the minor radius of the elliptical arch. (Fig. 11). The structural analysis of these sections must obviously be done in the design development stage. This will be covered later in this paper.

DESIGN FLEXIBILITY: FIVE PLAN VIEWS

Fig. 12. STRAIGHT

Fig. 13. BENT

Fig. 14. SIDE BY SIDE

Our first application of the BERNOLD System was for a straight barrel shell with a radius of 13 ft. and a length of 54 ft. (Fig. 12). Among other applications are the bent barrel (Fig. 13) and the side by side grouping (Fig. 14).

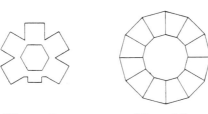

**Fig. 15.
INTERSECTING
CLUSTER**

**Fig. 16.
DIAGONAL
CLUSTER**

In Fig. 15 the intersecting corners of the plates will have to be cut or burned off in the field. These will essentially be wasted. Therefore, there may be barrel shell designs such as Fig. 16 where wastage will have to be seriously considered as a design limitation.

There may be situations where the BERNOLD Plates may be used in conjunction with wood forms, the wood forms used on a one time or even reusable basis. There may be design situations where the complexity of the barrel shell application limits the use of the plates altogether.

In any design development, the architect must keep in mind that he he is dealing with a rigid curved plate roughly four ft. square that may be joined at any of its four edges. Further flexibility can be had by increasing the lap of the plates.

BERNOLD SYSTEM DETAILS

Footings

The foot of the barrel shell will be subjected to two different loading conditions. When the concrete is applied, there will be stress outward at the base. When the lower two-thirds of the earth is placed against the shell, the stress will be inward. For this reason the reinforcing of the joint between the shell and the footing is critical. Two ways to handle this stress seem reasonable.

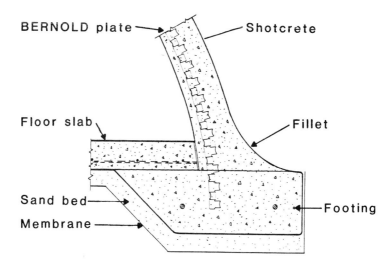

Fig. 17. PLATE STARTER COURSE

Plate Starter Course (Fig. 17)

In this detail the BERNOLD Plates are set into the footing form and held in alignment by a wood brace anchored across the footing. Footing plates are made the same size and placed with the same lap required for the remainder of the shell. In some instances the cross sectional shear area may require doubling the plates at this point. (Structural engineering calculations of this stress will be covered later.) These plates should be half height. The next plates in the rings are alternated half height and full height to fit the pinning system. These short plates on each side of the barrel shell arch will provide a means to adjust the standard plate module to the specific radius dimension of the required shell.

The concrete fillet can be designed to spread the shearing load of the shell on the footing under good soil bearing capacity. If the footing width is large to accommodate a low soil bearing capacity, there will also be cross sectional bending moment which may require crossing reinforcement over the longitudinal steel.

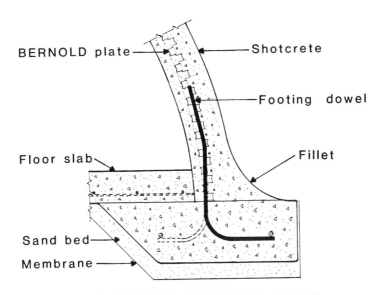

BERNOLD plate — Shotcrete

Footing dowel

Floor slab

Fillet

Sand bed

Membrane

Fig. 18. DOWELED FOOTING JOINT

Doweled Footing Joint (Fig. 18)

The designer may choose to use a section of the BERNOLD Plate as a jig by which footing dowels may be accurately placed. Accuracy of the placement of the dowel is not needed in the dowel to dowel distance as the open wave of the plate will allow considerable lateral tolerance. There will be a necessity to keep cumulative error out of the placement. For this reason two or more plate sections, lapped in the normal manner, should be used for dowel location.

Because of the openness of the plate wave, there will probably be no requirement for radius bends in the 30d extension of the dowel above the footing. For the shortest radius allowed, the dowel can be field bent after the footing has set and before the shell plates are placed.

This dowel footing will also allow minor adjustments for plate radius modulus differences. The area of the dowel steel should be the same or greater than the plate cross sectional area carried. Horizontal shear will probably be the controlling factor. However, the dowels should be checked for bending moment also.

Base Chord Closures (Fig. 19)

The necessity for base chord lateral restraint is a matter for the structural engineer to decide. If the lateral (shear) bearing capacity of the soil is matched with the vertical face area of the footing, no base chord reinforcing may be required. It is generally accepted that the horizontal component of the compressive loading of the top third of the shell will be taken by compression against the bottom two-thirds loading. But if the engineer is going to provide for negative bending moments in the shell foot, then this worry wart architect is going to feel more secure if there are some steel cross ties through or under the floor slab.

All that is required is to extend at every fourth or fifth foundation dowel, a cross tie dowel (Fig. 19) into the floor slab plane. When the slab mesh is placed, a single rod is tied across between the extended cross tie dowels.

286

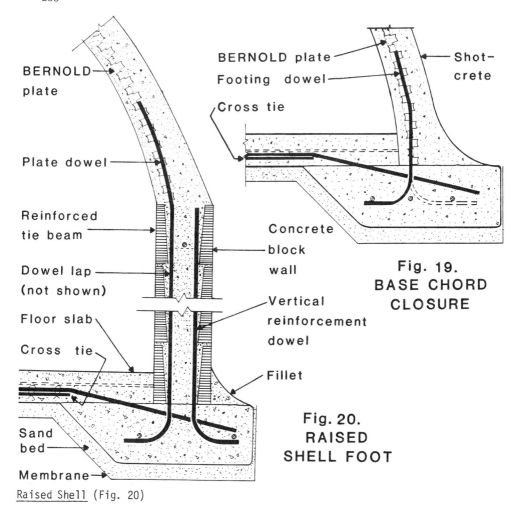

BERNOLD plate

Plate dowel

Reinforced tie beam

Dowel lap (not shown)

Floor slab

Cross tie

Sand bed

Membrane

BERNOLD plate

Footing dowel

Cross tie

Concrete block wall

Vertical reinforcement dowel

Fillet

Shot-crete

**Fig. 19.
BASE CHORD
CLOSURE**

**Fig. 20.
RAISED
SHELL FOOT**

Raised Shell (Fig. 20)

The floor space next to the barrel shell, when the curved barrel is brought to the floor line is hard to design in the least, and wasted in most cases.

To provide for right angle space for a programmed distance above the floor a block or cast foot may be added to the shell. In this instance the block wall is laid with the shell dowel set in the tie beam. The eight inch block modulus fits reasonably well with plate wave spacing. As when the dowels are placed in the footing, a double length of BERNOLD Plate should be used as a spacing jig.

Because the shell foot wall will first receive the inward thrust of the lower two-thirds of the overburden, the steel is kept on the outside face. The plate dowels are kept on the inside face to resist the outward thrust of the shell. The proper number can be continued on around to lap with the cross tie dowels cast in the footing for the base chord closures.

When the shotcrete is applied, it is thickened at the shell foot to fit the top of the wall. Note that a concrete fillet has been placed between the wall and the footing. This will assist in the membrane application.

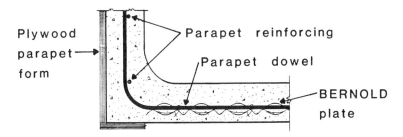

Fig. 21. PARAPET REINFORCING

Parapets (Fig. 21)

The aesthetic, structural, soil holding and safety necessity for barrel shell parapets is a decision that each designer must make for himself.

The detailing is quite simple. Remember parapets are retaining walls. The bending moment must be carried back into the shell. In that the BERNOLD Sheet cannot
be economically formed to a double curve, reinforcing bars are best suited to the
task. They are tied into the BERNOLD Plates between the weaves in sizes and locations determined by the structural engineer (Fig. 4). A temporary plywood form
with external bracing is placed in front of them to receive the shotcrete. Experience tells me that this form should be cut to the final profile required for the
parapet. This will enable the shell finisher to work more quickly and reduce the
patching required later.

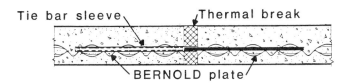

Fig. 22. THERMAL BREAK IN SHELL

Thermal Breaks (Fig. 22)

Remembering that the function of the thermal break is to slow down the movement of
heat to or from the exposed portion of the barrel shell into the sheltered portion
of the shell, the placing of this insulation becomes important. When the shell is
placed over solid wood forms, the thermal break is relatively easy. With the
BERNOLD Plate it is a little trickier. In the BERNOLD Application there is no
easy way to hold the insulation in place, so cross steel tie rods are threaded
through the insulation and tied into the BERNOLD weave on each side. The insulating sleeve is placed on the exposed side to cut down on the increased heat transmission property of the steel.

One additional suggestion not shown in the detail above is the idea of using two
pieces of the interior stucco bead shown in Fig. 27 to hold the thermal break insulation in place. Even if the insulation should slip down below the finished
inside surface, it can be cut flush and can be covered by the top plate of the
end wall.

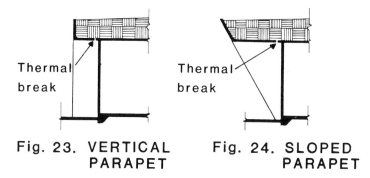

Fig. 23. VERTICAL PARAPET   **Fig. 24. SLOPED PARAPET**

Vertical or Angled Overhangs (Fig. 23 and Fig. 24)

In our first design with the BERNOLD Plate the shell extension beyond the exterior walls of the structure was parallel to the vertical walls. (Fig. 22) This meant that there was little horizontal thrust component which had to be overcome. The thermal reinforcing detail described was sufficient. Its only function was to keep the two portions of the shell in alignment.

Subsequent designs, however, will require that the projection be greater at the top of the shell than at the shell foot (Fig. 24). This will introduce an overturning moment into the projected overhang. To overcome this moment the thermal break crossing tie rods will need to be designed in order that the sleeved ends may make a mechanical connection to the concrete in the projected overhang.

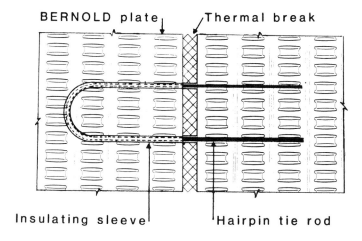

Fig. 25. TIES WITH HORIZONTAL TENSION

To provide this insulated anchorage where the thermal break ties are put into tension, a hair pin configuration is used. The p.v.c. pipe sleeve can be installed before the hair pin is bent. The unsleeved ends are threaded through the insulation and tied between the weaves of the inside plate.

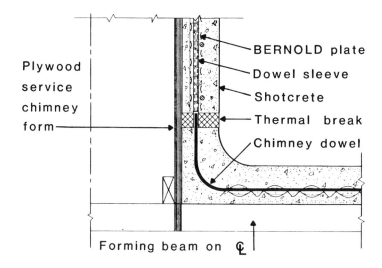

Fig. 26. SERVICE CHIMNEY
WITH THERMAL BREAK

Service Chimneys (Fig. 26)

The service chimney, used to take piping and ductowrk out of the shell above grade, is essentially a four sided or circular parapet wall. It may, however, require a thermal break. This requirement must be decided by the mechanical engineer. If the internal heat gain or loss through the sheltered envelope is sufficient to compensate for the exposure, the service chimney's "nose bleed" may be added to the thermal load of the exposed envelope. Otherwise an R-6 rigid insulation band should be inserted at the lowest point of the service chimney.

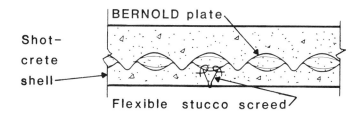

Fig. 27. INTERIOR STUCCO SCREED

Interior Stucco Screed (Fig. 27)

This screed of U. S. Gypsum's No. 4-A, Flexible Corner Bead may not be needed, providing the shotcrete contractor has skilled stucco technicians. The problem of uniform radius for the interior surface is more than a mechanical measurement. It is an aesthetic judgment. Where you are not sure of the skill and judgment available, the flexible corner bead may reduce the number and severity of the punch list items.

The bead is wired or attached with a "pig ringer" to the inside of the BERNOLD Plate as shown in Fig. 27. The bead is run parallel to the "V" ribs. They are placed on a center to center distance of approximately three feet. The distance should be checked against the size darby used in local stucco work. The flexible bead will provide a screed for the darby work. This should greatly increase the chance for an acceptable uniform interior shell surface. Because the beads will show in the finished surface, the placement of the screed lines should be uniform for the length of the shell or they may be placed to form a pattern to fit wall or ceiling lines or even light fixture openings. In any event, it is strongly suggested that they be detailed and located on a reflected ceiling plan.

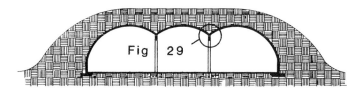

Fig. 28. BEARING INTERSECTIONS

Bearing Intersections (Fig. 28)

Some architectural programs are best met with parallel barrel shell applications where the shells do not complete the 180 degree radius. The shells may rest on interior bearing walls or on post and beam framing. In these applications the compressive loads are not delivered in a vertical mode. The horizontal components therefore must be restrained or balanced and eventually converted to vertical components or balanced by compressive loads against the soil or some sort of buttressing details to resist the overturning of side wall conditions.

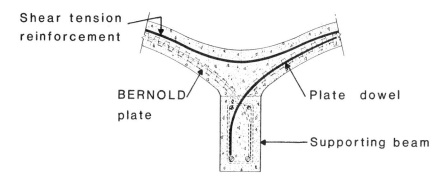

Fig. 29. BEARING REINFORCEMENT

Horizontal Transfer Details (Fig. 29)

The BERNOLD Plates lend themselves readily to this configuration. All that is required is to provide one set of dowels for each shell anchorage. Shell shearing loads at this juncture must be accurately calculated at the juncture. These loads may require the addition of fillet reinforcing to resist the vertical shear at and near the bearing point. To accomplish this, continuous reinforcing chairs may be wired to the plate to secure the proper location of the steel. When the shotcrete is applied, a fillet may be built-up at the intersection. This filleted joint will also more readily accommodate the membrane application.

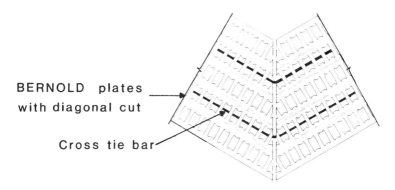

BERNOLD plates
with diagonal cut

Cross tie bar

Fig. 30. GROIN PLATE INTERSECTION

Groined Vaulting (Fig. 30)

Where the architectural programs require the intersection of·barrel shells of
equal or differing radii, the plates may be field burned or shop sheared to the
proper angle. As erection proceeds, the intersection may be held in place by pre-
bent ties extending 30 bar diameters either side of the intersection. The bars
are wired in place in the usual manner. Should a cantilever beam section be required in the intersection to support extended projection, additional steel may be
placed in the filleted area at the groin (Fig. 31). The depth requirement of the
tensile aspects of these stresses are readily accommodated by the thickened fillet
section. These details will apply to right angle intersections or intersections
of an obtuse nature.

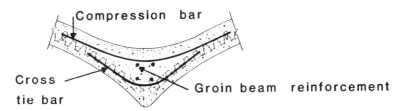

Compression bar

Cross
tie bar

Groin beam reinforcement

Fig. 31. GROIN REINFORCEMENT

Where angular intersections extend fully across the shell (equal radius intersections), the structural engineer should recognize the increased stiffening affect
of the intersection. Where the stress analysis of the forces acting in such a
location do not lend themselves to easy evaluation, it may be recognized that in
most instances the shell is not weakened by such intersections. Computer programs do exist (I am told) which will provide analysis where doubt may exceed
readily available "handbook" analysis.

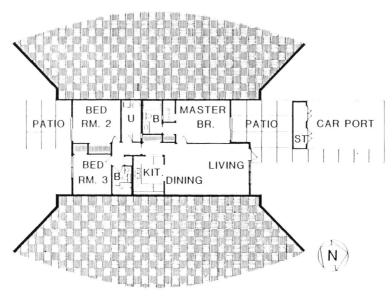

Fig. 32. FLOOR PLAN

THE DELTONA HOUSE

With the cooperation of the U.S. Gypsum Company's Metal Products Division, a proto-
type application of the BERNOLD Plates has been designed. Hopefully by the time
this paper is delivered, photographs and a T.V. tape will be available to show
field application of the BERNOLD System and the application details shown above.

Floor Plan (Fig. 32)

This 1100 square foot plan was developed to meet the medium to low income housing
requirements of the starting family. Its earth sheltered operational savings will
be reflected in the increased capacity of the family to meet the present high
interest rates.

The three bedroom plan will allow for moderate family growth. The low maintenace
characteristics will allow trade-up at some future date without major expenses.

The barrel shell parapet is extended with angular retaining wall to accommodate
the flat and rather narrow 80 foot wide lot.

Earth Sheltering

The ratio of exposed envelope area to sheltered envelope area is 1 to 3.5. The
average "U" factor for the exposed envelope is .08. For the sheltered envelope
the "U" factor is 0.8. Therefore, as long as the average sheltering depth is sufficient to keep a ΔT_s of one half degree Fahrenheit between the soil temperature
and the internal design temperature, the system will balance up to an exposed ΔT_e
of 17° Fahrenheit. This means it will operate at a summer inside temperature of
about 76° Fahreinheit. With a three quarter degree winter difference in ΔT_s, the
winter inside temperature will be about 69° Fahrenheit minimum.

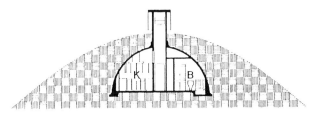

Fig. 33. CROSS SECTION

Cross Section (Fig. 33)

The cross sectional profile of this structure differs from our first design (The Montverde House) in that from a point three feet four inches above the floor line the interior wall of the shell is at right angles to the floor. Although this reduces the available floor area about 2.5%, the profile developed more readily accommodates the placement of furniture and equipment. Not shown in this section, the bath tub and shower stall have been lowered 16 inches. This provides a four foot eight inch height at the wall intersection. The enlarged shell foot width will permit the installation of conduit and an increased negative moment when the shell is covered.

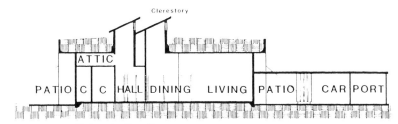

Fig. 34. TRANSVERSE SECTION

Transverse Section (Fig. 34)

The location of the thermal breaks and the skylites is shown in the transverse section.

Because of the limited floor area and the ratio of exposed to sheltered envelope, no atrium as such was used. Two small skylites were combined with the service chimney to reduce construction costs and provide interior light to the center hall and kitchen. All ducting to the exterior will go through this center service chimney.

Construction

This house was specifically designed so that the owners might do as much of the construction as possible. The BERNOLD Plate system will permit them to do all the work except the shotcrete work on the shell and those trades which are required by local code to be performed by licensed technicians.

SHELL ENGINEERING

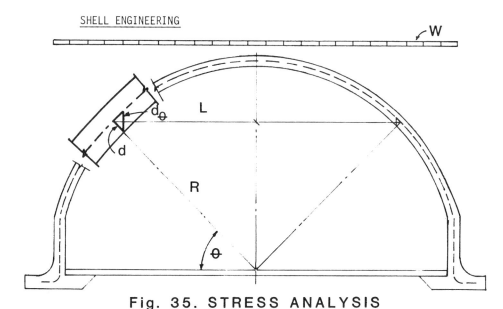

Fig. 35. STRESS ANALYSIS

The thickness of shell for a shear failure is a vertical plane. Any other angle of failure results in a greater thickness of shell and is less critical. The vertical depth is a function of the size of the angle θ.

$$d\,\theta \;=\; \frac{d}{\sin\theta} \qquad \text{(1)}$$

where d is the thickness of the shell measured to the center of the steel reinforcing.

as θ increases the shearing thickness decreases until at $\pi/2$ the thickness for shear is equal to the slab thickness. At the same time the shear force V_1 is decreasing as θ increases from 0 to $\pi/2$.

$$V = W\,L \quad \text{where } L = R\,\cos\theta \qquad \text{(2)}$$
$$V = W\,R\,\cos\theta \qquad \text{(3)}$$

$$\nu = \frac{V}{d} \;=\; \frac{W\,R\,\cos\theta}{d\,/\,\sin\theta} \;=\; \frac{W\,R\,\cos\theta\,\sin\theta}{d} \qquad \text{(4)}$$

Where ν is the shear stress.

At the section of critical shear there is a compression force which is approximately the hoop compression force.

$$N_u = P\,R \qquad \text{(5)}$$

The allowable shear stress is,

$$\nu = 2\left[1 + \frac{N_u}{2000\,Ag}\right]\sqrt{f'}\;c\;(0.85)\quad 11.3.1.2 \qquad \begin{array}{c}\text{A C I}\\ \text{11.3.1.2}\end{array} \qquad \text{(6)}$$

Equations (4), (5), and (6) result in

$$\frac{W\,R\,\cos\theta\,\sin\theta}{d} = 2\left[1 + \frac{W\,R}{2000\,Ag}\right]\sqrt{f'}\;c\;(0.85) \qquad \text{(7)}$$

Or with Ag = d x 1, solving for d,

$$d = \frac{W \ R}{1.7 \ \sqrt{f'} \ c} \left[Cos \ \theta \ Sin \ \theta \ - \ \frac{0.85 \ \sqrt{f'c}}{1000} \right] \qquad (8)$$

For a soil depth of 3.0 feet, the applied loads are:

W Live Load	=	50 p.s.f. x 1.7	= 85 p.s.f.
W Soil	=	3 x 120 x 1.4	= 504 p.s.f.
W Concrete	=	75 x 1.4	= 105 p.s.f.
W Total	=		694 p.s.f.
W Total	=	694/144	= 4.82 p.s.i.

For

R	=	13.667 feet
f' c	=	5000 p.s.i.

$$d = \frac{4.82 \ x \ 13.667 \ x \ 12}{1.7 \ \sqrt{5000}} \left[0.5 \ - \ \frac{0.85 \ \sqrt{5000}}{1000} \right] = \ 2.89 \ in. \qquad (9)$$

In the above equation, $\theta = 45^{\circ}$ because this angle produces the maximum required thickness.

Since d is the distance from the reinforcing to the face of the concrete, the required shell thickness is 2 x 2.89 = 5.78 in. The factor of safety is typical of concrete design.

This analysis assumes that shear failure is the critical failure mode. This has been demonstrated in Japan[1] to be the case. Bending of the shell is substantially prevented by the restraint of the soil against displacement. The allowable concrete compression is,

$$f' c \cdot t \ = \ 5000 \cdot (t) \qquad (10)$$

The applied load is W R.

$$W \ R \ = \ 4.82 \ x \ 13.667 \ x \ 12 \ = \ 790 \ lbs. \qquad (11)$$

Equating (10) and (11)

$$t \ = \ \frac{790}{5000} \ = \ 0.158 \ in. \ < \ 5.78 \ in.$$

THEREFORE, COMPRESSION IS NOT CRITICAL.

The above procedure results in shell thicknesses which have been successfully used in practical applications. Sizing the shell for bending stresses results in very thick walls and is too conservative.

The above shear formulas are based on standard ACI methods and should meet building code requirements.

---

[1] Bernold Loading Test, conducted by Public Works Research Institute, Ministry of Construction, Tokyo, Japan.

CONCLUSIONS

Cost

At this time we have detailed the BERNOLD Plate in two residential projects. Neither have been bid or contracted. Based on the information provided by U. S. Gypsum, the cost of the erected plates covered with five inches of shotcrete will be about $6.00 per square foot of shell area (not floor area). Relating this to the contracted cost of the one bar reinforced shotcrete over wood form project (bid a year ago and updated for inflation), the cost would be about the same. This method required that the wood forms be used three times.

Placement Time

It appears that the erection time for the BERNOLD Plate will be about half that for the wood forms, and the total shell can be erected and shotcreted at one time rather than the three separate operations required for the reusable wood forms. If the total project were wood formed at one time, our estimate is that the cost would increase to about $9.00 per square foot of shell in place.

Cold Joints

When reusable wood forms are used, the cold joint between each casting becomes a forming, reinforcing, and in our case, a before membrane waterproofing problem. The BERNOLD Plate resolves all of these problems.

Combination Forming

In instances where the shape of the BERNOLD Plate limits the design, they may be combined with other forming. We have used this technique to form the elliptical parapet visor on the residence shown at the start of this paper. The BERNOLD Plate stops at the thermal break. This provides an effective detail and a great deal of design freedom.

ACKNOWLEDGMENT

We express our appreciation to the Metal Products Division of U. S. Gypsum Company for their assistance in providing technical drawings and a grant to do the research required for this paper.

I also express my appreciation to John Szabo of U. S. Gypsum Company, Metal Products Division, who worked with my affiliated engineer, Richard B. Richardson, P.E., to develop the engineering stress analysis. It should be noted that U. S. Gypsum is endeavoring to put together a thorough research program which will provide computer assisted design of possibly more economical shell structures.

TECHNICAL INFORMATION

For further information on BERNOLD Sheet characteristics, contact U. S. Gypsum Company/Metal Products Division, 101 South Wacker Drive, Chicago, IL 60606.

GEOTECHNICAL ASPECTS OF SITE SELECTION AND EVALUATION
FOR EARTH SHELTERED-TYPE HOUSING

C. Dale Elifrits* and Nolan B. Aughenbaugh**

*Assistant Professor, Geological Engineering

**Professor, Geological Engineering

Department of Geological Engineering
University of Missouri-Rolla, Rolla, Missouri 6540]

ABSTRACT

A house built in contact with earth materials i.e. soil, rock or combinations
thereof, is uniquely influenced by its geologic environment, compared to the basement of a conventional house. The liveability of such earth sheltered-type homes
must be greatly improved over that of conventional basements if these dwellings
are to compete with above surface homes. The geological aspects of a site contribute much to the success or failure of earth sheltered-type houses. This paper
discusses the various techniques of site selection and site evaluation, with particular attention focused on the aspects of foundation materials, excavation at
the site, drainage and seepage, accessibility, construction planning, and use of
passive solar energy. A check list for quick site comparison/evaluation from a
geotechnical standpoint is provided.

INTRODUCTION

Geotechnical aspects of site selection and development for earth sheltered housing
involves site geology. Site geology is defined as the rocks, soils and land forms
which comprise the proposed building site. These are the natural factors which
influence site characteristics e.g. drainage, stability, excavation, foundation
and landscaping at the location. The natural and disturbed conditions of these
factors at the site are of great importance in evaluating the desirability of a
location for an earth sheltered structure. It cannot be overemphasized that a
man-induced disturbance of the natural geologic conditions interrupts the equi-
librium of the ongoing processes which can adversely affect a structure if not
recognized and accommodated in design prior to construction. Thus the site
geology cannot be ignored in the design and construction of an earth sheltered
building, to assume successful utilization of a location.

Lack of consideration of these natural geological conditions or assumed ability to
drastically modify natural conditions in order to develop a site frequently results
in large costs to the developer/owner and possible failure or substandard performance of the structure after completion. Examples of these undesirable results
include water problems around and within the structure due to poorly planned
drainage or a siting of the structure in a location where drainage was naturally
poor; failure of water proofing to perform as anticipated due to makeup of backfill
materials preventing ample drainage away from structure walls; large cost overrun
in excavation due to higher than anticipated rock to soil ratio in the excavation

297

zone. Through proper examination of the site in terms of its geologic composition and integration of the resulting information into site acceptance and design, these problems can be avoided and a successful project completed on schedule and within cost estimates/limitations.

SITE GEOLOGY FACTORS

DRAINAGE

Good building practices dictate that a structure should not be located in a natural runoff course and that the altered drainage should be away from the actual foundation zone. Since an earth sheltered structure is going to be placed in the conventional structure foundation zone or below this level, the necessity of designing for drainage in the natural undisturbed state to conform to the structure is probably the most important geotechnical factor to assure success of the project.

Areas of poor drainage, where water saturated zones are near the surface, should be considered as poor sites. Ideal sites are hillsides that do not receive excessive amounts of runoff water from an upgrade watershed during precipitation events and upon which no side slope seeps occur.

Subsurface water flow should not occur through the area of the proposed excavation. Sites with side hill seeps during wet weather or on which small streams flow at elevations above the foundation elevation should be avoided in order to lessen the chances of water movement into the building zone. Failure to avoid locations which are naturally wet produces costly problems and detracts from the desirability of the finished project. Remedial measures designed to overcome poor natural drainage must be fail safe- a nearly impossible objective for a residential construction project.

In addition to the site being well drained, a source of granular material which is free draining for backfill must be available at or near the site. As a matter of economics, this material most desirably should come from the excavation. Other sources e.g. quarries, gravel pits, stream beds might be acceptable provided the granular material is clean and free of fines and organic materials. This free draining backfill must then be placed in such a manner as to (1) assure adequate protection from water born fines which will reduce its ability to facilitate drainage and (2) allow gravity drainage away from the foundation and subgrade zones of the structure. With proper design and placement of the structure and as shown in Figures 1 and 2, its attendent drainage system reduces the likelihood of water pressure on the floorslabs and wall waterproofing, thus improving the chances of attaining the designed performance of the waterproofing during unusually wet periods. Reliance on waterproofing to substitute for adequate drainage is bad practice. Waterproofing is the second line of defense after drainage.

FOUNDATION/EXCAVATION CHARACTERISTICS

Areas with thick soils are excellent for earth shelter-type development, provided proper consideration is given to the soil's properties. Many factors should be investigated before construction is started. Among these are the soil's bearing capacity, the shrink-swell potential, the moisture content and permeability, its erodability and its thickness, if the distance to bedrock is not known.

Initial evaluation of most of these parameters can be made by looking at existing structures in the area. Local Soil Conservation Service (U.S.D.A.) offices have soils maps and are most helpful in providing data for their area. Their service is free of charge. One major factor must be considered in the soils evaluation -- an earth sheltered structure most likely will have foundation loadings much greater than conventional structures. Thus observation of conventional structures with respect to foundations and the soils must be tempered. A soils engineer should be consulted for testing and final evaluation before design is completed.

Areas which have very little soil cover over bedrock also are acceptable for earth sheltered development. Data that should be determined for these sites include thickness of soil zone; type of surface at the bedrock/soil innerface e.g. smooth bedding plane, rough, weathered bedrock, sink holes; the orientation of the bedrock surface and the soil and bedrock makeup.

Much of this initial evaluation phase can be easily made from observation of nearby road cuts, creek valleys and excavations. Also Soil Conservation Service or State Geological Survey records and personnel can provide assistance.

The identification of the location and condition of the soil/rock interface is most important. The soil/rock interface is of major consideration in excavation cost estimates. Local contractors can provide estimates of the costs of rock excavation versus soil excavation. This interface is also very important for foundation design; the footing of the structure must be placed on either all rock or all soil to insure against excessive differential settlements. Preferably rock should be the footing support zone. An uneven bedrock surface may be accommodated by piers and grade beams to bridge soil zones.

Strength and stability of the rock should be assessed. Most rock materials have strength values well in excess of the anticipated footing loading. However, open joints (fractures) and weathered bedding planes might reduce the rock mass strength and influence its long term stability. Therefore, their influence on the excavation and the stability of the site must be considered in the design.

SLOPE CONSIDERATIONS

Hillsides are the most desirable settings into which to place earth shelter structures. However, it must be noted that sites without slopes into which to place the structure should not be rejected for that reason only.

Should a sloping site be used, it should be one which is stable and shows no signs of creep or other downward movements. Such signs might be tilted trees, jumbled debris at the base of the slope or scarps at the top area of the slope. Design of the excavation and subsequent backfilling should be considered when looking at a slope. Geometric layout should be arranged to avoid steepening the existing slope angles with backfills. The manner in which runoff of water occurs on the natural slope should be noted to avoid a modification of the pattern with the placement of the structure.

Vegetation which is natural to the existing topography should be noted and used in final landscaping where possible. The natural stabilizing materials e.g. leaf mulch, rock cobbles which have held the undisturbed slope in place should be used in landscaping and slope stabilization on backfills, berms and tops.

ACCESSIBILITY FACTORS

The site should be situated such that access might be gained for construction without major disruption to the natural setting and without temporary bridging or other site improvements that are not required for final occupancy of the dwelling.

Care should be taken with respect to the entry of concrete trucks and their movement to the location where forms must be filled. Steep side hill slopes must be avoided with such trucks for safety and maneuverability considerations. Proper site planning should result in borrow from the excavation being placed out of the way of these trucks and in locations where it might be used advantageously during concrete placement e.g. ramp to elevate trucks above higher wall forms.

Concrete placement with a crain or pump as well as placement of other heavy members such as roof trusses or precast concrete beams requires space to swing the crane and space to position trucks with the objects within reaching distance of structure and crane. These areas should be accessible as described above and require little damage to trees or other natural features.

Access to utilities or positioning of well, septic tank etc. should be considered as the site is initially viewed. These items should not be isolated by the development because access for service during their use is essential. Driveway, garage and other layouts should be considered when initially viewing the site with respect to their impact on the natural setting, their service during all seasons and their cost of construction and maintenance.

SETTING

Setting is of great importance and should be considered with respect to the solar gain possibilities, the prevailing wind directions and the general manner in which the site and proposed structure will interact. South facing is desirable for solar heating at higher latitudes. However, in hotter climates the setting of the home should be one which allows natural ventilation using prevailing winds if possible. For temperate zones where both cold winters and hot summers occur, the setting should include both solar heating and natural ventilation.

The proposed project's relationship to adjacent structures must be considered. Privacy and security of the structure are influenced by its visibility. These considerations can be well accommodated by innovativeness in design and planning, but the physical limitations of the landforms at the site may override general visibility, therefore, requiring innovative landscaping to correct.

Figure 1 shows the almost ideal condition of access to the building site from an established road. Note that the garage was not constructed until the house was completed. However, the security problem presented by the lack of visibility of the structure is notable and must be addressed by the owner.

ON SITE EVALUATION PROCEDURES

Before a site is pruchased with the intent of constructing an earth sheltered building a few, quick, simple and inexpensive evaluations should be completed. Even the small expense of an option to purchase with rights to drill or trench on the site will often payoff in cost savings later in the project. The following is a series of evaluative procedures which can be followed, beginning with the least expensive and progressing to more complex as might be required, depending on the site and the anticipated development.

TOPOGRAPHIC SETTING

This characteristic can be quickly qualified at any site by either walking the site, studying a topographic map or examining aerial photographs. The overall topographic setting influencing the accessibility and siting should be assessed with respect to the desired design of the home and land use.

Vegetation and landforms often give clues to the rock/soil types and, by comparison to sites of known characteristics in the area, can be used to quickly evaluate the quality of the site. Healthy vegetation would indicate a desirable root zone which implies soil development with a desirable moisture handling capacity. This would most likely be interpreted as a site where excavation and landscaping would not be major problems.

Gentle slopes with good cover would indicate a stable area, providing no signs of sliding or slumping were present. Steep slopes with little cover would indicate

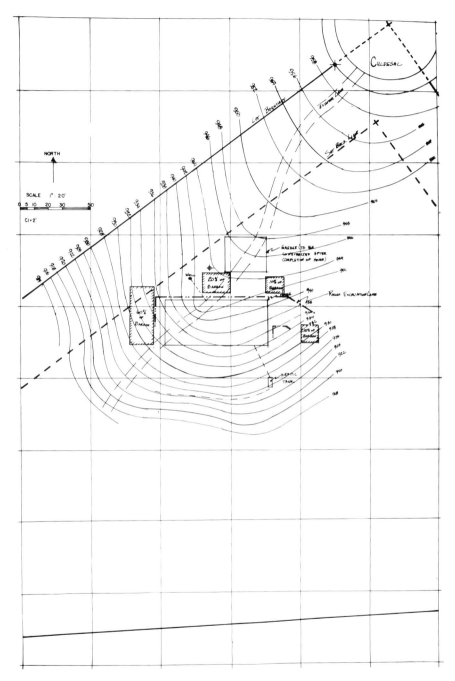

Figure 1. Site plot on a two foot contour interval
topographic map of the lot.

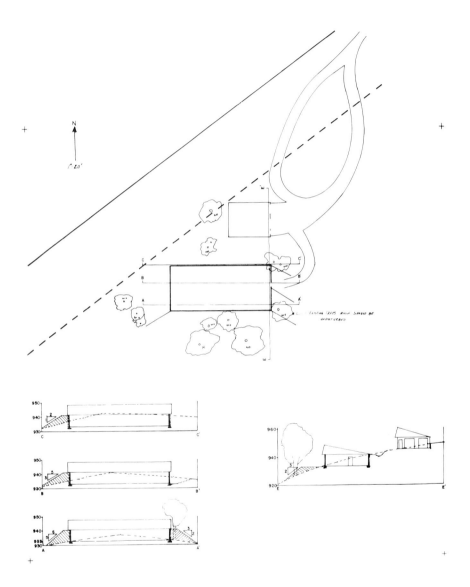

Figure 2.   Section Views of Proposed Devleopment Shown
            in Figure 1.

either rock materials or an area prone to slope failure --either of which would be less desirable conditions. Changes in slope on a site frequently are related to a change in soil or rock type. These features may be quickly noted for purpose of identifying the best location for the structure or identifying the zones of simi-lar materials around the site. Geologic and soils maps should be consulted from which soil and rock unit names for the area can be identified. These names aid in communication when requesting test data or information about the area.

ROAD CUTS, GULLIES, NEIGHBORING EXCAVATIONS

These features provide direct observation of the geologic materials below the sur-face without any expense to the observer, Figure 3. An additional source of sub-surface information which is usually available is well logs for logged well bores in the area. State Geological Survey records or U.S.G.S. Water Resources offices have these on file. There is no charge for their use.

By observing gullies, road cuts, etc. the following information may be gathered directly:
a) Soil type(s) and depths in the area
b) Rock type(s) in the area
c) Location of soil/rock interface
d) Slope characteristics of the soil/rock in the area
e) Zones which are wet or tend to produce or hold water
f) Excavation characteristics of the earth materials
g) Erodability of the materials and the success of stabilization by natural or manmade processes.

By observing neighboring excavations and projects the probable foundation, drainage, and landscaping systems required on the site in question might be observed. It must be kept in mind that if a conventional structure is observed, the foundation loading for a proposed earth sheltered structure will be greater. Contact with local contractors and engineers will frequently yield valuable site data without additional expense and time.

Once such observations have been made and knowledge of the probable materials at the site is gained, their engineering properties can be estimated. In the case of rock materials, identification of the rock at the site from geologic maps, well logs, and road cuts will enable the prospective developer to consult local engineer-ing offices or state Geological Survey Offices for results of tests which have been run on the samples of the rock unit. Similarly, soils can be evaluated for their engineering properties. From these data the desirability of the site from the standpoint of bearing capacities, drainage through onsite materials, slope stabilization and landscaping consideration may be made. All of the foregoing analysis procedures should be accomplished without expense to the developer except in terms of time. By use of these data, as might be summarized in author's check-list form (see the Appendix) many sites can be compared geotechnically. The value of these simple observations which may be correlated around the hillside or across a forty acre field to a yet-to-be-exposed site cannot be overemphasized in the initial examination of a potential building lcoation.

BACKHOE TRENCH

Once a proposed site has been identified as desirable, from the above, or more site specific data is required, the most revealing yet least expensive onsite explora-tion can be done by digging a series of trenches or pits with a backhoe. A sketch map or topographic map should be used as a base map of the site. Owner permission should be secured since the natural setting of the lot will be disturbed. Pit locations should be plotted on the map for reference. However, if the excavations are backfilled properly, this is only a temporary disturbance.

304

Soil zone with natural slope angle

Soil/rock interface Rock with natural slope angle, Note uneven weathered rock surface and variation in slope due to variation in rock in this exposure of Jefferson City Formation, a cherty dolomite at this site.

Figure 3. Example of a road cut where 25 feet of exposed material
may be observed. Note slope angles, soil/rock surface
and natural stabilizations.

Trenching in this manner yields information such as the following. Note that locations of trenches should be in probable excavation limits.
a) Actual observation of the materials present at the location of the proposed excavation,
b) Depth to soil/rock interface at the locations on the site if the depth is within the machine's reach,
c) Actual foundation zone condition,
d) Excavation qualities of the materials,
e) Samples of all materials,
f) Moisture conditions, (If required the pit may be protected and water movement be observed as it might occur around a building placed in the excavated zone.)
g) Slope stability and backfill qualities of the material,
h) Samples (for use) in laboratory testing for design calculations.

Thus, the trenching provides the best opportunity to observe materials in place, identify and quantify/qualify them as exposed and take samples which yield valuable data. Also, the relative cost of the activity with respect to the information gained is low i.e., $50 to $75 for the machine and operator in the Rolla, Missouri area for three, four to ten-foot-deep pits.

DRILLING

Where subsurface data is required and backhoe trenching is not practical or allowed drilling test holes will yield similar information. Frequently the cost of drilling is greater than that of trenching with a backhoe. Additionally, the drilling procedure allows only pinpoint sampling of the subsurface without the convienence of direct observation of the exposed materials. However, the following information can be obtained by drilling:
a) Depth to soil/rock interface,
b) Samples of the materials,
c) The foundation zone conditions, although they may not be visually examined as in the backhoe trench,
d) Moisture conditions, with some ability to identify probable wet zones as in the backhoe trench (At further expense, instruments may be installed in the bore hole to collect moisture data over time),
e) Excavation qualities of the materials (to be interpreted from the drilling rate and samples).
Drilling is a widely used exploration procedure and local drillers are available in most areas. Costs in the Rolla, Missouri average $50 per logged hole augered to refusal, i.e. to rock which an auger will not penetrate. The information gained from four holes would be less than that from two backhoe pits, assuming that all will penetrate the foundation zone. However, even though the drill holes cost much more, they disturb the site to only a small degree. The detail, quality and quantity of information desired, as well as economics and site disturbance, should be guidelines for choice of subsurface exploration method.

HYDROLOGIC CONSIDERATIONS

As mentioned in the preceeding sections of this paper, water movement over the site at the surface and through the subsurface materials must be considered. The following items should be mapped with respect to their spatial and topographic relationship to the foundation and backfill zones of the structure's proposed location.
a) Surface runoff courses,
b) Side hill or other seeps, both wet and dry weather, if possible,
c) Placement of utilities e.g. trenched water/sewer lines, electrical entrance which will have a chance to bring water to the structure,
e) If municipal water supply and waste treatment are not available, the relationship of these facilities (well and septic system) to each other and to the

structure with respect to proper protection of each.

The base map mentioned in the backhoe trenching section should be used to note these data. The locations of seeps, changes in slope, vegetation changes and the data gathered in digging on the site may then be plotted on the map. Rapid correlations among these various materials and conditions present may be established for a final evaluation and for design procedures.

MISCELLANEOUS CONSIDERATIONS

With the expanding of development into areas where municipal services, i.e. water, sewer, fire and security are not readily available, these aspects of the site must be considered. Also state, county and township codes should be checked for possible restrictions relating to all phases of the development and construction.

Of these considerations the water source and waste water treatment have direct geologic relationships. A well should be designed by a competent, active hydrologist. The slight cost of such professional input and the possible additional cost of the designed well will be repaid many times over in the quality of water and service of the well. Lack of such design and construction of a well frequently results in poor quality water, inadequate supply of water, and unusually high pump and system maintenance costs.

Waste treatment should be within code using the best professional judgement for the area. Experience has shown that aeriation septic tanks work well provided a method of discharge of the effluent is available e.g. a gravel packed lateral field in well drained soils. Again a contact with local or state water protection or health agencies should be made in order to evaluate requirements for the area. Local codes or development restrictions should also be known.

Interaction of construction procedures and the site should be examined. Access to the site during all seasons, erosion and drainage control during construction should be given consideration. A chart such as Malcom Wells has produced in Figure 5 might help in evaluating the environmental aspects of the project. These factors will influence final landscaping and thus must be anticipated for that design.

CONCLUSIONS

By following these few simple, geotechnical procedures, one can evaluate a site for an earth sheltered structure. Once such evaluations are complete and plans made, with the use of professional help where needed, a successful project can be completed.

The use of these ideas should aid in the development of the following basic input information which will help the designer, contractor and owner to be pleased with the results. In summary the following facts should be noted:
a) Develop a complete site evaluation that includes:
   1. Drainage and moisture relationships
   2. Foundation zone/excavation information
   3. Structure-site relationships
b) Use available maps, observations etc. to plot all information and plans on a contour map (two foot contour interval) of the site
c) Use professional geotechnical input
d) Plan and design your home and landscaping to be as compatible as possible to the natural conditions.

THE ABSOLUTELY CONSTANT INCONTESTABLE STABLE
ARCHITECTURAL VALUE SCALE

First published by Progressive Architecture in 1971

Subject to evaluation:

	-100 always	-75 usually	-50 sometimes	-25 seldom	+25 seldom	+50 sometimes	+75 usually	+100 always	
destroys pure air									creates pure air
destroys pure water									creates pure water
wastes rainwater									stores rainwater
produces no food									produces its own food
destroys rich soil									creates rich soil
wastes solar energy									uses solar energy
stores no solar energy									stores solar energy
destroys silence									creates silence
dumps its wastes unused									consumes its own wastes
needs cleaning and repair									maintains itself
disregards nature's cycles									matches nature's cycles
destroys wildlife habitat									provides wildlife habitat
destroys human habitat									provides human habitat
intensifies local weather									moderates local weather
is ugly									is beautiful

Negative score out of a possible 1500	Positive score out of a possible 1500

final score:

Figure 5. The Absolutely Constant Incontestable Stable
Architectural Value Scale by Malcolm Wells

308

APPENDIX

EARTH SHLETERED BUILDING SITE-GEOLOGICAL EVALUATION CHECKLIST

	SITE 1	SITE 2	SITE 3
Site Topography			
1. Direction of slope(s)			
2. Steepness of slope(s)			
3. Changes in slope angles			
4. Land area draining over the site			
5. Distance to nearest wet zone			
a. Horizontally			
b. Vertically			
6. Relief adequate for gravity drainage from foundation elevation?			
7. Flat or benched areas suitable for anticipated use?			
8. Slope stability-signs of slides, creep, severe erosion, etc.			
9. Access to the actual building area			
a. From existing roads, highways			
b. From existing electrical service			
c. From existing water, sewer or for development of such on site			
d. For construction activities			
10 Time of year of evaluation			
Site Geology			
1. Soils evaluations			
a. Soil thickness			
b. Soil composition, name-from SCS maps			
c. Soil moisture conditions, drainage			
d. Soil slope stability			
e. Soil quality as a plant support medium; nature of soil horizons			
f. Soil excavation character			
g. Soil engineering properties			
2. Rock			
a. Kind of rock e.g. sandstone, limestone, granite			
b. Name of rock formation-from geology map, S.C.S. information, well logs			
c. Rock character e.g. fractured, massive, karst.			
d. Rock permeability-how well drained?			
e. Excavation methods anticipated			
f. Rock engineering properties			
3. Soil/Rock Interface			
a. Depth to the interface			
b. Type of surface at the interface			
c. Probable orientation, continuity of the rock surface			
d. Probable thickness of highly weathered rock which should be removed.			

	SITE 1	SITE 2	SITE 3
e. Water movement at the interface			
4. Potential of use of the materials on the site as construction materials for the project.			
Hydrology			
1. Depth to saturated zone (water table)			
2. Probable direction and flow rate in near surface materials as shown by seeps, springs, etc.			
3. Highest wet weather water bearing zone with respect to foundation			
4. Cost estimate for designed well, if required as water supply			
5. Septic lateral field requirements for the site conditions and to meet codes where no common treatment facility is available.			
Miscellaneous			
1. Adjacent land use			
2. Distance to			
a. Interstate or heavily used highway			
b. Railroads			
c. Other noise and vibration sources			
d. Quarries, industrial development			
e. Power transmission lines			
3. Relationship to expanding urban development			
4. Check of zoning and codes.			

CONSTRUCTION PROBLEMS IN THE MINNESOTA DEMONSTRATION
PROGRAM

R. Sterling* and M. Tingerthal**

*Underground Space Center, University of Minnesota,
Minneapolis, Mn.
**Minnesota Housing Finance Agency, St. Paul, Mn.

ABSTRACT

This paper will use the actual construction experience of seven projects completed
under the Minnesota Solar/Earth Sheltered Demonstration Housing Program as a basis
to describe the building costs and construction problems experienced in earth
sheltered construction.

KEYWORDS

Earth sheltered housing; building costs; construction problems.

INTRODUCTION

The purpose of this paper is to provide details of problems encountered during the
construction of seven earth sheltered residential buildings. It will also provide
actual construction cost information about the same seven projects.

The subject projects were constructed between August, 1978 and June, 1980. All
were built primarily by experienced contractors, none of whom had previous experience with earth sheltered construction. Each of the structures was professionally
designed and engineered. Architectural supervision of construction was provided
in all cases to a greater or lesser degree. Project locations include urban, suburban, rural and rural town sites.

As detailed later, both project costs and the number and degree of construction
difficulties vary widely among the projects. The reasons for these variations are
not always clear, but many factors appear to contribute to the relative success of
the projects. These include site, location, quality of design, type of design,
contractor qualifications and experience, contractor-architect relations, and construction materials selected. Wherever possible, the authors have endeavored to
indicate the cause or contributing factor of a particular construction problem or
unusually high cost.

311

THE MINNESOTA DEMONSTRATION PROGRAM

In 1977, the Minnesota state legislature appropriated $500,000 for the purpose of funding a demonstration program to explore the feasibility of earth sheltered housing construction. The Minnesota Housing Finance Agency (MHFA) was assigned to oversee the program and the Underground Space Center was engaged to design and perform the monitoring functions of the program.

Provided with little guidance by the enabling legislation, the MHFA set forth to design a program which would address the major problems faced by earth sheltered construction. Preliminary research done by MHFA had indicated that there were, theoretically, no technical barriers to sound and acceptable earth sheltered construction. The study also indicated that high costs would not be a deterrent to successful construction. Lack of public acceptance and lack of availability of data on the energy performance of earth sheltered structures were identified by the study as the two major obstacles that the demonstration program should be designed to address.

This led to a program design which included, as its major components, maximum possible public exposure to earth sheltered structures and substantial energy performance monitoring. The resulting program, reduced somewhat in size from the original design, saw the completion of seven residential projects, including three houses owned by the State and located in State Parks and four projects which were built and sold on a speculative basis.

The demonstration program has been initially successful in meeting its two major goals. Over 40,000 people have toured the seven structures and were provided with an opportunity to view an earth sheltered house. Energy performance data is being collected from all of the houses by the Underground Space Center and will continue to be collected and analyzed until 1985. In addition, these houses have provided valuable information about construction costs, construction problems, post-construction problems, and financial and market acceptance.

For reference in the discussion of costs and construction problems, the following are brief descriptions of the demonstration houses:

Seward West Townhouses (12 units)
Location: Minneapolis, Minnesota (urban)
Description: Two-story; 1,180 sq. ft. average per unit; reinforced concrete block
               walls and precast concrete roof and intermediate floor.

Willmar House
Location: Willmar, Minnesota (rural town)
Description: Two-story; 2,100 sq. ft. plus attached garage, reinforced poured
               concrete walls, laminated wood beam and wood roof decking, precast
               concrete intermediate floor.

Whitewater State Park House
Location: St. Charles, Minnesota (rural)
Description: Two-story; 2,200 sq. ft. plus attached garage; reinforced concrete
               block walls, precast concrete roof and intermediate floor.

Burnsville House
Location: Burnsville, Minnesota (suburban)
Description: Two-story; 2,000 sq. ft. plus attached garage and storage; reinforced concrete block walls, precast concrete roof and intermediate
               floor.

Wild River State Park House
Location: Almelund, Minnesota (rural)
Description: Two-story; 1,920 sq. ft.; reinforced concrete block walls, wood
              beam and deck roof, wood joist intermediate floor.

Camden State Park House
Location: Lynd, Minnesota (rural)
Description: One-story; 1,640 sq. ft.; reinforced concrete block walls; precast
              concrete roof.

Waseca House
Location: Waseca, Minnesota (rural town)
Description: One-story; 1,300 sq. ft.; reinforced concrete block walls; precast
              concrete roof.

CONSTRUCTION PROBLEMS

As a part of the demonstration program procedures, contractors were asked to document the problems encountered during construction. The amount of detail provided
by the contractors varies widely and much of the information was never obtained in
written form. Furthermore, the information received is certainly not exhaustive.
Therefore, the number of problems which are indicated to have occurred for a particular structure does not necessarily indicate an overall difficulty greater or
less than for any other house. The apparent frequency of problems in a given
project is partially a reflection of the completeness of the information submitted
by the builder. Because of these reporting inconsistencies, the information will
not be presented in a comparative manner. The specific problems will be grouped
under several general categories and will be explained in that context. Many of
these problems are certainly also encountered in non-earth sheltered construction,
but they are presented here because the consequences are often far more severe
when they occur with earth sheltered construction.

Site, Sitework and Backfilling.

Many of the problems in this category arise from the fact that most homebuilding
contractors are not accustomed to the relatively deep excavations and careful
backfilling procedures which are required for earth sheltered construction. This
lack of familiarity can cause problems in two major ways. Either the procedure is
performed incorrectly, resulting in damage to the structure, or the amount of time
needed to perform the required procedure is underestimated during the bidding process, resulting in eventual cost overruns. Several examples will illustrate this
point.

Site re-grading at the Camden house. The Camden house is located on a large,
treeless, slightly rolling site. The specifications called for extensive regrading of the site to provide for drainage away from the house. Due to complications
with some site work which was to be accomplished by parties other than the contractor, this regrading was not done immediately. The contractor proceeded with
construction of the house despite the fact that the regrading had not occurred.
During a subsequent heavy rain, the house was flooded, resulting in minor damage.

Soil retention at the Waseca house. The Waseca house is located on a fairly nar-
row lot in a residential subdivision. The builder failed to take measures to retain the soil on his property between the time that the house was backfilled and
when it was sodded. Heavy rains caused substantial drainage onto a neighboring
lot and the builder was forced to pay damages to the neighboring property owner.

This is a less severe problem in larger sites, but failure to take proper soil retention measures can result in damage to waterproofing.

Backfilling at the Burnsville house. The site of the Burnsville house proved to be difficult for several reasons. It is a heavily wooded site and the house is essentially built into a small ridge which runs east-west through the house. The south side is the open facade, with major penetrations on the north for an entryway and garage. Because of the tightness of the site, it was difficult for large equipment to maneuver. This, along with the fact that the roof is quite steeply pitched, resulted in the backfilling of the roof virtually by hand. Obviously, this time-consuming process was extremely costly and brought about other time delays. One additional site-related problem at the Burnsville house was the presence of large boulders within the soil. These boulders were not only difficult and costly to move, they also displaced a significant amount of soil, resulting in the need to truck soil in from another site.

Waterproofing

The majority of waterproofing problems appear to occur because of improper application of the waterproofing material or because of inappropriate material choices. Leaks occurred most frequently around roof projections (skylights, chimneys, vent stacks, etc.). While most of the demonstration houses have experienced some minor leaking problems, two houses will serve as examples of this problem.

Leakage at the Whitewater house. The most severe construction problem encountered in the demonstration program is wall and roof water leakage at the Whitewater house. It has not yet been determined if the leakage is the result of inappropriate materials or improper application of materials. However, this is the only house in the demonstration program which used a built-up, coal-tar bitumen and tarred felt waterproofing system. It is strongly suspected that the materials simply could not perform adequately.

Leakage at the Seward townhouses. All of the units in the Seward complex have experienced some leakage around roof projections. It appears that this is the result of improper application procedures in using sprayed-on bentonite. All roof projections have been excavated and additional waterproofing has been applied. Leakage in most of the units has been stopped.

Structural

Competent engineering should eliminate structural problems during and after construction. The proper level of engineering was not always present in the projects, however, and engineering errors and incorrect interpretations of drawings by some builders did occur in the program and resulted in extremely costly problems. Two examples demonstrate this point.

Rear wall at the Wild River house. The Wild River house is the only house in the demonstration project which uses a wood system for both the roof and the intermediate floor. This means the overall structure gains less strength from the roof and the intermediate floor than is the case where these elements are concrete. In this case, the rear wall had little assistance from the roof and intermediate floor in resisting the lateral earth pressure. Following backfilling, it was discovered that the rear wall had deflected. It was necessary to remove the backfill and anchor the rear wall to a "dead man" which was cast to the rear of the structure. This is presumed to have been a problem of not obtaining the proper engineering analysis since the engineering design for this situation is not complicated.

Roof supports at the Waseca house. The drawings for the Waseca house were mis-
understood by the builder and interior support columns were placed too far apart
to adequately support the roof planks. When the roof planks were installed, there
was a major deflection. The roof planks had to be removed and an additional beam
installed to adequately support the roof.

Retaining Walls

A discussion of retaining walls should perhaps be included in the structural section, but they seem sufficiently problematic as to warrant a separate section.
Problems with retaining walls seem to occur for several reasons. One major reason
is inadequate structural design, where retaining walls are not designed to withstand the actual amount of pressure. Another major reason is poor design and
building siting, which treats retaining walls as an afterthought rather than as
an integral part of the building. This can result in retaining walls which are
both costly and massive, and can lead to dangerous or unwise reduction of the size
of the retaining walls to reduce cost. Some examples will illustrate these
points.

Retaining walls at the Waseca house. The Waseca house contains four massive retaining walls which extend as high as 14 feet. During construction, and after a
heavy rain, one of these walls collapsed under the weight of the wet soil. This
wall has been reconstructed on the old footings and is being carefully watched
for any sign of movement.

Retaining walls at the Whitewater house. The original design of the Whitewater
house called for a straight elevational two-story design with retaining walls extending from the upper front corners of the structure. These walls would have
been 16 feet high. When bids were received, the bid for the retaining walls was
extremely expensive and, consequently, the decision was made to reduce the size
of the retaining walls. This decision also resulted in a reduction of the amount
of earth cover on the walls of the structure. While the data is not yet available
to verify this fact, it appears that the energy efficiency of the building will
suffer significantly because of this decision, since the wall insulation was not
increased to compensate for the loss of earth cover.

BUILDING COSTS

In addition to the examination of construction problems, an important part of the
documentation process was the recording of the construction costs for each house
in the program. The contractors for each project were required to submit a construction cost statement giving a breakdown of the cost items in the project. It
was hoped that this exercise would help to quantify what were the expensive items
in earth sheltered house construction and how these costs varied from project to
project. There are several reasons why the information gathered is not entirely
satisfactory in this regard. First, despite the use of a standard cost form, the
individual costing practices of the different contractors resulted in costs being
lumped into some items with no costs being shown in others. Second, the houses
represent a wide range of contractual situations with almost none of these being
in the general marketplace of construction contracts. For example, three of the
projects were State Park houses in rural areas requiring a state contract for construction. The small number of local contractors, and the complexities of a state
bidding package were two reasons why these projects tended to be more expensive
than the private sector projects. In the speculative projects, the different
situations ranged from twelve townhouse units developed by a neighborhood non-
profit group to a Vocational school project and the first earth sheltered project

of a small design and construction firm.

Despite all the limits on the usefulness of the information, the authors do believe it is worth presenting and commenting on the collected costs for future
reference.   The costs have been organized into three tables. Table 1 shows the
actual cost per house (or hypothetical "average" townhouse unit) broken down into
a reasonable number of categories to allow some discussion of partial costs while
minimizing the problem of the differences in costing practices. Sub-totals of
cost are shown for the rough structure, finish structure, mechanical systems,
structure and mechanical, site work, overhead and total building costs. The notes
to the table explain some of the more obvious discrepancies between the cost tabu-
lations, but all the figures must be viewed with some caution as to whether all
the costs relating to that item are in fact included in the figure shown.

TABLE 1 ACTUAL COSTS	Spec. Townhouses Seward West, Minneapolis, MN 1140 sq. ft. (1)	Spec. House Willmar, MN 2100 sq. ft.	State Park House Whitewater State Park 2200 sq. ft.	Spec. House Burnsville, MN 2000 sq. ft.	State Park House Wild River State Park 1920 sq. ft.	State Park House Camden State Park 1640 sq. ft.	Spec. House Waseca, MN 1300 sq. ft.
1. Concrete & Masonry & Metals	$15,070	$15,900	$29,646	$17,430	$15,720	$25,587	$39,052
2. Waterproofing & insulation	1,417 (10)	1,220 (11)		12,722	350	15,300	9,240
3. Roofing, siding, doors, windows, rough carpentry	5,667	19,300	21,502	29,637	42,020	23,570	31,162
A. Sub-total Rough Structure	22,154	38,420	51,148	59,789	58,090	66,457	79,454
4. Finish carpentry, cabinets, doors, frames, etc.	5,311	3,379	7,085	2,500	11,000		16,819
5. Skylights	188						
6. Roll down shutters	1,217					960	
7. Plaster/Drywall	3,083	746		4,775	2,500		3,793
6. Tile/Carpeting/Flooring		4,483	2,000	906	1,800	2,600	6,615
7. Painting & Decorating	2,166	3,020	3,626	3,265	4,800	3,100	2,015
8. Furnishings	554			857	1,000		1,391
9. Specialties		1,951	2,012	653	500	175	3,750
B. Sub-total Finish Structures	12,519	13,579	14,723	12,956	21,600	6,835	34,383
10. Plumbing	4,542	4,040		2,790	3,700		6,098
11. HVAC	1,976	1,944	20,405 (3)	1,792	5,900	18,080 (3)	
12. Solar system	9,147	- -	(3)	-		(3)	
13. Electrical	2,369	2,864	3,985	2,590	2,500	4,454	12,093
C. Sub-total Mechanical Systems	18,034	8,848	24,390	7,172	12,100	22,534	18,191
D. Sub-total Structure	52,707	60,847	90,261	79,917	91,790	95,826	132,028
14. Acc. buildings/Garage	$ 3,667	$ (4)	$ (4)	$ (4)	$ (9)	$ (9)	$ (4)
15. Earth work	2,200	2,938	3,400	5,667	2,500	4,800	3,926
16. Site utilities		1,337		5,802	4,000		
17. Site improvements	3,161	5,052	2,223	5,538			3,708
E. Sub-total Site work	9,028	9,327	5,623	17,007	6,500	4,800	7,634
F. Sub-total Struct. & Site	61,735	70,174	95,884	96,924	98,290	100,626	139,662
18. Overhead, Admin. & Bond	3,583		3,157	3,000	5,960	5,000	
19. Surveys & Soil borings	216	1,447				256	
20. Taxes & Constr. Interest	1,314	25					
21. Architects fees	3,942	4,040	(5)	(5)	(5)		5,161
22. Miscellaneous fees	2,342	6,056 (2)			500	292	338
G. Sub-total Overhead costs	13,397	11,568	3,157	3,000	6,460	5,548	5,499
H. TOTAL BUILDING	$75,132	$81,742	$99,041	$99,924	$104,750	$106,174	$145,161
23. Land	1	12,888	-	(6)	-	-	6,750
GRAND TOTAL	$75,133	$94,630	$99,041	$ (6)	$104,750	$106,174	$151,911
SELLING PRICE	2 BR 66,600 3 BR 76,600	81,500	(8)	110,000 (7)	(8)	(8)	85,000 (12)

NOTES:

1. Unit size and costs averaged as 1/12 of total project.
2. Includes open house labor of $1,400.
3. HVAC system includes solar system.
4. Included elsewhere in costs.
5. Not included in these cost figures.
6. Not available.

7. Takes into account $17,000 grant but this grant covered open house costs and other special costs of the program.
8. Not applicable.
9. Garage not included.
10. Waterproofing cost only listed.
11. Insulation cost only listed.
12. Asking price, unit not yet sold.

Table 2 presents the same cost information limited only to the sub-total categories and presented in terms of the cost per square foot of living space for the structure. This is a frequent but imperfect measure of the expense of a building type and of the building components that are affected by the square footage.

Table 3 reviews the sub-total categories of costs in terms of a percentage of the total building costs (item H in table 1, i.e. excluding land costs). This serves to highlight the differences in the relative expense of the building components among the different projects.

TABLE 2 COSTS PER SQUARE FOOT	Spec. Townhouses Seward West, Minneapolis, MN 1140 sq. ft.	Spec. House Willmar, MN 2100 sq. ft.	State Park House Whitewater State Park 2200 sq. ft.	Spec. House Burnsville, MN (1) 2000 sq. ft.	State Park House Wild River State Park 1920 sq. ft.	State Park House Camden State Park 1640 sq. ft.	Spec. House Waseca, MN 1300 sq. ft.
A. Rough Structure	$19.43	$18.30	$23.25	$29.89	$30.26	$40.52	$61.12
B. Finish Structure	10.98	6.47	6.69	6.48	11.25	4.17	26.45
C. Mechanical Systems	15.82	4.21	11.09	3.59	6.30	13.74	13.99
E. Site Work	7.92	4.44	2.56	8.50	3.39	2.93	5.87
G. Overhead Costs	11.75	5.51	1.44	1.50	3.36	3.38	4.23
Sub-total Structure	46.23	28.97	41.03	39.96	47.81	58.43	101.56
Sub-total Structure & Site	54.15	33.42	43.58	48.46	51.19	61.36	107.43
TOTAL BUILDING	$65.91	$38.92	$45.02	$49.96	$54.56	$64.74	$111.66

NOTE: (1) 2000 sq. ft. living space only used for cost/sq.ft.

TABLE 3

COMPONENTS AS A PERCENTAGE OF TOTAL BUILDING COSTS

1. Rough Structure	29%	47%	52%	60%	55%	63%	55%
2. Finish Structure	17	17	15	13	21	6	24
3. Mechanical Systems	24	11	24	7	12	21	13
4. Site Work	12	11	6	17	6	5	5
5. Overhead Costs	18	14	3	3	6	5	4
Sub-total Structure (1-3)	70	75	91	80	88	90	91
Sub-total Structure & Site (1-4)	82	86	97	97	94	95	96
TOTAL BUILDING	100%	100%	100%	100%	100%	100%	100%

With the costs presented in these three tables, it is possible to draw some limited conclusions about the construction costs of the different projects:

Seward West townhouses. The structural costs, particularly the rough structural costs, are low on all three counts i.e. cost, cost per square foot and percentage of total. The rough structure costs are those most affected by the savings in building twelve units at one time and hence, bear out the predictions of lower costs in multiple developments. Because the units are self-contained, the multiple construction does not affect the finishing and mechanical system costs as greatly as it does the rough structure costs, and here the small size of the unit gives a penalty in terms of the cost per square foot. This is natural for most small structures because the basic systems and amenities must be provided regardless of the size. The costs for this building may be the most rigorously assessed since a large general contractor with a permanent cost estimating staff was involved. The mechanical system costs on the building were high since an active solar system was included. It is quite probable, in the authors' opinion, that units built without the solar system would have energy bills of only $50-$100 per year more than units built with the system (due to very low energy requirements of the earth sheltered/passive solar design elements). Since the $9,147 shown against the solar system probably does not include all the building modifications and costs necessary to add the solar system, the payback period is extensive and the units might have been more marketable if the price had been dropped $10,000 and the building designed only with the ability to add solar panels later if necessary.

Other cost items which can be commented on are the external roll down shutter systems which probably have their installation costs hidden elsewhere but still appear to represent a reasonable investment for the shading, privacy, security and thermal insulation they provide. The skylight costs are misleading since the costs of the structure to support the skylight and the cost of making the opening in the roof slab is omitted. The site work costs are probably higher than they would have been on a slightly larger site due to the extensive retaining walls required to make rapid grade changes. The overhead costs are the highest of any of the projects but are more clearly identified on this project because of its nature and the involvement of several different institutions.

Willmar house. With the house in Willmar being constructed by a Vocational-Technical school it is not clear exactly how representative the construction costs would be of a typical contractor's bid. Subcontract prices would obviously be similar but labor rates for the main contract work would probably be higher than in this case. The house has the lowest cost per square foot of any of the structures and is most notable perhaps for the absence of any unusually high costs. Mechanical costs and site work were kept low and there were no special shuttering or active solar features.

Whitewater State Park house. Some of the general cost effects of building in the State Parks have already been discussed. The Whitewater house is the largest house in the program but only the third most expensive in initial cost. Savings come from leaving a portion of the lower floor unfinished, the small costs used for waterproofing and insulation and a low cost for site work from the elimination of the need for significant retaining walls. The mechanical system costs, however, were the highest in the program reflecting the use of a large active solar system with collectors located on the south facade of the building. Savings in cost were also made by leaving concrete block exposed for the side walls on the exterior of the house. Some of the cost savings in this project will adversely affect its energy performance i.e. the low level of insulation used (despite the fact that insulation is far more cost-effective than an active solar system). More attention to exterior finish materials would also probably have been a wise investment. The project does include an earth covered garage which is obviously more costly than an exposed garage but provides an attractive north side to the project.

Burnsville house. This structure has 2,000 square feet of living space, a 300 square foot storage loft and a 500 square foot earth covered garage built into the structure. Only the 2,000 square feet of living space are used in the cost per square foot figures which penalizes this project slightly compared to the other projects which must provide storage space within the living area. This house is expensively finished on the exterior and is the most photographed house besides the townhouse units. The structure costs are very reasonable considering the materials used, including expensive waterproofing, extensive insulation and thermal breaks, balconies, etc. Mechanical costs for the project were low, reflecting the passive design strategy and grouping of systems into a mechanical core. The unusually high costs of this project were in the site work. No site investigation was done, the excavation ran into large boulders and the cost of excavating and backfilling the two-story house on a tight site was underestimated. Site costs were also increased because expensive site utilities were required and the design included expensive retaining walls and landscaping.

Wild River State Park house. The Wild River House is well finished both inside and outside and this is reflected in the high costs for item 3, Table 1 and for finish structures (sub-total B). No garage is included in the costs shown in the tables and hence this reflects lower comparative costs than should really be assessed in relation to the other projects. Offsetting against this is the considerable amenity of a wrap around balcony providing an extension of the living area in the summertime. Waterproofing and insulation costs in item 2, Table 1 are clearly distributed elsewhere in the cost table.

Camden State Park house. The vagaries of bidding an unusual project in a rural area were clearly demonstrated with this project. A design was originally completed for Maplewood State Park and a quotation had been given for an 1,770 square foot house with garage for $95,000 on that site in northern Minnesota. However, due to the needs of the state park system, the location of the house was moved at the last minute to Camden State Park. The house was redesigned for the new site but retained many of the same features. The cost, when initially bid in this Southwestern Minnesota area, jumped by $35,000 to $130,000. Reducing the square footage of the house to bring the cost into line with the other houses was not efficient in terms of the reduction in the contractor's bid costs. This combination of circumstances gives this project, although simple in structural form, the second highest cost of all the projects (especially in cost per square foot). The particularly high costs are present in the rough structure cost and in the mechanical system costs (including an active domestic hot water solar system). The costs as listed do not include the cost of a garage.

Waseca house. This project was by far the most expensive in the demonstration program and, since it was also the smallest project, its cost per square foot was almost double most of the other projects. No individual reasons for this high cost are apparent from the figures, since the percentages for each portion of the cost are similar to several of the other buildings. The project does not include an active solar system and the site was relatively flat and open. Part of the high cost could be traced to the extensive exterior retaining walls necessitated by high berms and a courtyard at either end of the structure. The wall that failed during construction necessitated replacement, which obviously increased the costs for this item. Another design feature contributing to a high cost is the high light monitor over the central atrium, adding substantial extra cost but no additional square footage. Even with these design features, however, it is difficult to understand why the costs should be so high for this house.

CONCLUSION

In this paper, the authors have presented a brief summary of construction costs and major construction problems experienced in the Minnesota demonstration program. The paper does not represent an exhaustive analysis of either the costs or problems, but merely highlights some of the more significant points discovered during the program. A review of the demonstration program as a whole reveals many problems in the design, marketing, construction and cost of the houses. This prevalence of problems is not viewed by the program designers as being particularly troublesome. One of the primary reasons for conducting a demonstration program is to identify those areas where problems occur.

By presenting this paper the authors hope to accomplish one of the major purposes of any demonstration program. That is, to document actual findings and present them publicly so that the issues raised can be discussed and analyzed. The materials presented here should also provide a reference for parties who intend to build or design earth sheltered residential structures in the future. With proper planning and execution, all of the problems presented here would have been avoided. None can be attributed to unavailable technology. Therefore, a review of these findings as part of a careful planning process should assist builders and designers in avoiding a repetition of these and similar problems in future projects.

A PRELIMINARY, EXPERIMENTAL, ENERGY PERFORMANCE ASSESSMENT OF FIVE HOUSES IN THE MHFA EARTH-SHELTERED HOUSING DEMONSTRATION PROGRAM

Louis F. Goldberg and Charles A. Lane
Underground Space Center
University of Minnesota
Minneapolis, Minnesota

ABSTRACT

Net electrical energy consumptions, component load distributions and heating efficiency comparisons are presented for five earth-sheltered houses located in Minnesota for the period of June 1980 to February 1981. The energy performance of two of the houses is detailed for a set of two contrasting ambient weather conditions. The results presented enable an assessment of the earth-sheltered housing concept to be made and demonstrate the effectivness of the 'Total Gaseous Internal Energy' as a quantitative earth-sheltered housing comparative indicator.

KEYWORDS

Earth-sheltered; experimental; energy.

INTRODUCTION

The Minnesota Housing Finance Agency (MHFA) Solar/Earth-Sheltered Housing Demonstration Program was defined in 1978[1] to demonstrate the earth-sheltered housing concept and specifically achieve the following goals:

- provide high visibility
- acquire reliable energy performance data
- demonstrate appropriate proven technologies and energy savings.

An important aspect of this program is to explicitly compare the energy performance of earth-sheltered houses with their conventional, above grade, low Thermal-Integrity-Factor (T.I.F.) counterparts. This enables specific conclusions about the viability of the earth-sheltered concept from an energy consumption perspective to be drawn. In order to facilitate this process, all the houses in the program use electrical energy only for all heating and other functions, even though this may distort energy cost comparisons by neglecting the effects of using currently cheaper energy sources such as natural gas.

The contents of this paper are devoted to the second aspect of the MHFA program, namely, the data acquisition, or monitoring segment.

Current Monitoring Program Status

At present, the monitoring program encompasses eight houses, seven of which are
earth-sheltered, the eighth being an above grade control house. All eight houses
are located in Minnesota, three (including the control house) in the Twin City
greater metropolitan area, the remainder being in rural areas. The monitoring
status of these houses, as of March 1981, is summarised in Table 1.

Table 1   Monitoring Program Status

House Location	Type	Transducers Installed	Data Capture System Installed	Data Capture System Operating	Manual Watt-hour Meter Readings
Burnsville	e-s	X	X	X	X
Camden	e-s	X	X		X
Lind Park	a-g	X			
Seward	e-s	X	X	X	X
Waseca	e-s	X			
Whitewater	e-s	X			
Wild River	e-s	X	X	X	X
Willmar	e-s	X	X		X

Note: e-s = earth-sheltered
        a-g = above grade

The perhaps glaring omissions in Table 1 are caused principally by the following
factors:

- the housing construction program has been behind schedule
- not all the houses have been continuously occupied
- construction related difficulties have impeded monitoring
- the instrumentation equipment has not fulfilled its reliability
  specification.

More specifically, the control house at Lind Park is still in the finishing stages,
the Waseca house is unoccupied and the Whitewater house is experiencing
construction related problems particularly with regard to moisture seepage. The
remaining houses have all suffered to a greater or lesser extent from monitoring
equipment reliability difficulties. These reliability problems are somewhat offset
by energy consumption data being manually gathered by the house residents, thus
providing a 'back-up' facility on arguably the most important aspect of the
monitoring program.

Monitoring Program Objectives

In terms of the composite MHFA program outlined, a set of specific objectives has
been established for fulfilling the energy performance aspect of the project.
These objectives have evolved as the monitoring program has progressed and are
currently somewhat more sophisticated and ambitious than those originally
expressed[1]. The houses are largely monitored on two different levels of transducer
saturation, the objectives increasing with the transducer saturation level.

The objectives for the Burnsville, Seward, Waseca, Wild River and Willmar houses
which have a low transducer saturation level are:

1. to determine the gross energy consumption and the relative contribution of the major component loads,

2. to establish the overall temperature distribution throughout the house and compare this with the ambient temperature, and,

3. to quantify the variation of relative humidity representative of the house.

In addition, where appropriate, other parameters and systems are also monitored. Examples are, the overall performance of the active solar heating system as well as the roof cover temperature profile of the Seward house. Hence, the transducers employed in these five houses are sufficient to measure temperature, energy consumption and internal relative humidity only.

The Camden and Whitewater houses have a higher level of transducer saturation. As as a consequence, they have in addition to the three objectives delineated above, the following appendages:

4. to ascertain the hot and cold water consumption,

5. to record ambient weather conditions (external humidity, solar insolation, wind speed and direction), and,

6. to measure heat fluxes through the concrete floor slab.

The control house at Lind Park is the most intensely instrumented house with a correspondingly ambitious set of objectives. In addition to objectives 1-5 (a rock storage bed under the wood joist floor replacing the concrete slab), the following have also been stipulated:

7. to obtain the envelope heat loss through the exterior walls,

8. to experimentally determine the overall exterior wall conductivity and the external convective heat transfer coefficients, the latter as a function of ambient weather parameters, and,

9. to establish detailed performance data on the active solar heating system.

These additional objectives enable not only a more comprehensive comparison to be made with the earth-sheltered houses, but also serve as a benchmark for the assessment of relevant computer simulation results.

The Camden, Whitewater and Lind Park houses are thus also equipped with weather stations, flowmeters and heat flux gauges where necessary to transduce the additional parameters.

INSTRUMENTATION

The instrumentation used in each house monitored consists of a data logger (or data capture device), cassette tape recorder and appropriate transducers. The data logger is a combined control, signal measurement and channel multiplexing unit manufactured by Automatic Hardware Systems, a St. Paul, Minnesota based electronics company. The unit utilises the ubiquitous 6502 microprocessor for control. The operating software is programmed in machine language which is stored in EPROM's (Erasable Programmable Read-Only-Memory). The multiplexer assembly uses integrated circuit devices exclusively. Operational programming is accomplished via a front panel keypad which enables each channel to be individually calibrated with a series of stepwise linear coordinates. RAM (Random Access Memory) contents are maintained during a mains power failure by nickel cadmium batteries. The data loggers are configured in 30, 74 and 108 channel versions. Analogue inputs accepted are 100mV, 10mV and 1V level dc voltages, the 10mV range being configured for differential input. Provision is also made for counting the positive going edge of square wave pulses and annunciating the status of several logic level inputs. Data is stored in unformatted mode on a standard Philips cassette via a converted audio tape recorder which accepts serial digital input.

Temperatures are transduced by Motorola MTS 102 reverse biased semi-conductor junction devices. An instrument package manufactured by Texas Electronics is used to measure relative humidity, insolation, wind speed and direction. Heat fluxes are measured with Thermonetics Corporation H12-18-5-P thermopile gauges. Electrical energy consumption is recorded by conventional watt-hour meters fitted with mechanically coupled pulse initiators calibrated so that one contact closure occurs approximately every 90 watt-hours. Water flow is measured by a similar arrangement with the pulse initiator which is driven by a nutating disc in the flow stream yielding one pulse every 0.1 gallons of flow.

The principal objective in selecting this instrumentation system was one of cost minimisation in order to spread the available budget over the requisite number of houses. With such a criterion, inevitable compromises on equipment reliability, accuracy and precision were made. The data logging equipment was initially promoted as offering a low cost means of attaining the reliability of commercial equipment costing an order of magnitude more. Unfortunately, time and experience have refuted this claim and highlighted the fundamental contradiction of the selection criterion, namely, reliability and low cost are not compatible in these circumstances.

Some six months after the installation of the data acquisition equipment as supplied by the manufacturer, it became apparent that the data capture reliability was inadequate to fulfil the designated program goals. Hence an equipment modification program initially encompassing only the 30 channel systems was initiated. The substantial part of this modification effort has been devoted to effectively decoupling the data logger from its surroundings so as to minimise potentially catastrophic failures caused by various forms of transient voltage surge. These modifications have resulted in a significant improvement in the 30 channel equipment reliability except in the case of remote regions (such as Wild River) which experience inordinately large supply line voltage transients. The equipment is incapable of operating in such an environment.

Even with these modifications, however, the overall equipment effectiveness assessed on the basis of the entire project, including all three channel options, has never exceeded 20 percent. This low value is significantly attributable to mechanical failure of the tape recorders, particularly in the 30 channel systems. The situation described above is being systematically rectified by the acquisition of replacement equipment with reliability as a fundamental specification, even though this entails some budgetary strain owing to the high cost of such equipment.

SCOPE OF THE PRESENTED DATA

In view of the program status described in the preceeding paragraphs it is presently feasible to present data for the Burnsville, Camden, Seward, Wild River and Willmar houses only. This data is divided into two catagories. The first describes the overall energy performance of all five houses and the second assesses in greater depth some of the detailed behaviour of the Seward and Wild River houses. The latter category is further restricted to the examination of two representative twenty-four hour periods encompassing opposed sets of ambient weather conditions for each of the houses.

The volume of data presently on file at the Underground Space Center is several orders of magnitude larger than the abbreviated sample presented here might suggest. However, this data will continue to be largely inaccessable for analysis purposes until the data base management system currently in the planning stages is fully implemented.

OVERALL ENERGY CONSUMPTION PERFORMANCE

The salient specifications of the five houses are presented in Table 2. From a heat transfer perspective, all the houses have a qualitatively comparable envelope fabrication consisting of concrete walls and roofs which are either of the poured, masonry block or precast plank types. The roof insulation consists of a layer of extruded polystyrene board (styrofoam™) ranging in thickness from 51mm to 127mm covered with an earth layer in the thickness range of 356mm to 610mm. The walls are all insulated to at least the 2.4m depth level while all the walls and roofs are waterproofed.

The averaged monthly energy consumption data reported in Table 3 is obtained from the watt-hour meter readings taken manually by the residents on a biweekly basis. This reading schedule is in general neither synchronised with the calendar nor meticulously regular, and hence, requires averaging and conversion into monthly intervals. The unfilled portions of the table indicate that readings within the particular monthly time span were not taken owing to either the residents being unavailable or to the meters being removed for calibration and/or replacement.

Table 3 shows the available component loads for each of the five houses as well as the net electrical consumption. An exception to this is the Burnsville house where the watt-hour meter in the main line was incorrectly installed and hence has not been functional (the problem is in the process of being rectified). Additional parameters reported (where appropriate) are:

- total heating load
- residual load, that is, the load component not specifically accounted for
- residual load ratio, that is, the residual load expressed as a proportion of the net load
- 'equipment ratio', which is defined as the parasitic or unavailable energy required to drive the direct/solar heating system.

As expected, particular values of the residual ratio appear smallest in Camden and largest in Seward, since these houses have the greatest and smallest energy transducer saturation respectively. A comparison between Wild River and Willmar, which are perhaps comparable in net energy consumption, shows that Wild River has a higher residual ratio in general than Willmar, attributable to the heating system air handler (whose load component is unavailable) in Wild River.

The heating energy performance of the five houses (where available) is summarised in Table 4. The Thermal-Integrity-Factor tabulation (the monthly heating degree-day data being extracted from reference 5) is presented purely for the sake of interest and for such comparative purposes as may interest the reader, since its value as a precise physical comparative parameter is questionable to the authors. Hence only the average monthly electrical heating energy per unit of heated area is considered in the following discussion.

326

TABLE 2 Salient House Specifications (extracted from reference 1)

Location Architect/ Developer	Earth-Shelter Type	Heated Space (ft2)	Building Materials	Waterproofing		Insulation		Roof Earth Cover
				Roof	Walls	Roof	Walls	
Burnsville; Carmody & Ellison; Design & Const.	elevational south facing; 1 single family	2000	reinforced concrete block walls; precast concrete roof	rubber membrane (60 mil butyl)	rubber membrane (60 mil butyl)	4"styrofoam	4" top 7'; 1" below 7'	14"
Camden State Park; Peter Pfister	elevational south facing; 1 single family	1800	reinforced block walls; precast concrete roof	butyl (60 mil)	butyl (60 mil)	5"styrofoam	4"styrofoam down to frost line; 1"under 1st 8' of floor slab	18"/ sand-bed
Seward (Minneapolis) Seward West Redesign,Inc.	elevational south facing; 1 of 12 single family townhouses	1056	reinforced concrete block walls; precast concrete roof	spray bentonize	spray bentonize	4"styrofoam	3" top 8'; none below 8'	18"
Wild River State Park; Clark Engler	elevational south facing; 1 single family	1920	reinforced block walls; wood roof	rubber membrane (butyl)	spray bentonize	2"styrofoam	2"styrofoam for top 8'; 1"styrofoam below 8'	18"/ sand-bed
Willmar Willmar Area Voc-Tech. Institute	elevational south facing; 1 single family	2100	poured concrete walls; precast upper floor; laminated rafters roof	rubber membrane (60 mil butyl, extends 12" down walls)	bentonite panels	4"styrofoam	4"styrofoam upper level; 1"styrofoam lower level	24": 4" gravel clay; 16" soil

TABLE 3 Averaged Monthly Energy Consumption

Load Type* (kW-h & %)	June 1980	July 1980	Aug 1980	Sept 1980	Oct 1980	Nov 1980	Dec 1980	Jan 1981	Feb 1981
BURNSVILLE									
Garage lights & sockets	9	7	27	20	33	--	--	--	27
Dryer	147	129	149	163	159	--	--	--	146
Water Heater	484	509	508	555	614	--	--	--	486
Heat Coil #1	0	0	0	0	14	--	--	--	174
Heat Coil #2 & fan	47	109	70	72	236	--	--	--	1465
Total heating system	47	109	70	72	250	X	X	X	1639
CAMDEN									
Outside lights	7	3	6	8	23	22	30	10	--
Washer	67	123	3	1	4	8	70	105	--
Range	30	27	27	21	49	42	27	52	--
Refrigerator	72	78	78	71	69	61	60	56	--
Solar unit	16	18	16	14	9	5	5	8	--
Water heater	102	121	124	135	175	259	254	265	--
Furnace air handler	23[+]	56[+]	25[+]	4[+]	10	26	84	70	--
5kw heat coil	0	0	0	0	108	299	959	810	--
10kw heat coil	0	0	0	0	129	342	1074	843	--
Main	436	564	427	447	809	1359	2813	2432	--
Residual	119	138	148	193	233	295	250	213	X
Residual ratio (%)	27.3	24.5	34.7	43.2	28.8	21.7	8.9	8.8	X
Total heat	0	0	0	0	237	641	2033	1653	X
Equipment ratio (%)	X	X	X	X	4.2	4.1	4.1	4.2	X
SEWARD									
Air handler	--	--	3	20	49	86	156	113	140
Furnace heat element	0	0	0	0	0	672	1619	1275	1363
Main	--	--	179	168	224	988	2007	1587	1677
Residual	X	X	176	148	175	230	232	199	174
Residual ratio (%)	X	X	98.3	88.1	78.1	23.3	11.6	12.5	10.4
Equipment ratio (%)	X	X	X	X	X	12.8	9.6	8.8	10.3
WILD RIVER									
Dryer	--	63	53	29	30	44	90	85	118
Water Heater	432	676	443	223	230	574	747	650	692
Heat Coil in duct 6	0	0	0	0	3	351	83	106	79
Heat Coil in duct 10	0	0	0	0	51	816	801	726	910
Main	--	1462	1291	1274	1317	3353	1850	2318	2620
Residual	X	723	795	1022	1003	1568	129	751	821
Residual ratio (%)	X	49.5	61.6	80.2	76.2	46.8	7.0	32.4	31.3
Total heat	0	0	0	0	54	1167	884	832	989
WILLMAR									
Outside lights	0	11	6	6	1	3	21	3	--
Cooktop & oven	32	23	32	31	26	40	50	40	--
Dryer	85	26	48	46	74	42	60	50	--
Water heater	322	309	302	283	329	367	379	378	--
Furnace	256	256	270	265	728	768	2435	2015	--
Main	1029	1110	945	980	1476	1636	3412	2954	--
Residual	334	485	287	349	318	416	467	468	X
Residual ratio (%)	32.5	43.7	30.4	35.6	21.5	25.4	13.7	15.8	X

*Units are kW-h unless otherwise indicated
X cannot be calculated
+ Circulating passive solar heated air

Two features of prominence may be noted. Firstly, Burnsville and Willmar used electrical heating energy from June to September 1980 inclusive while the other houses did not. Secondly, during the core of the heating season, December 1980 and January 1981, Wild River used a significantly lower amount of heating energy per unit of heated area than did Camden, Seward or Willmar. In February 1981, Wild River again required a noticeably lower amount of heating energy than did Burnsville or Seward. This difference, on average by a factor of two, is perhaps a clear indication of the relative efficiency of Wild River. This superiority may be resultant from a multiplicity of interactive factors two of which are certainly the relative contributions of active and/or passive solar energy and coupled thermal mass. These effects may also be argued to be responsible for the electrical heating energy consumption of Burnsville and Willmar during the 1980 cooling season. However, factors such as high thermostat settings, malfunctioning control systems or consistently gross watt-hour meter reading errors could also conceivably be the explanation. The wall insulation differences do not appear to account for these phenomena, particlarly as Wild River has generally less insulation than the other houses.

Again, referring to Table 3, the equipment ratios show the effectiveness of the Camden heating system design with a value of 4% when compared with the 10% average for Seward. Other trends in the data are self evident, particularly with regard to the component load distributions. Distributions for the relative component loads of Camden, Wild River and Willmar are shown in Figures 1, 2 and 3 respectively. These figures reveal that both Wild River and Willmar used on the order of twice the net amount of energy for approximately the same period of time than did Camden. In contrast though, the total relative heating load of Wild River is 25%, while those of Camden and Willmar are 45% and 52% respectively. This again emphasizes the relative efficiency of Wild River from a different perspective and indicates that overall energy consumption should be used with discretion when comparing the energy performance of earth-sheltered houses.

The primary significance of the energy consumption data would be in its comparison with equivalent data for conventional above grade housing. As the control house is not being monitored, the comparison cannot yet be made within the context of the MHFA program, and hence, will be one of the major priorities in the program's evolution.

TABLE 4   Heating Energy Performance

Performance Parameter	House	June 1980	July 1980	Aug 1980	Sept 1980	Oct 1980	Nov 1980	Dec 1980	Jan 1981	Feb 1981
Average Monthly	Burnsville	0.3	0.6	0.4	0.4	1.3	--	--	--	8.8
Elec. Heating	Camden	0	0	0	0	1.4	3.8	12.2	9.9	--
Energy/Unit of	Seward	0	0	0	0	0	6.8	16.5	13.0	13.9
Heated Area	Wild River	0	0	0	0	0.3	6.5	5.0	4.7	5.5
($kW-h/m^2$)	Willmar	1.3	1.3	1.4	1.4	3.7	3.9	12.5	10.3	--
Thermal-Integ-	Burnsville	--	--	--	0.65	0.84	--	--	--	2.03
rity-Factor	Camden	0	0	0	0	0.89	1.20	2.65	1.92	--
($Btu/ft^2/$	Seward	0	0	0	0	0	2.14	3.60	2.53	3.19
heating	Wild River	0	0	0	0	0.19	2.05	1.08	0.91	1.27
degree-day)	Willmar	--	--	--	2.28	2.34	1.23	2.72	2.01	--

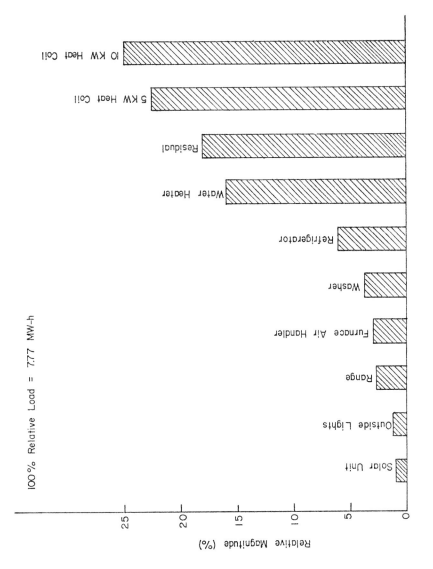

Figure I.   Actual Relative Energy Consumption, Camden House,
From 6/18/80 To 1/15/81.

330

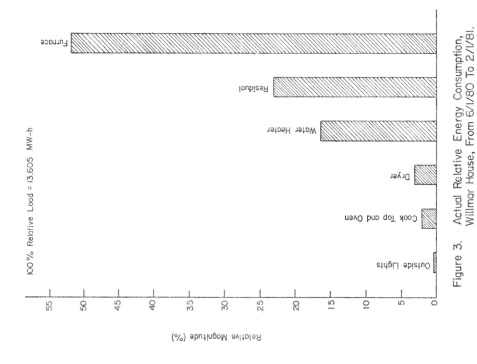

100% Relative Load = 13.605 MW-h

Relative Magnitude (%)

Figure 3. Actual Relative Energy Consumption, Willmar House, From 6/1/80 To 2/1/81.

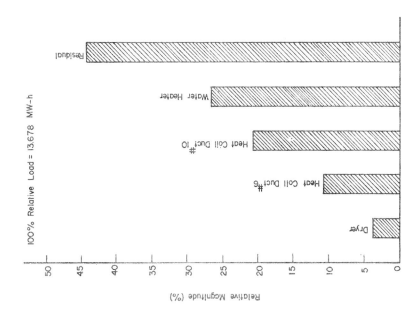

100% Relative Load = 13.678 MW-h

Relative Magnitude (%)

Figure 2. Actual Relative Energy Consumption, Wild River State Park House, From 7/11/80 To 2/15/81.

SEWARD AND WILD RIVER PERFORMANCE ASPECTS

The primary performance parameter used in the performance analysis is that of the 'Total Gaseous Internal Energy' (TGIE) which consists of the sum of the internal energies of the air and water vapour within the building envelope. The quality of the water vapour within the envelope is essentially unity, thus implying that the water vapour is saturated. This permits the internal energy of the water vapour to be determined from a psychrometric analysis requiring relative humidity, moist air (or dry bulb) temperature and barometric pressure only as input parameters[3].

In order to implement this analysis, each house is partitioned into a series of temperature zones consistent with the number and location of the internal temperature transducers (Figures 4 and 5). In the case of Seward, the relative humidity is assumed uniform throughout the house at the value measured at the transducer location (see Figure 4). In view of the paucity of internal temperature transducers in Seward (3 only) the error hereby introduced is proportionately small compared with the larger error of gross simplification of the temperature gradients within the house. In the case of Wild River, a series of relative humidity distribution tests were undertaken to establish the relative humidity gradients throughout the house. These results have been translated into mean correction ratios which serve to adjust the humidity measured at the transducer location to the appropriate temperature zone value. The temperature zone details for both houses are summarised in Table 5.

TABLE 5   Temperature Zone Details

House	Zone[a]	Net Gaseous[b] Volume (m^3)	Temperature[a] Transducer Reading	Relative Humidity Correction Ratio
Seward	A	51.7	①	1
	B	102.9	②	1
	C	13.8	(② + ③)/2	1
	D	95.9	③	1
Wild River	A	47.0	①	1.38
	B	95.8	(① + ②)/2	1.35
	C	52.5	②	1.30
	D	48.1	③	1.40
	E	101.8	(③ + ④)/2	1.33
	F	22.8	④	1.33
	G	28.8	(⑤ + ⑥)/2	1.33
	H	15.9	⑥	1.33
	I	5.9	(① + ② + ③ + ④)/4	1.40

Notes: a. See figures 4 and 5.
        b. Excludes 'solid' volume of furniture, etc.

It is interesting to note that in the case of Wild River, the relative humidity transducer is generally 30 to 40% pessimistic which has been verified to be attributable to the transducer location (Figure 5) and not to poor calibration. Further, there is a notable relative humidity distribution throughout the house, both of these factors requiring further research to determine their effect (if any) on overall energy performance.

332

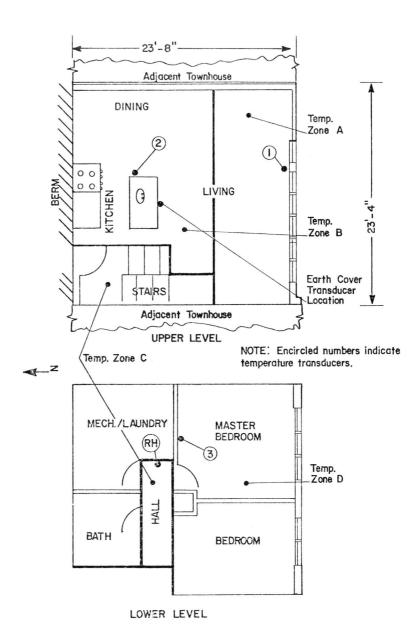

FIGURE 4. SEWARD TEMPERATURE ZONE
LAYOUT

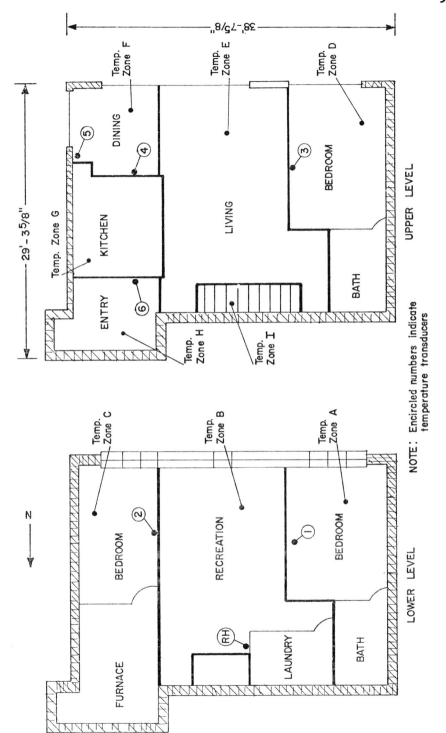

NOTE: Encircled numbers indicate temperature transducers

FIGURE 5. WILD RIVER TEMPERATURE ZONE LAYOUT

As may be seen from Figures 4 and 5, the architectural layout of the houses is similar, apart from the fact that Wild River is a single entity while Seward is one of the center units in a row of twelve townhouses.[2] Hence, Seward has a smaller earth-shelter coupling per unit volume than does Wild River. Seward, with a total temperature zone volume of 264.3m^3 is smaller than Wild River with a corresponding volume of 418.6m^3.

In addition to the TGIE, shown in the energy response profiles of Figures 7, 10, 13 and 15, the other parameters reported common to both houses are (Figures 6, 7, 9, 10, 12, 14):

  - mean internal temperature
  - external temperature
  - upper-lower level temperature difference
  - upper level south-north temperature difference
  - relative humidity at the transducer location
  - electrical heating energy.

Earth cover temperature profiles (Figures 8 and 11) and solar system performance parameters (Figures 6 and 9) are also shown for Seward.

The Wild River data is presented for a warm day (9/5/80) and a cold day (12/15/80). In the case of Seward, the data is for a sunny day with 88% of possible sunshine[4] (11/25/80) and for a cloudy day with 2% of possible sunshine[4] (12/15/80). Insolation conditions for Wild River are not available from the National Weather Service. Such data can only yield high order transient response performance. Steady-state and 'gradual' response analyses will only be feasible when bulk data processing becomes a reality, as mentioned previously.

In scanning these figures, it must be noted that the energy data is shown in an unconventional manner, being given in the units of 'energy (W-h or MJ) per 15 minutes'. This is in effect a power parameter and has the physical significance of being the amount of energy used within a given 15 minute interval. Hence, the area under these curves represents the cumulative energy consumption. This reporting format is a necessary consequence of the manner of energy measurement and the data sampling rate. This latter factor together with the accuracy and precision of the energy readings accounts for the linear and discontinuous nature of the energy plots.

Measurement Precision and Accuracy

The accuracy of the temperature measurements has been assessed at $\pm$ 0.5°C at the calibration point, with a system precision of $\pm$ 1 in the least significant digit of a 4 digit precision number, that is, $\pm$ 1°C in 1000°C or 0.001°C in 1°C. Each transducer is individually calibrated by means of an accompanying variable resistor, thus theoretically yielding transducer uniformity. However, owing to the non-linear temporal effects of doping diffusion as well as variable resistor carbon track expansion, the transducers do drift off calibration at seemingly random rates. Hence, wherever possible, the data is presented in difference or mean form in order to maximise the interpretative yield. This is facilitated by the transducer characteristic of linear junction resistance change with temperature, even though the calibration suffers non-linear degradation.

The relative humidity transducer accuracy is quoted by the manufacturer as being $\pm$ 2% of the signal conditioner output value (typically $\pm$ 1% relative humidity), while the system precision is identical with that of the temperature transducers.

The energy readings have a precision of $\pm$ 1 pulse (corresponding on average to $\pm$ 90 W-h) which implies an accuracy of +1/-0 pulses owing to the monotonically increasing pseudo-digital nature of the energy readings.

The accuracy, in particular, of all these measurements is a legitimate source of concern and mitigates against quantitatively precise conclusions. However, conclusions on qualitative trends are still valid and hence the readings are sufficient to fulfil the overall program goals.

Data Sampling

Each analogue transducer, that is, all except the watt-hour meters, is sampled every 15 seconds. These readings are accumulated and averaged over a 15 minute period. Similarly, the number of pulses registered by each watt-hour meter are aggregated over a 15 minute period. This averaged or aggregated data is then stored in RAM for later output to the cassette recorder, 4 data sets being output simultaneously. After each 15 minute period therefore, all the accumulator registers are reset to zero. This sampling process has the dual effect of damping out short term fluctuations within the accumulation period as well as considerably reducing the volume of data produced. No assessment of the distortions introduced by this process can yet be made, although this will be done with the more sophisticated replacement equipment currently on order.

Discussion of Results

On a macroscopic or system basis, the data underscores two features of earth-sheltered housing. Firstly, that the energy performance of each house is unique, and secondly, that within the earth-shelter context, the houses are particularly responsive to ambient weather conditions and insolation. The TGIE variation for Seward on cloudy and sunny days is 0.15% and 0.4% respectively, the variation for Wild River on cold and warm days being 0.8% and 1.29% respectively. These variations are the maximum experienced and are expressed as proportions of the daily mean values. Extending the comparison to the indoor temperature distributions, Seward again reveals consistently lower temperature differences than does Wild River in both the vertical and horizontal planes, this behaviour occurring for both facets of the ambient condition sets. These comparisons substantiate the energy performance individuality of the houses.

The coupling between the external temperature, determinable electrical heating input energy and TGIE as described in Figures 7, 10, 13 and 15 may be interpreted as defining the energy response of the houses. This response is specific to each house and hence offers the potential of being used as the basis for a comparative assessment of the relative merits of different earth-shelter configurations. The energy response profiles have the following common features:

- a close correlation between the external temperature and the TGIE on warm or sunny days (Figures 7 and 13)
- a poor correlation between the external temperature and the TGIE on cold or cloudy days (Figures 10 and 15)
- a close temporal coupling between the addition of electrical heating energy and a rise in the TGIE.

An assessment of the relative magnitudes of the electrical energy input within a 15 minute period and the corresponding rise in the TGIE reveals that only a small fraction (of order 0.5%) of the input energy is translated into a TGIE increment. The definitive reasons for this large discrepancy are as yet unclear, although the range of possible explanations would include ventilation effects, temperature stratification, infiltration, parasitic conduction leakage as well as systematic influences such as the poor placement and/or response of the temperature transducers, data sampling and averaging techniques, or indeed, any combination of all of these factors. As the discrepancy is universal throughout the data, the explanation is more likely to be systematic in nature, the solution to the dilemma thus dependent on the acquisition of more sophisticated instrumentation.

Figures 7 and 13 show that the thermal response of Wild River is more than twice as fast as that of Seward, that is, for Seward the TGIE rise lags the external temperature rise to a much greater degree than is the case for Wild River. As these phenomena appear to be largely dependent on solar effects, an examination of the active solar system performance of Seward is necessary, since unfortuntely, this performance mitigates against any simplistic explanation of the observed phenomena being adequate.

A study of the three right hand profiles of Figures 6 and 9 shows that apparently the active solar system is behaving abnormally and is circulating internally heated air through the collectors in the reverse direction. The effect is more clearly illustrated in Figure 6 which indicates that between 4h00 and 8h00, the large excursions in the electrical heating energy are mirrored in the solar collector temperature rises which in turn correspond with the excursions in the air handler energy. The effect is repeated between 18h00 and 20h00. It should also be noted that between 12h00 and 16h00 the system seems to behave normally, that is, no electrical heating input corresponding with a 14°C collector temperature rise and air handler operation. The anomalous solar system behaviour is obvious in Figure 9, since on a cloudy day (-4°C mean external temperature), a 12°C solar collector rise is impossible, particularly in the presence of continuous forced convective heat transfer. This behaviour may be explained in terms of a control system malfunction causing a motorised damper to operate in a manner inverted from that intended. In system terms, the effect is to remove heat from the house which would account for the contradiction shown in Figure 10, namely, an increase in external temperature occurring simultaneously with a decrease in the TGIE. This contradiction is absent in Figure 15, where, for Wild River, the increase in external temperature is tracked by the increase in TGIE, although with a much expanded temporal lag when compared with Figure 13.

Hence Figures 13 and 15, in the absence of the complicating factors of Seward, show two of the features of Wild River which may account for its previously discussed relative energy efficiency. They are:

- a high thermal response to insolation
- effective thermal mass coupling, shown firstly by the tendency of the TGIE to decrease at a lower rate than the external temperature after 17h00 in Figure 13, and by the continuous TGIE increase being contemporaneous with the gradual drop in external temperature after 16h00 in Figure 15.

Returning to the internal temperature distributions, the higher thermal response of Wild River is again evident in Figure 12 from the correspondence between the increase in the external and the south-north temperature difference rises. This correspondence is carried through to the upper-lower temperature difference profile, although the effect is relatively much smaller from an amplitude perspective. In Figure 14, on a cold day, the south-north temperature rise oscillates about a gradually rising mean, while the upper-lower temperature difference increases continuously. These stratifications are probably closely related to the ventilation system configuration which indicates the significance of this aspect of earth-sheltered housing design. Figure 14 also shows that the electrical heating input energy perturbations are reflected in all the internal temperature profiles as well as in the relative humidity profile, the effect being largely absent in Figure 12. The same perturbation reflection also occurs in the corresponding profiles for Seward (Figures 6 and 9). The smaller variations in temperature differences shown by the Seward profiles (of order 1°C, compared with the 7°C peak experienced by Wild River under similar conditions) suggests that either the ventilation system of Seward is more effective in distributing the available heating energy, or that the lower thermal response of Seward predominates in determining temperature stratification. In both Seward and Wild River, the south-north temperature difference tends to be positive, while the upper-lower difference is positive for Wild River and negative for Seward. This may be caused by the unique characteristics of the ventilation and/or natural convection systems of each house. Alternatively, it may be a function of the passive solar response of the house, whereby Seward receives and stores insolation predominantly on the lower level, while Wild River does likewise on the upper level.

The relative humidity generally follows the trends of the internal temperature profiles discussed above which is psychrometrically consistent. Hence, no further elaboration is warranted, suffice to note that the humidity levels at the transducer locations are moderate and do not vary by more than 8% at most.

Finally, consideration of Figures 8 and 11 gives qualitative insight into the behaviour of the earth cover on Seward. It should be noted at the outset that the grass growing on the cover is not well established, leaving large patches of bare earth. Thus, the resultant surface heat transfer effects establish the context of the observations made. In view of the instrumentation accuracy inadequacies discussed earlier, quantitative conclusions are not justified at all. Figure 8 indicates that the diurnal external temperature rise percolated about 127mm into the cover, the soil at lower depths remaining practically isothermal. This is most probably caused by the location of a freezing front at approximately 191mm above the soil separation layer, the moisture below the front being unfrozen and therefore capable of draining away. This results in a sharp thermal conductivity difference across the freezing front, the high conductivity in the frozen segment promoting the propagation of external temperature variations. This has important implications in the design of the earth sheltering (berms, covers, etc.), since it appears that moisture is a dominant heat transfer regulator. Hence, the indications are that the maintenance of a dry soil cocoon around the house should be a key facet of the overall design philosophy, particularly in cold climates.

Figure 11 is notable for the speed of propagation of the heating front through the earth cover, the peak temperature on the soil separation layer being reached approximately 8 hours after the external temperature peak. The damping of the temperature rise increases with depth. This is also consistently explained by the high conductivity of the moisture impregnated earth cover, since the entire depth is below freezing. The damping effect is consequently caused by a moisture gradient, the moisture content decreasing with soil depth.

From Figure 11 it becomes apparent that over the 24 hour period shown, the average heat leakage through the roof slab is increased as a result of the earth cover in comparison with the leakage that would occur without the cover. Even though this is relevant in this specific transient context only, it does lead to speculation on a more effective earth cover design which would seek to create a dry soil layer adjacent to the roof as previously mentioned. This may be achieved by inserting a moisture barrier of polyethylene sheet at 2/3 of the cover depth below the surface, the depth of the cover being increased to about 70cm.

PROGNOSIS AND CONCLUSIONS

Once the quantum leap from above grade to earth-sheltered housing has been made and the associated energy benefits thereby reaped, a different set of design parameters to optimise these benefits is predicated. The 'Total Gaseous Internal Energy' may be used as the earth-shelter system comparative quantifier, the optimisation parameters then being thermal mass, thermal mass coupling, passive and active solar response and earth contact configuration.

The effects of earth contact configuration and thermal mass coupling seem to be distinct, implying that there is a definite energy envelope surrounding the earth-sheltered house within which alone the thermal mass is active. There is evidence to suggest that the earth surrounding the house does not automatically form part of the thermal mass, but rather acts as a limiting boundary condition. In this context, the boundary performance may be optimised and the thermal mass increased if the earth immediately adjacent to the building is moisture free. Thus the external thermal insulation location and quantity become a function of the boundary condition optimisation.

These hypotheses are contextually constrained to the heating season of a cold, Minnesota type climate. Generalising to include the cooling season as well as other climates, leads to speculation that in order to optimise earth-sheltered housing performance in terms of the parameters highlighted by the results, a variable conductivity earth contact insulation system is required. This system would ideally need to be controllable over the full diathermal to adiabatic range. It should also offer potential for linkage to the thermal mass to enhance the beneficial effects observed, particularly in the Wild River house.

Specific conclusions pertinent to the MHFA program are:

1. The program objectives defined for the Burnsville, Seward, Wild River and Willmar houses have been fulfilled.

2. Poor instrumentation reliability and transducer inaccuracies are largely responsible for the failure to meet the other designated objectives, especially those quantitative in nature.

3. The qualitative information yield has exceeded the program's defined expectations.

4. Sufficient data has been produced to enable the energy benefits of earth-sheltered housing to be assessed.

The results produced have, on reflection, raised more questions than they have answered and have cast a controversial light into the recesses of detailed earth-sheltered housing behaviour. These growth symptoms augur well for the future application and research of earth-shelter technology.

ACKNOWLEDGEMENTS

This paper and the results described herein have been produced with funding provided by the Minnesota Housing Finance Agency, contract No. 34000-00895-01. The authors would like to express their gratitude to the MHFA as well as to the Underground Space Center staff, who have made invaluable contributions to the research program as a whole. Particular thanks are due to Suzanne Swain and Liz Seykora for their meticulous efforts in producing the graphics presented.

REFERENCES

1. Shank B. The Earth-Sheltered Housing Demonstration Project - Minnesota Housing Finance Agency. Underground Space, Vol. 3, No. 5, pp. 259-268. Pergamon Press, 1979, United Kingdom.

2. Underground Space Center (Ahrens, D., T. Ellison and R. Sterling). Earth Sheltered Homes, Plans and Designs. Van Nostrand Reinholdt, 1981, New York.

3. Rogers, G.F.C. and Y.R. Mayhew. Engineering Thermodynamics Work and Heat Transfer. Longman, 1967, London.

4. National Weather Service, NOAA. Surface Weather Observations. Numbers 27796 and 31881, 1980.

5. McQuiston, F.C. and J.D. Parker. Heating, Ventilating and Air Conditioning Analysis and Design. John Wiley, 1977, New York.

340

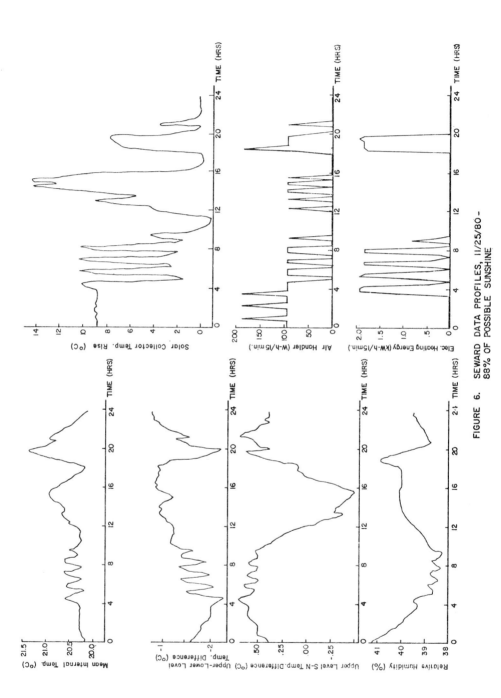

FIGURE 6. SEWARD DATA PROFILES, 11/25/80 – 88% OF POSSIBLE SUNSHINE

341

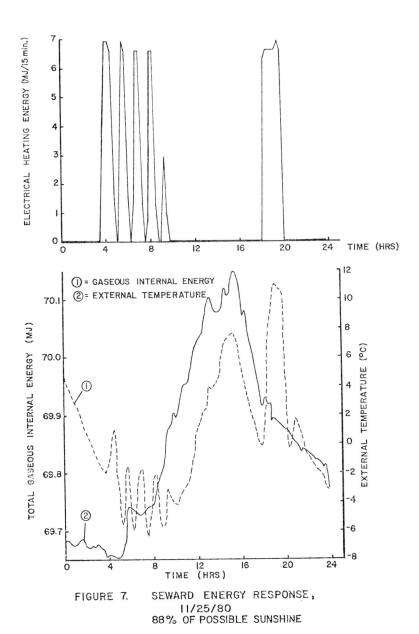

FIGURE 7. SEWARD ENERGY RESPONSE,
11/25/80
88% OF POSSIBLE SUNSHINE

342

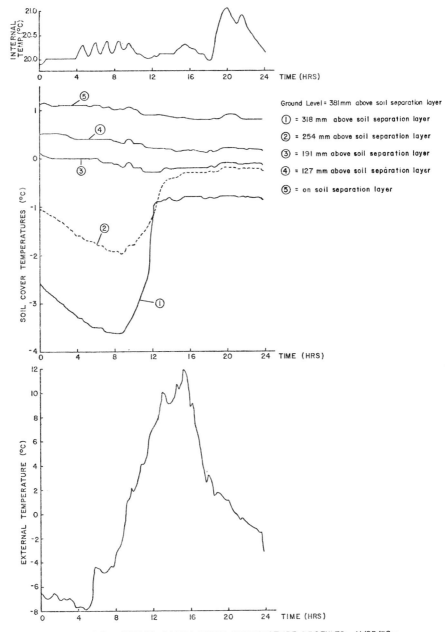

FIGURE 8. SEWARD EARTH COVER TEMPERATURE PROFILES, 11/25/80–
88% OF POSSIBLE SUNSHINE

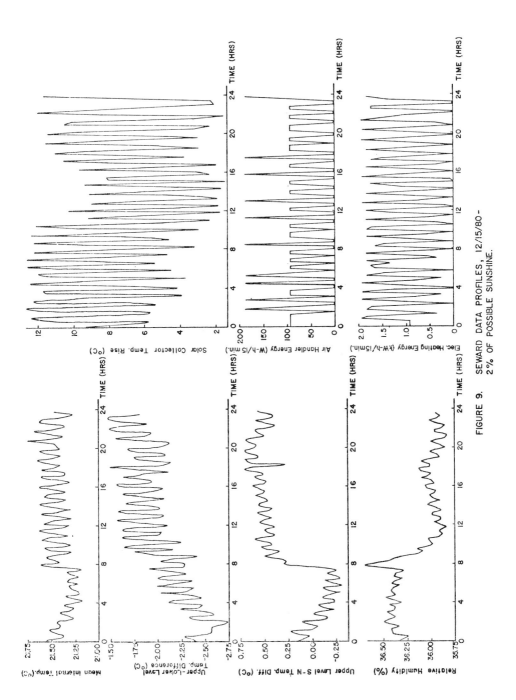

FIGURE 9. SEWARD DATA PROFILES, 12/15/80 – 2% OF POSSIBLE SUNSHINE.

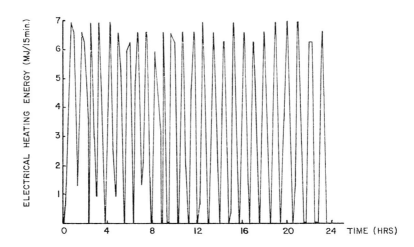

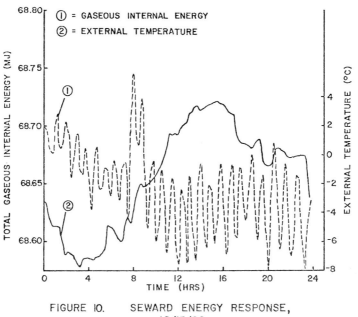

FIGURE 10.   SEWARD ENERGY RESPONSE,
12/15/80
2% OF POSSIBLE SUNSHINE

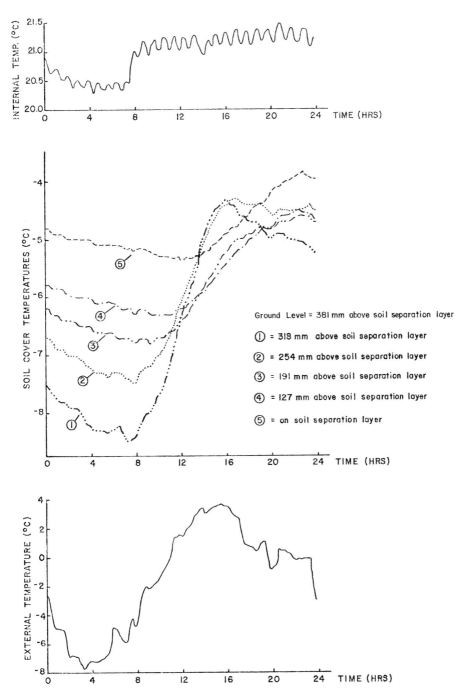

FIGURE II.   SEWARD EARTH COVER TEMPERATURE PROFILES, 12/15/80
2% OF POSSIBLE SUNSHINE.

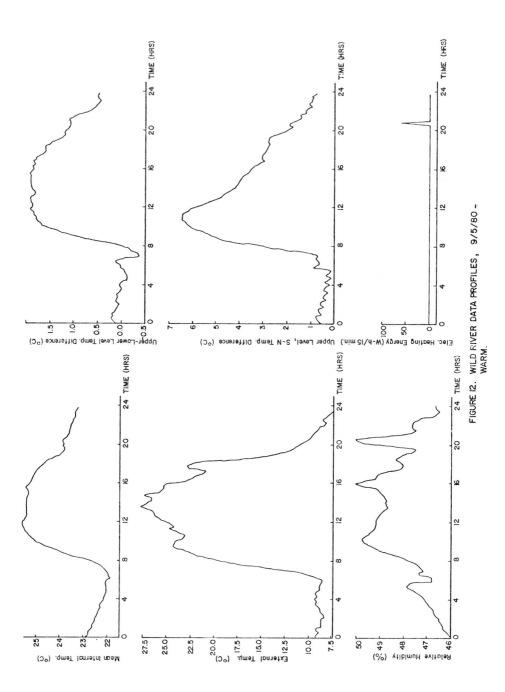

FIGURE 12. WILD RIVER DATA PROFILES, 9/5/80 – WARM.

347

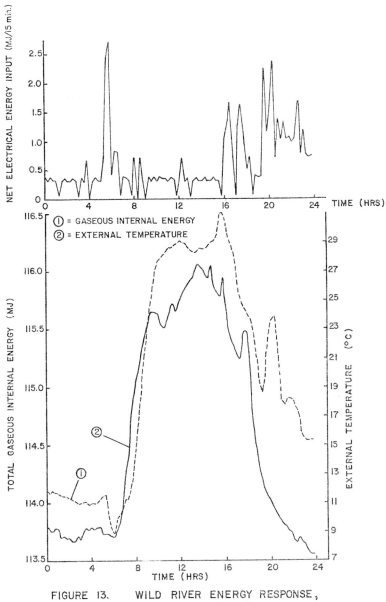

FIGURE 13.   WILD RIVER ENERGY RESPONSE,
9/5/80 - WARM

348

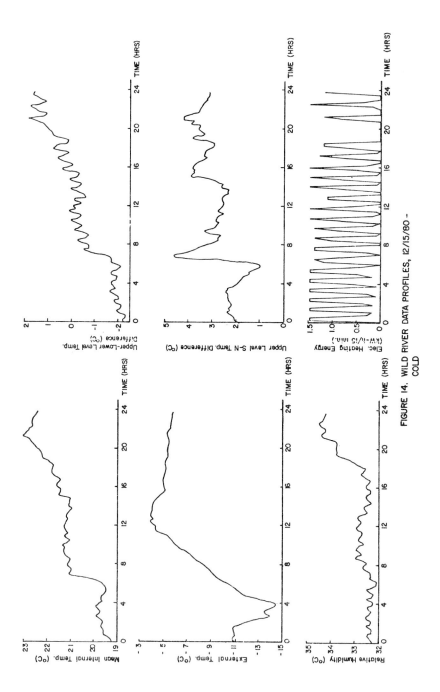

FIGURE 14. WILD RIVER DATA PROFILES, 12/15/80 – COLD

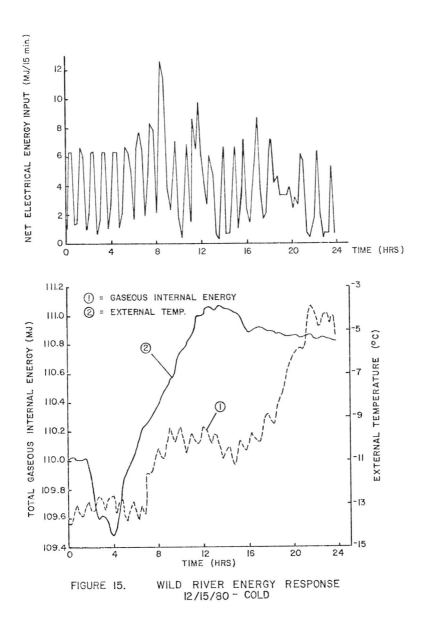

FIGURE 15.   WILD RIVER ENERGY RESPONSE
12/15/80 – COLD

GOING UNDERGROUND: AN OWNER/BUILDER'S EXPERIENCE

Robert B. Scott
Scott Earth Homes
P.O. Box 1753
Coeur d'Alene, Idaho 83814
208-664-4423

ABSTRACT

This paper details the construction process and products utilized in my underground home located in Northern Idaho. Specifically detailed are some of the
problems and solutions I encountered to assist other potential owner/builders in
avoiding some forseeable pitfalls. In addition, I have included some temperature
charting data compiled over a four month period, December 1980 through March 1981,
which helps evaluate the home's energy performance. Also included is a cost
breakdown. Although these figures will vary throughout the country the breakdown
offers some guidelines for a novice owner/builder.

KEYWORDS

Design and engineering; financing; site work; footings and walls; roof system;
waterproofing and insulation systems; backfilling; plumbing, electrical and
mechanical systems; framing; interior decoration; energy performance.

INTRODUCTION

My home is named "The Eagle's Nest" for its unique site high on a hill overlooking
Mica Bay on Lake Coeur d'Alene and the valley south towards Worley. The home is
earth sheltered with a passive solar, southern orientation taking full advantage
of the spectacular view and the sun's natural heat.

Construction features cast-in-place concrete walls and floor with prestressed
concrete planks on the roof. The waterproofing system is an elastomeric, sprayed
in place Elastall 900 membrane. The insulation is sprayed in place urethane foam
to an R30. The roof and rear walls are covered with a minimum of 18 inches of
earth. The majority of the south wall area is glazed in thermopane glass offering
solar gain and abundant natural lighting. The house has 1735 square feet of floor
space in addition to 708 square feet in the earth-sheltered garage that can
easily be converted to living area.

THE CONSTRUCTION PROCESS

Design and Engineering

Design criteria for my house was first and foremost that it should be completely earth-sheltered, semi-recessed into the hillside. I wanted to utilize passive solar space heating in conjunction with an air-tight wood stove. After considering the materials available to me in the North Idaho area, I decided to use pre-stress concrete plank roofing on a cast-in-place wall system.

My wife and I wanted the house to be ecologically sound and to fit into its rural mountain top setting in an unobtrusive manner. We spent months designing the right arrangement for the floor plan. Since we had no children, we set about designing a home for a childless couple to live and play in.

I wanted to keep the basic design as simple as possible to aid in the construction process and to reduce costs. In addition, I wanted a house that would lend itself to expansion in case we were ever able to have children, or eventually wanted to sell it and maintain a reasonable resale value.

Our final plan was a single level, one bedroom, two bath house with an office that could serve as a second bedroom or guest room. We put a cast-in-place spa at floor level on the south side of the third potential bedroom which we were using as a den/atrium. The garage was completely earth covered to allow for expansion of living space in the form of two additional bedrooms, a bathroom, and a small rumpus room for children. All of the necessary plumbing for this third bath was in place before the garage slab was poured.

We incorporated an airlock entry between the main house and the garage - an essential component for energy efficient living in the North Idaho climate.

In Kootenai County, where the house was built, the building department demands that all earth-sheltered homes submit plans that have been engineered by a state licensed engineer.

In trying to find a qualified engineer, we encountered some degree of difficulty locating a person with expertise in cast-in-place and pre-stressed concrete systems who was willing to take the time to do a relatively small residential job.

Finally we located an engineer who could and would do the job but found that we had to go into very detailed explanations informing him of the nature of earth sheltered construction and eventually bought him a copy of Earth Sheltered Housing Design. The cost of the engineering was double what we anticipated it to be. I feel the reason for the increased cost was the newness of the underground housing concept in our area.

I was very impressed by the fact that the engineer did not over-design the house so as to be able to take a direct hit. He also was innovative in designing a cast-in-place, heavily reinforced concrete beam to span an eight foot opening in our central load-bearing wall, which saved us considerable expense in engineering and fabricating a wide-flange steel beam to span the opening.

353

Financing

Due to the fact that we were innovators in the field of underground housing in our area, our initial contacts with potential lenders were frustrating and non-fruitful. Two banks said they weren't even interested in talking to us when we telephoned them. The third banker said he would look at the proposal.

We got together our plans, bids from subcontractors, our financial statement and our cost breakdown and went down a few minutes early for the appointment. We were told that the bank officer had gone home sick that day. The manager appeared to speak to us, briefly glanced at the plans and told us his bank wasn't interested in financing this type of project (meaning earth-sheltered).

Finally, we approached a local, fairly conservative, savings and loan. After our meeting with the loan officer, the loan committee approved our loan.

Retrospectively, we feel that the main reason the loan was approved by the committee is because of our presentation. We had compiled detailed cost breakdowns, specification sheets, and financial data. In addition, we had inundated the committee with full color product brochures for our major products - especially the prestress plank system, and the waterproofing and insulation four stage system. We tried to anticipate any questions or reservations the committee might raise.

The negative aspect of the good news was that the lender wanted 14% interest with a 60% loan to value ratio. At that same time the lender would loan 80-90% loan to value at $11\frac{1}{2}\%$ interest on a "normal" conventional house.

Site Work

Before purchasing the ten acres for our home, we drilled a well to determine if we could get a sufficient water supply. This was an important contingency clause in our offer to purchase the land. We hit water in sufficient quantities at 160 feet.

For the excavation of the building site I used a Case 1150, the largest machine that was economically feasible for use on the site. The alternative was to use a smaller machine at a lower hourly rate but with much less earth moving capacity. The operator of the 1150 was a former engineer named Dan Rosa, who had chosen to lead a less stressful lifestyle. He was invaluable in offering practical advice throughout the construction process. He suggested that we peel back and store the rich topsoil for use in the final topcoating during the backfill process. He used the subsequent layers of lower quality soil to build two ramps alongside the excavation that would enable us to gain elevation for concrete trucks to chute the concrete into the footing and low wall pours. This did away with the need to rent a concrete pump on two separate occasions which saved me hundreds of dollars in rental time.

We encountered solid granite two-thirds of the way down the hole which necessitated blasting. I was able to locate a local blaster who allowed me to drill most of the holes myself which saved me considerable money and him considerable time. The laws in this area pertaining to blasting are very lax regarding permits and insurance which permitted this portion of the construction to be trouble free. This might not be the case in many other areas of the country.

Rosa did an excellent job of digging a square, level hole that made the back-filling process go quite smoothly. He also put in roads that allowed heavily loaded concrete trucks easy access to two points adjacent to the building site for pours.

I had a smaller earth moving machine on the site, a Case 450, to do ditching and earth moving in relatively inaccessible areas. I found on an hour to hour basis, it was much cheaper to use the larger machine.

Footings and Walls

When we had cleared the rubble after the blasting we discovered that I had a spring producing five gallons per minute directly under the site of my living room footings which immediately turned my beautiful excavation into one large swimming pool. We hand dug a trench around the inside perimeter of all the footings and an additional trench around the outside perimeter of all the footings, placed perforated drain tile in all of the trenches and then put a six inch layer of pea gravel across the entire excavation. It worked perfectly removing all standing water. We hand dug the footings to undisturbed soil and formed and poured them.

We placed two strands of #4 rebar lengthways in the footings and short pieces of #3 bar across the footings. I used a 2 x 4 and placed it in the wet concrete immediately after pouring the footings to create a keyway for the walls.

After the keyway was completed we put the vertical rebar for the walls in place using #4 bar at 16 inches on center placement. We poured the slabs for the 1000 gallon spa and the 1000 gallon heat storage tank at the same time as we did the footings.

We formed and poured the six inch frost foundation walls next and then waterproofed and insulated them after a two week curing period. The waterproofing compound was elastomeric trowel-on-grade Elastall 900 by United Coatings of Spokane, Washington. Sheet insulation was placed two inches thick against the low walls and partially backfilled to be retained in place.

High wall pour – near disaster. We formed and poured the high walls after the low walls were backfilled. These consisted of an 84 foot back wall, eight feet high, and the load-bearing shear walls, that graduated from eight feet in the back to nine feet in the front, and were 28 feet in length. There were two reasons for the shallow pitch from the front to the rear. The first was to drain any standing water from the roof and the second was to allow more light to enter the interior of the house.

The #4 rebar that had been placed in the center of the footings was pulled to within two inches of the inside surface of the high walls and then tied with #3 bar horizontally on 18 inch centers. At this time I placed 1 x 4 inch nailing strips horizontally on two foot centers inside the wall surface of the forms. These strips later became nailers for sheetrock and interior wall finish.

All conduit and plumbing were placed in the wall forms and prepared for the pour. Three weeks prior to the wall pour I had discussed the use of 3/4 inch exposed aggregate in the wall with the local concrete company. They told me that it would be no problem pumping the mix and recommended that I put a retarder on the forms and sandblast the wall after stripping the forms. The morning of the pour one of the company sales representatives informed me that he had just learned that they could not pump the exposed aggregate mix I had ordered.

There was some dispute about whether or not I could try pumping the mix. Discussion with the operator of the concrete pump followed and he said I would be charged $75 per hour for up to eight hours to clean out his machine if pumping failed. I subsequently decided I had no alternative but to try to use

There was some dispute about whether or not I could try pumping the mix. Discussion with the operator of the concrete pump followed and he said I would be charged $75 per hour for up to eight hours to clean out his machine if pumping failed. I subsequently decided I had no alternative but to try to use chutes to place the concrete. At one point during the pour we had six 12-foot long chutes hanging off a truck when they broke, collapsing down on the wall. We had to dump six yards of concrete over the bank that had gone off while we were trying to put the wall forms back in place. All in all a very disheartening and expensive experience that I believe might have been avoided if the representatives from the concrete company had done their homework.

A couple of days after the marathon 12-hour wall pour, we poured the slabs throughout the house. I had a small amount of difficulty finish troweling the slab in the areas where the conduit protruded through the concrete. We allowed the walls and slab to cure by draping the whole area with black visqueen and sprinkling it with water twice daily for two weeks.

Roof System

We used a total of 21 prestressed concrete planks - 4' x 28' x 1' thick - with three hollow cores running the length of each piece. While the planks were still at the prestress yard I took a 200 pound high pressure air hose and blew the dust out of selected cores to be used for the heat ducting system. The cores sized out adequately for air movement at little extra cost other than the concrete spray sealant applied at the prestress yard after dusting. An advantage of this ducting system is that heat loss from the system is retained in the roof's concrete thermal mass.

The planks were loaded and transported to the site on 40-foot tractor trailers. This was a relatively expensive proposition as the nearest source of the planks was over 40 miles away from the construction site. The transport up the mountain road to my building site was remarkable. The operator of the trucking company had a mid-sized bulldozer pull the trucks one mile up the road backwards, by dragging the rear end of the trailers around the horseshoe turn near the top of the hill.

Installing the planks. We used a 40 ton self-propelled, hydraulic crane to place the planks. I supervised this operation myself, learning as I went. The first five planks took us two hours and 40 minutes to put in place. The last five planks took less than 20 minutes. I learned fast.

Placement of the planks went relatively easy with no major mishaps but the transport and erection costs were expensive due to the inaccessibility and remoteness of my building site.

I welded the tie plates between the planks myself saving considerable money in the cost of a rigger. We bent the rebar protruding up from the high walls into the hollow cores and grouted the rebar into the cores and all of the spaces between the planks with a 3/8 inch grout mix.

Waterproofing and Insulation Systems

Due to the fact that we had running water behind our walls and under our floors from the spring unleashed by the blasting, I chose to use Elastall 900 waterproofing, an elastomeric urethane coating, instead of Bentonite. The system uses a four stage application method. First the concrete was sealed with a product called Uni-Tile Sealer to prevent concrete dusting and to give it a moisture barrier. Secondly, United Coatings arranged for an approved applicator to spray on a 45 mil coat of Elastall 900 on all exterior wall and roof surfaces to form a uniform, seamless waterproofing membrane from the front of the roof to the bottom of the footings.

The third step was the application of a sprayed on, high density, urethane foam insulation with an R-value of 7.14 per inch and a density rating of 2.5 to 3 pounds per square inch. We sprayed the foam one inch thick at the footings creating a fillet at the intersection of the walls and the footings. We increased the thickness up the wall to three inches at the top of the wall and to four inches on the roof.

Finding a foam applicator was a relatively difficult task as most of them are only equipped to do extremely large commercial jobs. United Coatings assisted me in locating a high quality applicator who agreed to spray the foam on my house between two larger jobs.

Finally, the fourth step, included recoating the entire house with Elastall 900 creating a watertight, seamless membrane around the entire house.

Backfilling

Prior to backfilling we wrapped the entire house in heavy mil visqueen. I was able to buy a couple hundred sheets of moisture-damaged particle board for less than $1 per sheet from a local lumber yard. I leaned these up against the walls and placed them on the roof over the visqueen to protect the waterproofing and insulation against possible puncture during the backfilling operation. I then put down a two inch layer of pea gravel with a three inch layer of 1½ inch round washed rock to promote good drainage, avoid ponding of water on the roof, and to prevent penetration by rodents seeking the warmth of the insulation.

I used a small earthmoving machine called a Bobcat to evenly spread the various layers of backfill. I then used 12-14 inches of the topsoil, previously set aside during the initial excavation, to backfill the roof and walls. Once again, Rosa (the earthmover) offered sound advice in the suggestion of backfilling the corners of the house first, then behind the load-bearing shear walls, then all the areas between. In the same vein, we built a ramp on to the roof and then distributed the earth radiating out from the ramp.

Rosa would bring the clean fill dirt to the edge of the roof. I would then use the Bobcat to spread and compact it. I might note the most dangerous situation I found myself in during the entire construction process occurred while I was in a tired but overconfident state of mind. I drove the Bobcat off the roof and wedged the machine between the back of the house and the bank. Luckily, no harm was done to me, the machine or the house. Rosa hooked a chain on to the blade of his Case 1150 and extracted the Bobcat and driver from an embarrasing situation.

Plumbing, Electrical and Mechanical Systems

The plumbing in an earth-sheltered house is very much the same as in a conventional house with the possible exceptions of joining the waste air vents together into one or two runs located in a hollow core of the prestressed planks. They can then be taken to an outside wall and run to the front of the house. This will avoid a forest of plastic pipes on your roof.

The electrical system in my home was very expensive due to the fact that the electrician was not innovative and was extremely reluctant to run wiring up in the hollow cores of the planks. The electrician insisted on running all major runs in conduit in the slab. Two possible alternatives to conduit, both less expensive, are to run all major runs in the hollow cores down into the wood frame partition walls or to use a major raceway in the slab.

The heating system is one of the areas where I think I read a few too many books. I wanted to try everything and I almost did. My heating system consists of an air-tight woodburning stove, backed up to the double sided 15-inch thick stone, center load-bearing wall. The stove has a stainless steel hot water jacket in the back of the fire box that is plumbed through a heating core to the 1000 gallon spa. From the spa the hot water is pumped through a filter to the 1000 gallon cast-in-place heat storage tank. There is 80 feet of coiled copper pipe that acts as a domestic hot water pre-heating heat exchanger passing through the heat storage tank. The tank is located under a six-inch slab floor surfaced with quarry tile. The tank is 4' x 4' x 8' and is sealed with another United Coatings product called Uni-Clad, an epoxy paint-on coating. Hot water is pumped from the tank through plumbing in the slab to a radiator in the forced-air furnace and back to the tank.

The forced-air furnace is a small electric model but gas, oil or even a heat pump could be used as it is primarily a back up to the wood stove (required by most lenders). Air is forced up past the heat coils and radiator by a two-speed fan. The return air source is directly above the wood stove which gives the furnace the ability to circulate the warm air produced by the stove through the ducting system, to the entire house. This system can also cool the house in the summer and provide ventilation as needed, by use of a damper to control fresh air intake.

In addition, I put 800 feet of black plastic pipe in my septic tank for a waste water heat recovery system. Eventually I decided that the waste heat would not be a high enough temperature to make it worthwhile to connect it to my system at this point. However, it is a future resource if energy costs continue to escalate, and I may at some point connect it in to the other systems.

Framing

I did most of the interior framing myself and used a few tricks to keep the prestressed planks from buckling the sheetrock as they settled.

First I laid out the top plate on to the bottom surface of the planks and shot them into place with a stud gun. Then I framed the interior partition walls one-half inch shorter than the opening size and nailed them into place using a spacer to keep the top plates separated the one-half inch. When we sheetrocked we nailed into the lower top plate allowing the roof planks to settle down one-half inch without exerting any downward pressure on the sheetrock or wood framing.

The most important aspect of the exterior framing was to build a cant into the back side of the parapet wall to allow for the earth movement as it expands and contracts in freezing weather.

One cost-saving method we incorporated was to frame the south wall of glass to accept a standard size thermopane sliding glass door pane instead of ordering a custom made window to fit a non-standard size opening.

Interior Decoration

The myriad of selections available in the field of interior design allow very wide latitude in satisfying personal preferences. We felt that light colors enhanced the modern look we wanted to achieve. We painted all walls and ceiling off-white for initial convenience and time saving. Light-colored wallpaper was utilized in the dining room and kitchen.

Floor coverings are important in underground houses because they are on a concrete slab. We used cushioned vinyl in the laundry room, kitchen and bathroom. The master bathroom featured ceramic tile for shower, counter top and floor. The carpeting we selected was the highest quality plush that would fit our budget. More important, however, was the 9/16 inch densified foam pad underneath the carpet which gives you a soft, comfortable feeling from the first step into the house.

Part of our wall finish utilizes smooth surfaced, random length, natural four inch cedar. The look is dramatic and because the wood is nailed directly to the nailers in the concrete wall leaving no dead air space behind, the effectiveness of the thermal mass in the wall is preserved.

The prestress plank roofing can be textured and painted as with any conventional ceiling.

Conclusion

In conclusion, I would say that building an earth-sheltered house with the materials, techniques and systems that I used required a new and different set of skills than normally used for conventional residential construction. Learning these skills, or helping subcontractors to utilize them, can be an expensive, time consuming and even dangerous experience.

As the concept of earth-sheltered construction becomes more widely accepted and the technology begins to become more readily available, the process of construction of underground homes will become more standardized, less experimental, and more accessible throughout the country.

EVALUATION OF UNDERGROUND LIVING

Energy Performance

After living in our house for almost one year, we have found it to be the most comfortable place we have ever resided. We anticipated and received a mild winter by North Idaho standards with average winter outdoor temperatures in the mid thirties. At no time did the house drop below the mid-50's even through there were long periods of time the house was unheated.

We decided this would be a rare opportunity to accurately gauge the amount of cord wood used to heat the house for one winter. As of this writing we have used less than one and a half cords with no other heat source.

As of the present time, we have not needed to use the elaborate backup heating system we installed.

The house has been totally dry with no humidity problem or failures of any stage of the waterproofing system.

The four tables following detail my temperature readings for the months of December 1980, and January through March 1981. All of the temperature readings were done between 9 and 10 o'clock each evening. The letter notations below the charts signify N for no fire in the stove prior to the temperature reading, and S meaning the sun was shining at least part of that day. Because of a heavy teaching commitment with night classes there are many nights when the stove was not used due to our arrival home in time to go to bed.

Overall, I am totally impressed with the performance of the house. It has performed above my expectations prior to construction and my energy savings will increase each year as heating costs rise throughout the country.

Psychological Performance

We find the house to be extremely light and bright in the daytime, even on cloudy or overcast days. In fact it is brighter than our former wood-frame house which offered similar exposures. In addition, it is a quite and tranquil environment. We are happy with our semi-remote location, the house design, and its performance.

When friends come to visit they always comment on how comfortable the environment is. They do not specify exactly what they mean but we feel that the feeling they notice is the constant warmth, even on the coldest of days, or the coolness, even on the hottest of summer days. Very few people realize they are underground unless you tell them beforehand.

We now share our home with our infant son, the biggest bonus of the construction process! It is reassuring to know that we have built a home that he and his son after him can take pride in owning and in which they can dwell comfortably.

TABLE 1 Temperature Charting, December 1980

	1	2	3	4	5	6	7	8	9	10	11	12	13	14	15	16	17	18	19	20	21	22	23	24	25	26	27	28	29	30	31
Living Room		—	VACATION		—						55	56	55	62	66	62	66	64	61	58	61	61	57	58	53	59	58	56	56	64	60
Outside													33	37	33	38	45	46	34	30	27	34	38	39	37	40	51	48	38	42	43
Laundry													59	62	63	70	66	68	66	62	61	63	63	59	58	56	67	62	63	66	65
Bedroom													60	60	62	62	68	64	68	66	61	63	62	61	60	58	65	62	63	68	65
													N	N	S	S	NS	N	N	N	N	N									

TABLE 2 Temperature Charting, January 1981

	1	2	3	4	5	6	7	8	9	10	11	12	13	14	15	16	17	18	19	20	21	22	23	24	25	26	27	28	29	30	31
Living Room	64	64	60	62	61	61	61	58	61	72	68	64	64	65	64	66	63	64	63	68	69	63	61	65	71	63	64	63	62	62	62
Outside	33	36	34	38	33	38	37	39	37	34	33	32	30	28	33	32	33	35	38	41	44	51	43	32	37	32	36	36	37	38	37
Laundry	62	64	66	62	62	62	60	59	60	66	68	64	64	66	64	65	64	65	63	66	67	63	61	65	71	64	63	62	63	63	62
Bedroom	64	65	65	63	63	63	62	62	61	62	68	68	63	64	66	65	66	65	65	64	69	64	63	64	68	63	65	66	68	63	64
	N	N	NS	S	NS	S	S	NS	N	NS	N																	N			

TABLE 3 Temperature Charting, February 1981

	1	2	3	4	5	6	7	8	9	10	11	12	13	14	15	16	17	18	19	20	21	22	23	24	25	26	27	28
Living Room	68	59	66	60	58	64	64	58	57	68	58	59	70	69	67	60	58	57	56	61	64	71	65	62	62	71	64	66
Outside	33	29	33	29	28	27	29	26	8	17	26	34	37	38	35	41	36	38	36	36	37	47	44	38	35	36	34	35
Laundry	69	61	63	59	57	64	65	59	60	66	58	60	65	66	66	59	60	59	58	60	63	68	66	64	64	66	65	65
Bedroom	68	61	66	63	59	63	63	60	58	64	60	60	63	66	65	60	59	59	59	62	62	66	65	63	63	67	64	67
		N	S		N						N					N	N	N	N			S	S	S	N			S

TABLE 4 Temperature Charting, March 1981

	1	2	3	4	5	6	7	8	9	10	11	12	13	14	15	16	17	18	19	20	21	22	23	24	25	26	27	28	29	30	31
Living Room	65	66	64	61	66	65	65	65	65	66	65	69	69	67	69	71	64	64	68	68	74	69	65	71	B		68	69	67	74	70
Outside	36	37	40	35	29	33	37	36	44	42	43	44	44	42	49	38	34	35	42	40	45	42	43	44	A	B	44	43	34	36	35
Laundry	66	67	64	62	63	65	62	64	64	64	65	62	65	66	69	72	65	64	68	66	70	68	63	69	Y		68	68	68	70	69
Bedroom	65	66	65	62	64	65	65	65	66	65	69	68	67	69	66	64	68	68	68	65	68	68	65	67	.		68	68	67	69	68
	NS	NS	N	N				NS	NS	NS	NS	NS	NS	S	S			N	S	NS	NS	NS	NS	NS			N				

TABLE 5 House Cost Breakdown

Item	Amount	Percentage of House Cost
Excavation, blasting, trenching, backfill, pea gravel, rock, landscape	8,479.05	10%
Drainage/Waterproofing low walls	640.51	1%
Footings	1,415.09	1%
Low Walls	2,584.17	3%
High Walls	8,544.71	10%
Roof-including installation & transportation	11,480.29	14%
Plumbing-Labor & Material	3,618.74	4%
Heating/Air Conditioning	1,359.99	1%
Electrical-Labor & Material	8,099.83	10%
Wall Finish-sheetrock, paint, trim	2,744.45	3%
Lumber/Cabinets/Framing	7,801.99	9%
Flooring, Laminate, Tile	2,720.68	3%
Insulation	5,667.66	7%
Waterproofing	1,312.00	1%
Doors/Windows	3,590.09	4%
Exterior Siding	558.05	1%
Concrete Flatwork	1,459.94	1%
Miscellaneous - includes labor, rental tools, permits, shed costs, etc.	6,999.11	8%
House	79,076.35	
Well/Septic/Cistern	7,301.76	9%
Land Cost	25,000.00	
	$111,378.11	

TERRA VISTA
HUD AWARD WINNING PASSIVE SOLAR EARTH SHELTERED HOUSE

Michael S. Milliner, President
M. S. Milliner Construction Company, Inc.
302-A East Patrick Street
Frederick, Maryland 21701

ABSTRACT

Through the Cycle 5 Residential Solar Demonstration Program, a design competition administered by HUD in 1979, partial design and construction funding for new passive solar homes was granted to 91 builders nationwide. One of the projects selected, Terra Vista, is an earth sheltered home utilizing direct and indirect solar gain as well as passive cooling. Construction of the home, located near Frederick, Maryland, commenced in late 1980, and was completed by spring, 1981.

This paper reviews various design considerations such as site, configuration, floor plan and solar features as they each relate to the livability and energy requirements of Terra Vista. Important construction features such as structure, thermal and moisture protection and glazing are also discussed.

KEY WORDS

Passive solar heating and cooling; earth sheltered housing, design and construction.

INTRODUCTION

In the fifth of a series of grant programs administered by the Department of Housing and Urban Development, a passive solar design competition was conducted in 1979. The objective of the program, entitled the Cycle 5 Residential Solar Demonstration Program, is to demonstrate projects which require a minimum of additional, non-renewable energy measured in terms of an "energy budget", and to educate homebuilders in the principles of passive solar design and construction techniques. Projects were evaluated as to the degree to which the solar design is compatible with the local climate, and provides a logical flow of heating and cooling to the home's interior.

In hopes of securing funding for an earth sheltered home, a team of architectural and solar designers joined forces with M. S. Milliner Construction, Inc., to prepare a preliminary design for submission in Step One of the design competition. From a total of 900 submissions, HUD selected 139 promising designs to receive a design grant for completion of design and working drawings. In Step Two of the competition, 91 finalists, including Terra Vista, were selected to receive partial funding for construction.

With partial federal funding, plans complete, and site selected, all that remained was to secure construction financing. Unfortunately, the record high interest rates of early 1980 made construction financing prohibitive. However, after securing an FHA insured permanent loan commitment, a construction loan was secured and construction commenced by fall, 1980. Despite the complexity and intensity of labor involved in such a building project, the construction process went extremely well, with completion by spring, 1981. The Open House required by the grant program is slated for late June, with occupancy available shortly thereafter.

In presenting the earth sheltered housing concept to the local citizenry, it was of critical importance that Terra Vista be of pleasing appearance and not terribly divergent from other more typical homes in terms of internal layout, natural light, and amenities. As such, the home possesses rather standard floor plan characteristics, and is mildly contemporary in external appearance. However, the primary design objective was to maximize energy efficiency through conservation and the utilization of solar energy. Therefore, virtually every aspect of the site, building configuration, floor plan and solar features was critically analyzed as to its affect on the flow of energy both externally and internally to the house. Construction elements such as structure, moisture and thermal protection and glazing systems were carefully considered as to the affect of each on the ability of the building envelope to collect, retain and store solar energy. The net result of the design, construction and solar features herein described is to allow Terra Vista to achieve an extremely low energy budget for both winter heating and summer cooling.

hud demonstration house
cycle 5 passive solar design award

m.s. milliner inc.

DESIGN

Site

The home is located in a small subdivision of custom homes near Frederick, Maryland. The 3 1/2 acre partially wooded site is ideal in every respect for the construction of an earth sheltered, passive solar home. Sloping gently (12%) to the south, the grounds adjacent to the home are densely planted with white pine on the east and north, with deciduous trees to the south and west. The sloping lot with dense foliage allows maximum protection from prevailing northwest winter winds, while providing moderation of air and ground temperatures in summer. The location mid- way up the side of a slope moderates temperatures under light wind conditions, pro- ducing convective up-slope breezes during the day and down-slope settling of cool air at night, while avoiding the extreme low temperatures experienced in valley bottom sites in winter or humid/fog conditions in summer.

The home was designed to require an absolute minimum of disturbance to the lot, with excavation planned to provide the required earth for berming and covering of the building envelope. Inasmuch as freedom from grounds maintenance was consider- ed essential, a natural groundcover with a meadow-like apperaance was determined to be desirable for the majority of areas immediate to the house. The groundcover selected is a mixture of wild flowers along with a tall fescue. Virginia Creeper is planted along the roof overhangs to create an easing affect at the wide fascias. Other shrubbery and plants are planted in high visibility areas such as the entrance walk and patio. Existing deciduous trees are located along the south side to provide shade for the greenhouse and patio.

Building Configuration

The requirement to minimize alterations to the existing topography, while effectuating a workable solar perspective, weighed heavily in the design of the building config- uration. The optimum solution was a structure elongated on the east-west axis with stepped up living levels along the north side. The maximum depth of excavation was thereby limited to 8', while creating a very low profile that blended in harmoniously with the natural terrain.

The roof of the lower living areas is nearly level, while the upper northern roof is raised to form a vaulted ceiling. Winter winds are thereby deflected, while providing an area for clerestory windows to extend the full length of the home.

Floor Plan

The image typically held by many individuals is that of a damp, dark and cramped underground house. The primary objective in designing the floor plan was to miti- gate those concerns by creating a spacious, airy and open layout with gracious quantities of natural light in all living areas. The result, with few exceptions, is a room arrangement quite typical of many modern homes and totaling 2,300 square feet of living space.

In order to minimize penetrations through the earth berming on the other exposures, which simplifies construction and maximizes energy efficiency, the entry is situated centrally on the south side. The air lock vestibule opens into a foyer which allows direct access to either the sleeping quarters or daytime living areas. Three bed- rooms are each provided with large closets and window areas which exceed all code

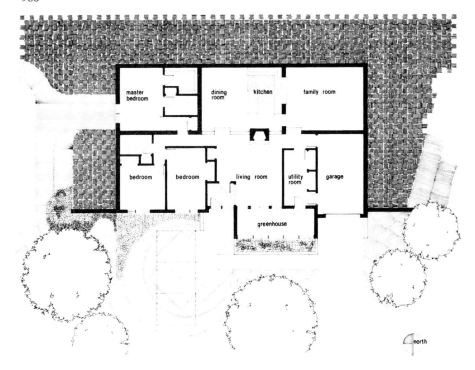

related egress and lighting requirements.

The daytime living areas are designed to allow a free-flowing traffic pattern and a
high level of visual openness. In the dining room or kitchen, which are located at
the rear of the home, a view is provided into the family room, through the clerestory
windows, and through the living room. From the living room, there is a view into
the clerestory lit kitchen and dining room, or through the greenhouse to the exterior.
The greenhouse, located next to the utility room to facilitate horticultural convenience, provides solar collection as well as an air lock to shield living room window
and door areas. A single car garage is located to the east, with access into the
utility room or a hall leading to the living areas. A patio with trellis cover is
situated to the southwest, with access from the entry walk.

                    CONSTRUCTION

Structure

The requirement to support heavy earth loads and provide a large quantity of thermal
mass indicated concrete and/or masonry as the logical structural elements. In considering availability, complexity of structure, and the critical quality control
required, it was determined that reinforced concrete block exterior and bearing walls
were most appropriate. Following the construction of steel reinforced concrete
footings with vertical steel dowels, conventional 12" concrete block walls were built
with Durowall reinforcing. After forming the required leveling caps, all masonry
walls were poured with concrete and reinforced with steel bars 16" o.c. vertically.

The roof is built of 8" deep prestressed concrete roof planks set in place by crane.

A 3" reinforced concrete topping provides the diaphragm action required by the structural design. The forms at roof eaves are built of treated lumber and left in place as the parapet structure with 1 x 5 cedar fascia. Floors consist of reinforced concrete slabs over plastic vapor barrier and crushed stone.

Moisture Protection

Berms and swales blend in well with the elevated and well-drained site, and are designed to create rapid runoff of water from the areas adjacent to the house. Corrugated plastic 6" drains and gravel are located at the base of the exterior walls as well as at the low side of both roofs. These drains are run underground and slope away from the home until they reach natural grade.

The waterproofing system determined most appropriate for this project was the Bentonize trowel applied system. The seamless application, flexibility in conforming to irregular configurations, and integration with vertical insulation provided the most convenient and economical approach.

Sidewalls were treated with 3/16" of Bentonite applied directly to the block, with additional thickness at roof joints and a small cove at the base. The foam insulation was imbedded directly in the freshly applied waterproofing with joints between sheets 'buttered' with 1/8" thick Bentonite. The insulation was thereby held firmly in place, providing the protection necessary for backfill and creating an additional waterproof barrier. Prior to the backfill operation, all sidewalls were covered with polyethylene for additional protection. All backfilled earth was firmly compacted in lifts of 8" to 10".

The lower roof slopes 6" to the south and was treated with 3/16" Bentonite applied to the concrete topping, however, polyethylene was imbedded directly in lieu of insulation. Following the polyethylene, foam insulation was laid horizontally on the roof, and then a second layer of polyethylene was loose laid over the insulation. A 3" layer of gravel was installed over the polyethylene, which provided a horizontal percolation bed to allow water to drain toward the low side where it is picked up by the 6" corrugated drain or a single interior roof drain. A 2" fiberglass silt screen is installed to protect the gravel and then 18" of earth cover is put in place by crane.

The upper vaulted roof is also protected with polyethylene imbedded in Bentonite, and then foam insulation. However, in order to prevent slippage due to the slope, the insulation is braced at the low side and small foam blocks are glued to the upper surface. An 18" earth layer is then installed directly on the insulation.

Thermal Protection

The need to conserve energy through cost effective levels of insulation and to provide a quantity of thermally treated mass commensurate with the level of passive solar potential, indicated varying quantities of foam insulation placed on the exterior of the building envelope. After much research into the effects of earth, moisture and frost, it was determined that the Dow Styrofoam brand high-density rigid insulation would be most appropriate for the subgrade conditions on this project. Sidewalls were treated with a 4" thickness to a depth of 4' below grade, 2" between 4' and 8' deep, and 1" below 8'. Roof areas were all treated with a 5" thickness. All exposed sidewalls are treated with 4" of beadboard insulation. The exterior finish is a fiberglass mesh reinforced and precolored stucco called Dryvit, and is applied directly to the beadboard. Soffit areas at roof overhangs are covered with 3" beadboard and also finished with Dryvit.

Careful attention was paid to thermal breaks throughout the building envelope. The retaining walls are, with the exception of below grade footings, structurally independent of the exterior walls and separated by 4" of foam insulation. Concrete floor slabs are thermally isolated from interior to exterior by providing an insulating break under door sills. Window and door openings in masonry are carefully wrapped with beadboard to prevent conduction losses.

Glazing

Terra Vista is one of four homes nationwide selected by 3M Corporation to demonstrate a new high transmission film ideally suited for multiple glazed solar apertures. The recently patented product is a 4-mil film of clear, UV-protected polyester, coated with two invisible layers of a specially developed anti-reflective material. The film transmits 93% of the sun's energy, as compared to 84% admitted by single pane glass. Multiple layers of film, therefore, transmit a greater percentage of solar energy while providing an insulating value equivalent to that of glass.

In this project, the optimum window design was determined to be two layers of glass with two layers of the 3M film suspended in between. The total R-value developed is 3.9, which is approximately twice that of thermopane glass. This system, in place 24 hours a day, can be shown to exceed the net efficiency of thermopane glass in place during sunlight hours and an R-6 movable insulation system in place during non-sunlight hours. The advantage to the occupant is obvious in terms of effort required to open and close an insulating device on all windows twice daily.

All fixed windows in Terra Vista are quadruple glazed while the operable bedroom windows and glass panels in doors are triple glazed (using one film between glass panes) All units, except the direct set greenhouse panels, were custom built with vinyl covered wood frames by Weather Shield Manufacturing, Inc.

HEATING AND COOLING

Load Calculations

The well insulated and earth protected mass of the building envelope creates an extremely stable interior environment year round, with practically no requirement for auxiliary heating or cooling. According to detailed computations required by the grant program, Terra Vista has a total annual heat load of 61.3×10^6 btu/year. Passive solar contribution is 46.2×10^6 btu/year which reduces the net load to 15.1×10^6 btu/year. The energy budget is computed as follows: (15.1×10^6 btu/year (5087 degree days/year)/(2,300 square feet of conditioned living space), which yields an auxiliary heating requirement of 1.3 btu/degree day/square foot.

The total annual cooling load is 33.9×10^6 btu/year. Cooling contributions from the earth contact, ventilation through an earth pipe, and a heat pump water heater are estimated to reduce the load by 50%. The moderating value of the massive structure will reduce the cooling load further, making the need for auxiliary cooling unlikely.

The domestic hot water heating load is 17.9×10^6 btu/year (more than required for auxiliary heating in the entire home). An air source heat pump water heater provided by Fedders Compreseor Corporation, is expected to provide 11.9×10^6 btu/year from interior ambient air or about 2/3 of the total. The air source may be solar heated household air. The adjustment between these sources will depend on the need, at any given time, for the cooling or dehumidification which results from the operation of the heat pump unit.

Passive Systems

The passive solar systems are integral to the design and efficiency of Terra Vista, creating a bright, sunny interior and providing 75% of the annual heating load. Direct gain is derived through 210 square feet of south facing vertical window area. Clerestory glazing is situated to allow direct sun contact on the rear north wall. Indirect gain is derived through 140 square feet of sloped glazing on the south side of the greenhouse. Heat transmission to the interior can be through open doors or the hollow cores of the roof planks.

Passive cooling, besides solar rejection features is provided by a 12" concrete earth pipe that originates at the furnace closet and slopes away from the house a distance of 90', terminating at a vertical air shaft. Buried 10' below grade, the pipe partially cools and dehumidifies summer ventilation air. High interior vents provide a chimney effect, exhausting any build-up of warm air.

Mechanical Systems

Auxiliary winter heat may be derived from one of two sources: a woodstove or heat pump. The woodstove is a free standing Buckstove Model 26000, centrally located to allow even distribution. The heat pump is a counterflow unit made by Carrier with 10 KW electric auxiliary heat. The ductwork is situated in a radial pattern, opening into a supply plenum under the furnace, and consisting of round steel coated with polyvinyl cloride. The closet in which the furnace is located functions as a return air plenum, providing a central return for daytime living areas and the master bedroom, and ducted returns for the other bedrooms.

A manually controlled exhaust fan is located on the exterior north side of the chimney and is used to assist ventilation if required. A thermostatically-controlled exhaust fan and ventilating louver provides ventilation through the greenhouse. Exhaust fans are also provided for both bathrooms. The range hood is custom built and utilizes a triple filter ventless exhaust which filters odors, smoke, and grease.

OPERATION OF SYSTEMS

Winter - Passive Mode

The stable earth cover moderates cold temperatures and reduces air infiltration. Direct gain apertures admit sunlight during the day, which through the greenhouse effect, is converted to thermal energy and trapped within the home. The insulated thermal mass of the structure absorbs and stores the heat, for release during non-sunlight hours.

Indirect gain within the greenhouse causes warm air to rise, passing through operable vents in the greenhouse ceiling and circulating through the concrete roof planks to the rear portions of the living areas. As greenhouse-heated air circulates throughout the house, it's heat is absorbed by the mass, causing the air to fall. The supply ductwork slopes downward from the registers to the plenum, except for the greenhouse supply which slopes away from the plenum. This continual drop in grade allows cooler floor level air to drain through the ductwork system back to the greenhouse where it replaces the warm rising air. One effect of this pattern of air movement is to warm the earth mass below the floor slab. The sliding glass door can also be opened for direct air transfer between living space and greenhouse. During non-sunlight hours, the stored heat within the mass is radiated into the living space to minimize temperature fluctuations throughout the 24-hour day.

winter day

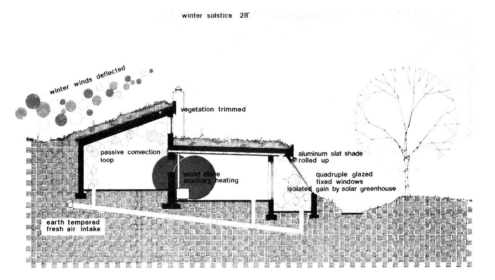

winter solstice 28°

winter winds deflected

vegetation trimmed

passive convection
loop

wood stove
auxiliary heating

aluminum slat shade
rolled up

quadruple glazed
fixed windows
isolated gain by solar greenhouse

earth tempered
fresh air intake

Internal heat sources such as lights, appliances and occupants is expected to provide a significant portion of the net annual heat load. Under extreme conditions,
however, when the solar and internal heat sources are insufficient, the woodstove
may be used for a short period to supply supplemental heat. The air circulating
stove can perform in a passive mode, or with fan assist. The damper on the earth
pipe may be partially opened to supply make up combustion air in lieu of air infiltration.

Winter - Mechanical Mode

During sunlight hours, the greenhouse convective loop may be assisted by the fan
in the air handler. In that case, room air taken into the return air plenum would
be supplied to the greenhouse and/or other living areas via the underslab ductwork.
By adjusting various return and supply registers, the air movement can be increased
or decreased through the greenhouse as required. Operation of the fan in the air
handler will also assist in even distribution of air heated by the woodstove. Only
under extreme conditions, and if use of the woodstove is not desired, would the
auxiliary electric heat be likely to be required.

Summer - Passive Mode

The stable earth cover moderates warm temperatures, while the thermal mass of the
structure prevents overheating. Evaporative cooling from the earth's surface has
a cooling effect on the home's interior. Roof overhangs are designed to shade vertical glass surfaces. An adjustable polypropylene shade blocks 80% of the sunlight
striking the greenhouse glazing. The vents in the ceiling of the greenhouse are
closed, while the insulating panels covering the vent and fan are removed.

summer day

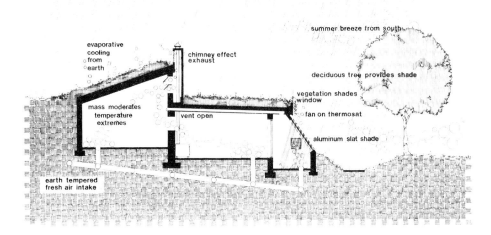

summer solstice 73°

summer breeze from south

evaporative
cooling
from
earth

chimney effect
exhaust

deciduous tree provides shade

vegetation shades
window

mass moderates
temperature
extremes

vent open

fan on thermosat

aluminum slat shade

earth tempered
fresh air intake

Large vents located as high as possible in the dining/kitchen area can exhaust stratified warm air into the framed chimney which has exterior vents on the high north side. The damper in the earth pipe is opened as required to allow replacement exterior ventilation air to be drawn underground and partially dehumidified and cooled. Windows and doors may be opened for direct ventilation. The thermal mass absorbs any excess daytime heat buildup and releases it at night to 'drive' the passive ventilation throughout the 24-hour day.

Summer - Mechanical Mode

The exhaust fan located high in the chimney can draw ventilation air through the home and exhaust it to the exterior. By closing all other return openings, ventilation air may be provided exclusively by the earth pipe. This air would actually be drawn through the ductwork system and introduced through the supply registers in each room. The fan in the air handler could provide a similar function. The fan in the greenhouse can be thermostatically operated to provide mechanical ventilation as required.

The heat pump water heater extracts heat from interior household air, with a concurrent dehumidifying effect. It is not expected that mechanical air conditioning will be required, however, the heat pump unit can provide this function if required

CONCLUSION

The Cycle 5 Program and construction of Terra Vista was an excellent learning experience in terms of the principles underlying passive solar and earth sheltered housing design and building. Attempts are presently underway to have the home monitored in the near future. At this time, it can only be said that the house performed as expected this past winter. Despite the extremely cold temperatures during December, January and February, the interior temperature never dropped below 55 degrees, with an average of 60 degrees throughout the winter.

Costs were indeed greater than that expected for a home above ground of similar size, configuration and amenities, however, the energy, environmental and longevity advantages are extensive. Public acceptance has thus far been much better than anticipated, with considerable interest growing in custom designed and built earth sheltered projects.

FERROCEMENT EARTH SHELTERED HOUSING PROJECTS

Loren C. Impson, President

Spatial Experiences, Denton, Texas

ABSTRACT

This paper explains the use of ferrocement for construction of simple earth sheltered houses. Practical information gained from construction projects in two earth sheltered communities is reported. Data relating to cost factors is also included.

KEYWORDS

Ferrocement; earth sheltered; economical housing; passive solar heating and cooling.

INTRODUCTION

Whitehawk and Rainbow Valley are two communities of individuals involved in applying alternative technologies for self-sufficiency. These communities were originated north of Denton, Texas by Robert and Ruth Foote. Whitehawk has been functioning for three years and encompasses 80 acres, half of which is used for private residences. The other half is held in common. Rainbow Valley consists of 220 acres with 120 acres held in common by the members for parks and food production. The primary goal of these two communities is self-sufficiency through economical housing, shared resources, and communal food production.

Three of the major costs of shelter are the initial expense of building materials and the ongoing expenses of heating and cooling. Earth sheltered designs were chosen for the projects described in this paper primarily due to their potential for reducing these costs. The mild Texas winters ordinarily require peak heating for about two months. With the earth sheltering techniques, proper solar orientation, ground cover, and passive design techniques, heating and cooling loads can be virtually eliminated. Earth sheltering in a normal year can keep the interior temperatures below 75° F. with two feet of earth and proper ground cover during peak cooling loads. The same structure can maintain 60° F. throughout the heating season. One of the structures at Whitehawk using these techniques reached a temperature of 80° F. when the outside temperature was over 100° F. for 42 consecutive days.

HISTORY

The earliest patent on ferrocement was granted in 1847 to a Frenchman, J.L. Lambot, for a method used in the construction of a ferrocement rowboat. An Italian engineer, Pier Luigi Nervi, is generally credited with popularizing the use of ferrocement in his work on roofs and boats in the 1940's. (Sabnis, 1979). The American Concrete Institute recently established the "549 Committee" to investigate the viability of uses for ferrocement. From their findings they have defined ferrocement as: "...a type of thin wall reinforced concrete construction where a usually hydraulic cement is reinforced with layers of continuous and relatively small diameter mesh. Mesh may be made of metallic materials or other suitable materials." (Batson 1979). This is a broad definition in which materials as diverse as cane and mud to polyethelene reinforced concrete may be included.

THE FERROCEMENT TECHNIQUE

The structures at Rainbow and Whitehawk utilize ferrocement for their shells. The detail is shown in Figure 1.

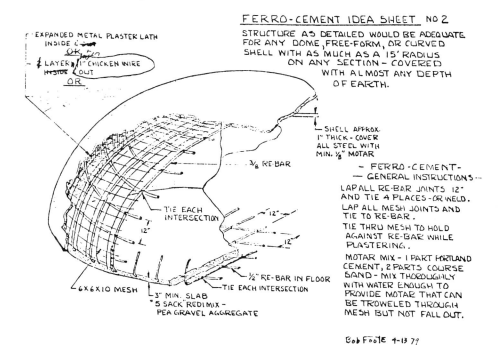

Fig. 1 Ferrocement detail (Foote 1979).

They are built from a layering system which includes expanded metal lathe, chicken wire or galvanized mesh; rebar; 6"x6" mesh (#6 or #10); and a high ratio of cement to sand, with an acrylic polymer for a cement admixture. Ferrocement lends itself to curvilinear shapes necessary for the membrane effect of a thin walled structure. A minimum of skills are necessary for this construction method. Although expensive equipment makes this method of construction more rapid, the ferrocement technique can be implemented with simple and inexpensive tools. The first ferrocement structure (consisting of over 550 sq. ft. of floor space) completed at Whitehawk cost a total of $3500 for materials and required 3000 man hours of construction time.

A building project utilizing ferrocement was directed by Jose Castro in Mexico. His project used ferrocement primarily for roofs. Through his research he determined that roof load requirements of 100 Kg/m^2 can be met with two bars 0.25'' in the edge of the roof, connected by two layers of chicken wire for spans up to six meters using a 1.1 meter rise. The resulting thickness is about 8mm. and weighs 22 Kg/m^2. Roofs using his design have supported uniformly distributed loads of 10.4 tons on a 6m x 6m dome (Castro 1979).

Since each house at Whitehawk and Rainbow Valley is, in a sense, a research project, experimental techniques are being used. The first structure at Whitehawk was constructed from #3 rebar on one foot centers with 6''x6'' wire mesh #10 at 50% overlap and an interior and exterior layer of galvanized 1/4'' wire mesh with a 50% overlap. The concrete shell has a rough troweled finish and is waterproofed with a bentonite clay compound. It is covered with a minimum of two feet of earth. Another structure at Whitehawk, using flatter roof lines, required the use of backbrace beams for structural integrity. The thinnest shell at Whitehawk is 5/8'' to 1'' thick. Its composite steel structure is comprised of two layers of one inch chicken wire, two layers of 6''x6'' #10 wire mesh, and two more layers of one inch chicken wire. There are three pieces of 3/8'' rebar in the structure, one at the base, one at 52° from the base, and one around the 10 ft. skylite in the center of the structure. The materials cost on this 44 ft. diameter domed shell was less than $3000. During the backfilling the backhoe was inadvertantly maneuvering on the backfill and the structure was noted to deflect internally, with no subsequent cracking.

The Spatial Experiences method of forming the skeleton of an earth sheltered ferrocement structure is based on I beams cut from 1/2'' plywood as the vertical members and 3/4'' fibreboard for the cross members. Each beam's curve is cut to provide for the size and shape of dome most frequently built with those forms. There is flexibility allowing for an increase in size using the same forms.

The construction process begins with the selection of a size and configuration for the structure. The method works equally well with domes, barrel vaults, as well as irregular floor plans. A slab is poured in the desired shape on undisturbed soil. The beams are raised and secured at the peak. Plywood 1/2'' thick is secured to the beams creating the skin of the forming method. This skin provides the base for one layer of expanded metal lathe, rebar as required for structural integrity, one layer of 6''x6'' #6, one layer of 6''x6'' #10, and two layers of 1'' chicken wire. These are stapled in place with inch crown roofing staples. A ratio of 2:1 sand to cement with 1/2 gallon of ''Duryl 79''[1] per yard is mixed to a plastic consistency and applied to the composite steel structure. The forms are removed from the interior, and the interior is plastered to specification. The structure is then externally waterproofed, covered with a smooth layer of earth for thermal mass, and insulated with styrofoam as required by climate. It is then backfilled with a minimum of two feet of earth for ground cover growth.

Due to the right angles resulting from traditional construction methods, massive amounts of concrete are required to bear the load of earth and vegetation on an underground shelter with safety tolerances. Less concrete is required using the ferrocement technique due to the curvalinear shape and the placement of the steel in the shell. The volume of steel in this method is double that in conventional poured in place concrete. Consequently the cost of concrete can be reduced as much as 87% in spite of the fact that a more expensive mix is used. This reduced cost offsets the increased costs of steel. The cost of poured in place concrete is contrasted with ferrocement construction in Table 1.

---

[1] A water dispersion of an acrylic polymer designed for modifying portland cement.

TABLE 1 Cost Comparison of Poured Concrete versus Ferrocement.

Poured in Place per sq. ft.		Ferrocement per sq.ft.	
Cement at $45/yd. 12" thick	$1.67	at $70/yd 1" thick	$0.22
Steel at 15¢/ft. #4 rebar 2'	.30	expanded metal lathe, 6"x6" wire mesh #6 and #10, two layers of 1" chicken wire	.46
Forms used 5 times 3/4" plywood	.20	beams and 1/2" plywood,5 uses	.10
Cost per square foot of wall area	$2.17		$0.78

Labor cost depends upon complexity of the shell and the concrete application method. Gunnite can be applied for $92/yd. (including mix) which would give a cost of 30¢/sq.ft. of wall area. Cement can be hand applied for 12¢/sq.ft. plus 21¢ for the mix, resulting in 33¢/sq.ft. of wall area. Both methods are satisfactory, only the length of time involved varies.

The cost break down on an 18 foot diameter dome built by Spatial Experiences at Rainbow Valley as a test module is given in Table 2.

TABLE 2 Cost Analysis

Steel; 6"x6" #10 wire mesh with 30% overlap, #3 rebar on 2' centers, 1" chicken wire with 30% overlap, expanded metal lathe, and staples.	$341
Forms with a life expectancy of 5 uses.	60
Cement, sand, and admixture.	331
Mortar mixer	83
Waterproofing	27
Interior paint	60
	$902

Labor allocations were:

4 laborers at 4 hours to erect forms	16 hours
2 laborers at 8 hours to staple steel	16
5 laborers at 12 hours to apply an exterior coat of cement	60
2 plasterers at 5 hours to apply the interior finish	10
2 laborers at 5 hours to seal exterior and paint interior	6
	108

This dome was constructed with laborers inexperienced in this construction technique. Larger structures can be built with an exponential reduction in both materials and labor. Excavation cost for this dome was $60. It was recovered for $250.

CONCLUSION

The benefits of earth sheltering are numerous. Through experience gained in these two projects the viability of a construction technique which has had little prominence for many years has been tested. These projects have shown that ferrocement construction makes housing in the $25-30/sq.ft. price range a possibility. Additional savings in fuel costs are achievable with the use of proper solar orientation and earth sheltering.

The disruption of ecological balances brought about by mis-application of some present day technologies, the high cost of transportation, high interest rates, and inflation provide the incentives for the style of living in these two communities. The owner builder approach, cash purchases, and the low cost of ferrocement have been combined to yield a stable base upon which to construct this type of housing. In spite of the current fear that home ownership will no longer be feasible for many American families, these communities have shown that through the innovative use of alternative technologies this traditional American value can still exist.

REFERENCES

Batson, G. B., G. M. Sabnis, and A. E. Naaman (1979). Survey of Mechanical Properties of Ferrocement as a Structural Material. Ferrocement--Materials and Applications, American Concrete Institute Publication Sp-61, Detroit, Michigan, 10-11.

Foote, R. (1979). "Ferrocement Idea Sheet", Whitehawk Press, Denton, Texas, 2.

Sabnis, G. M. (1979). Ferrocement--Past and Present. Ferrocement--Materials and Applications, American Concrete Institute Publication Sp-61, Detroit, Michigan, 3-4

CASE HISTORY: EARTH SHELTERED HOUSE IN VERSAILLES, MISSOURI

John D. Simmons,* Jerry O. Newman,** Robert E. Harrison,***

*Research Engineer, **Research Leader,
***Agricultural Engineer

Rural Housing Research Unit, USDA-SEA-AR
Clemson, South Carolina

ABSTRACT

Solar energy and earth embankment are two design considerations for reducing the energy input to heating a residence. This report describes how three supplemental heating systems interact with solar energy as a primary heat source. The three supplemental heat systems were (1) electric resistance, (2) heat pump, and (3) wood stove. The electric heating elements were used to calibrate the house energy use as a function of ambient outdoor temperature (degree day). Then, following a schedule developed by the Rural Housing Research Unit, a 16-week testing period was divided into four one-week cycles. One week of each cycle represented the operation of each of four modes in which different heating systems were used. The house thermostat was set at 70°F. It was determined that more energy was necessary to heat the house on cloudy days than on sunny days. The electric heating elements and the heat pump kept the house close to the desired temperature. However, they were the most expensive to operate. The solar system and the wood stove, although less expensive to operate, created significant temperature swings above 70°F in the house. This was due to their uncontrollability, and had these systems been more controllable, these economic benefits would have been many times greater.

KEYWORDS

Solar heat; earth embankment; attic collection; calibration technique; supplemental heating systems; heat pump; wood stove.

INTRODUCTION

The combination of earth embankment with a solar energy source provides a logical approach for energy conservation in residential housing. Earth embankment is particularly valuable because it moderates the environment adjacent to the structure by reducing the temperature drop and by decreasing the potential for air infiltration through the perimeter walls. But, the demand for below-grade structures has been suppressed because of the common image that portrays basements and below-grade spaces as dark, cold, and uncomfortable. Likewise, solar energy is abundant, and cost-effective solar systems are being developed for residences, but

379

the high initial cost has been a serious deterrent to its widespread acceptance and its general use in individual homes.

The Rural Housing Research Unit (RHRU) has been successful in developing solar systems that are cost effective, and significant progress has been made toward overcoming the sociological and psychological deterrents to living below grade. A popular house design known as the "Solar Attic" was developed several years ago by RHRU. It features "The Solar Attic System,"[1] a low temperature (110°F), air cavity solar system which is integrated into the structure of the roof and attic. The house was also characterized by a simple air handling system and a bed of crushed stone was used for storage of solar energy. Another popular house developed by RHRU engineers was the Solar-Earth house that features earth-embanked walls and solar heating. This house has a skewed roof made of conventional building materials, one exposed wall facing south, and a high-temperature, flat-plate solar collector used for both space and water heating.

The solar attic system is a cost-effective, low technology, low temperature solar system with an efficiency close to that of many highly sophisticated systems, but the high heat loss on successive cloudy days results in rapid dispersement of the available energy. The earth-embanked house has a low energy requirement but its high temperature solar collector was designed for water heating and, therefore, is not the best solar collector for space heating. The earth-sheltered, solar attic house is a combination of these two concepts and is characterized by thermal stability and energy efficiency that provides individuals with a comfortable, economical, and practical house.

The first earth-sheltered, solar-attic house was the product of a cooperative effort of RHRU scientists and a private citizen in Versailles, Missouri, who was constructing his own residence/office combination. In return for RHRU´s counseling, this home owner allowed and assisted RHRU in the monitoring of temperature and relative humidity, and he provided meticulously collected data on energy use. He also operated his multi-unit heating system during a 16-week period according to a schedule designed by RHRU for the purpose of identifying energy requirements of the house and how each of his four heating systems contributed to the overall house performance.

This report describes:

1. The design and construction features of the house and the various components of the heating system;

2. The winter performance of the house;

3. The contribution and performance of the individual components of the heating system.

DESCRIPTION OF RESIDENCE

The earth-embanked, solar-attic residence is owned by Mr. Bill Mason, a photographer in Versailles, Missouri. It is a single level, 1800 ft² structure which serves as his 1260 ft² two-bedroom private residence and his 540 ft² photographic studio (Fig. 1). Figure 2 is a floor plan of the house, and Fig. 3 is a cross-sectional view perpendicular to the south wall of the house.

The house is 60 ft long and 30 ft wide and is built into a south facing slope on a rural wooded lot. Three sides are earth embanked and the exposed wall which faces south contains 51 ft² of glazing. The exposed wall is shaded by a 10 ft overhang that covers the front porch or patio. The earth-embanked walls are pressure treated plywood foundation panels designed according to procedures outlined by

Fig. 1 Photograph of the house

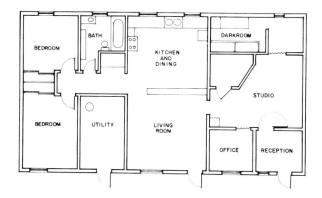

Fig. 2 Floor plan of the house

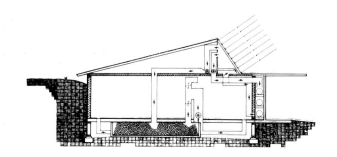

Fig. 3   Cross section of the house

Newman and Godbey.[2]   This lightweight panelized foundation system was constructed onsite and it was installed on an 8 x 16-in crushed stone footing. It consisted of two separate panel systems, one 2.5 ft tall which formed the crawlspace and a second 7.5 ft tall which enclosed the living area. The break in continuity of the foundation wall at the floor line made the fastening of the floor line connections a critical criteria in the structural design. The panels were hand assembled and their frames were locked together by overlapping, telescoping upper and lower plates. The roof and the floor systems are long, deep, horizontal beams designed to give lateral support to the foundation panels and secure them in place. Extra nailing or glue nailing of both the roof and floor sheathing is encouraged to insure adequate rigidity to support the foundation and its load. Secure fastening of the foundation panels to both the floor and roof structure is necessary to hold them in place and secure fastening of the floor and roof to the end wall panels is necessary to prevent differential movement.

The roof is a skewed truss with a short steep 55° south slope and a long shallow 3-in-1 north slope. A 6 ft vertical partition along the skewed ridge 2-1/2 ft from the front edge of the roof separates the attic into two sections. The south portion is a void-type solar collector and the north section is a well-insulated, conventional attic.   In the south or solar attic portion, a 1/2 in plywood deck was installed 18 in above the house ceiling to serve as the energy absorber plate and to create an air space between the attic and the house. This configuration forms a double attic cavity. This double attic configuration is a basic part of a solar attic system, and it has been found to be effective in reducing heat gain through the house ceiling during the summer months. This void-type solar collector utilizes windows in the south facing roof slope to pass sunlight directly into the attic space where it warms the attic. The warmed attic area acts as a thermal dam that often eliminates heat loss through the ceiling and thus provides a passive solar contribution to the house's energy needs. The solar attic is made active by circulating the warm attic air into the living area or through a rock bed energy storage within or under the house.

Air from the solar collector is also used to preheat water for the electric water heater. The hottest air from the attic collector is passed over and around three 42-gal galvanized storage tanks which are located in a well-insulated vertical column on the front of the house. A photograph of the solar water heating system is shown in Fig. 4.

Fig. 4 The water storage tanks

Since the solar system often supplies more heat than the house needs, a 480 ft^3 crushed rock, thermal storage was constructed in the crawlspace. Hot air from the collector is passed through the rock and there it gives up its energy; when the house needs energy, air from the house is passed through the rock bed to retrieve stored energy. The solar attic is the primary home heating system, but the sun does not shine all of the time and it is not generally practical to store solar energy for more than 1 or 2 days. Therefore, supplemental heating is needed to keep the house warm when there is no solar energy available from either the collector or the solar storage.

In this house, there were three supplemental heating systems. The heat pump was selected as the the basic supplemental heating system because of its acclaimed potential to supply heat economically to the house. The second supplement heating system was a bank of electrical resistance heating elements. This supplemental system could be used at will but basically, it provided energy when the heat pump was not adequate for a quick warm up or temperatures were so low that the heat pump did not operate efficiently. The third supplemental system was a wood burning stove that was utilized intermittently.

Good drainage is essential to earth-embanked housing. In this house, perimeter drains carry water away from the base of the house and a gently-sloped grade (for 12 ft on all sides) directs surface water away from the house. Gutters and downspouts collect rain water that falls on the roof and carry it away from the building.

INSTRUMENTATION

The house was instrumented for the purpose of determining energy consumption during a winter's heating season. Individual power meters were installed on the electric heating elements, the heat pump, the water heater, and the total house input. Outside temperature was determined with a conventional mercury thermometer and cloud cover was recorded by the owner. Weather bureau data provided the back up on both. Inside temperature along with relative humidity was recorded by a hygro-thermograph. The owner also recorded the amount of wood consumed by the wood burning stove.

TEST PROCEDURE

RHRU designed a 16-week testing schedule to study 4 distinct modes of operation, each of which was operated for 4 separate one-week periods. The test period started on November 5, 1979 and lasted until February 24, 1980.

Mode I: The house was heated by electric heating elements only and all other systems were not used. Since all the energy input to the electric heating elements plus nearly all of the electric energy used by lights and household equipment was converted to usable heat, there was no other heat source in the house. The total electric use was considered the total energy required and was used as a basis for calibrating the house energy requirement. The purpose of Mode I's operation was to calibrate the house heat loss as a function of outside temperature in degree days and to provide a comparsion for more economical heating equipment. The resulting heat loss rate was used as a basis for evaluating the contribution of (1) the solar system, (2) the heat pump, and (3) the wood stove.

Mode II: The house was heated by the solar system and the electric heating elements were used as the supplemental energy source. This mode was designed to demonstrate the ability of the solar system to heat the house as a function of available solar energy and as a function of the outside temperature (degree days).

Mode III: The house was heated by the solar system, and the heat pump was the primary supplemental system. However, the elecrtric heating elements remained functional, operating only during extreme periods when extra heat was required to maintain house temmperature. This test series provided the data needed to compare the cost of operating an electrical resistance supplemental system to the cost of operating a heat pump as a supplemental system for a solar system.

Mode IV: The house was heated by solar and the electric heating elements were the primary supplemental heating system. However, the wood stove was burned intermittently to reduce the amount of supplemental electric energy required. This series gives insight into the potential savings that can be accrued from intermittent wood stove operation as a supplement to the solar and electric resistance heating system. While the test was in progress, the house temperature was maintained as close to 70°F as possible. Also during this time, the owner kept records of weather conditions including the outside ambient temperature and cloud cover. Weather bureau data was used to supplement the owner's observations.

RESULTS

Table I includes the degree days for each test period, the total power used by the house, the amount of energy used per degree day, the percent of cloud cover for each week of each mode of operation, and the total for all operations in each mode.

Table II shows the energy used by each source of heat in the house and its percent of the total heat supplied during the period under each mode of operation.

Table III shows the percent of the heating energy that is supplied by each of the four heating systems during the total period under each mode of operation.

TABLE 1   Comparison of energy use of the house under various weather conditions and modes of operation

	Start Date Week of	Deg Day (F°)	Total House Elem (kWh)	Rate of Energy Use (kWh/dd)	Cloud Cover (%)
Mode I	11-5	188	580	3.09	85
	12-3	178	580	3.26	15
	12-31	241	910	3.82	85
	1-28	357	1120	3.14	55
	Total	964	3190	3.31	60
	Cloudy Days	217	800	3.69	100
	Sunny Days	747	2390	3.12	0
Mode II	11-12	147	380	2.59	45
	12-10	217	620	2.86	45
	1-7	231	710	3.07	50
	2-4	273	850	3.11	100
	Total	868	2560	2.95	60
Mode III	11-19	105	320	3.05	85
	12-17	147	400	2.72	42
	1-14	168	450	2.68	70
	2-11	266	710	2.67	50
	Total	686	1880	2.74	62
Mode IV	11-26	245	470	1.92	30
	12-24	189	350	1.85	100
	1-21	217	440	2.03	70
	2-18	161	270	1.68	60
	Total	812	1530	1.88	65

The electric heating elements deliver all of the energy they consum to the house.

The wood stove, on the other hand, looses large amounts of heat through the chimney and about 50% of the energy was assumed to go up the chimney. Manufacturer's literature indicates the COP of the heat pump was 1.65. This means that for each kWh of electricity consumed by the heat pump more than 1-1/2 times that amount of energy was delivered to the house in the form of heat.

TABLE 2   Summary of electric energy consumption of the house for selected modes of operation

	Heat Pump	Electrical		Total House
		Elems	Misc	
	kWh (%)	kWh (%)	kWh (%)	kWh (%)
Mode I	0 (0)	2349 (73)	851 (27)	3220 (100)
Mode II	0 (0)	1549 (62)	1011 (39)	2560 (100)
Mode III	684 (36)	316 (17)	880 (47)	1880 (100)
Mode IV	0 (0)	658 (43)	872 (57)	530 (100)

TABLE 3   Summary of individual heating system contribution to total house performance during selected modes of operation

	Solar Input	Heat Pump Input	Electrical Resistance Input			Wood Stove Input	Total Heat Input
			Elems	Misc	Total		
	%	%	%	%	%	%	%
Mode I	0	0	74	27	100	0	100
Mode II	22	0	48	30	78	0	100
Mode III	16	31	14	39	53	0	100
Mode IV	22	0	22	28	50	28	100

Mode I: Data collected during the 4 weeks when the electric heating elements were the only source of heat for the house is summarized in Table 1. The average energy requirement for this 4-week test period was 3.31 kilowatt-hour/degree day (kWh/dd) with weekly averages varying from 3.09 kWh to 3.82 kWh/dd.

The passive solar contribution or the total useful energy that the house receives from the sun without active solar collection was determined by comparing the energy requirements of the house on cloudy days to the house energy requirements on sunny days.

Observing only the cloudy days when direct solar radiation was quite low (Table 1), the average energy requirement was 3.69 kWh/dd and on the sunny days, the average energy requirement of the house was 3.12 kWh/dd. Therefore, on the average sunny day, the house required 0.57 kWh/dd less energy than it required on cloudy days. Thus, direct radiation falling on the house provides a passive solar contribution which reduced the energy requirement of the house on sunny days.

Mode II: Data collected during Mode II, when the solar system was actively used and the electric heating elements were providing the supplemental space heating energy, is shown in Tables 1, 2, and 3.

The energy requirement of the house was 2.95 kWh/dd or 0.36/ kWh/dd less than that required when electric energy only was used to heat the house. As will be seen later, house temperatures were usually above 70°F. This occurred during and for several hours following a solar collection period.

Mode III: Data collected during Mode II when the heat pump was the primary supplemental system is shown in Tables 1, 2, and 3.

Mode IV: Data collected during Mode IV when the wood stove was being used intermittently and the electric heating elements were being used to supplement the solar system is shown in Tables 1, 2, and 3.

When the wood stove was used intermittently to reduce the electric energy requirement of the house, the house used 1.88 kWh/dd electric energy to maintain the house temperature. This is 43% or 1.43 kWh/dd less than that required for the 4 weeks of Mode I period when the house was heated by the electric heating elements. The house temperature remained at a temperature higher than the 70° F setting during a major portion of the time when the wood stove was being used.

The house requires more energy per degree day on cloudy days and since its energy requirement increases as the number of degree days increase, both degree days and amount of sunshine during each mode were compared to determine if test conditions were similar during all of the test periods. Table 1 shows the degree days and the amount of cloud cover during each week of each test, and it shows the total and average during each mode of operation.

From Table 1, the total degree days for Modes I, II, and IV fell within a 20% range varying between 812 to 964. Thus, it can be assumed that tests for Modes I, II, and IV were conducted under similar thermal weather conditions but during Mode III, the degree days were 20% less, but during the other three test periods the test modes were quite uniform within the 60-65% range.

When the available solar energy is considered as indicated by percent cloud cover in Table 1, there is considerable variation from week-to-week. The average percent of cloudy weather during all of the test modes were quite uniform within the 60-65% range.

Figure 5 is a series of graphs of house temperatures for a representative week extracted from each of the four modes of operation. The first is a graph of the house temperature cycle when it was heated by the electric heating elements only. There is evidence that the temperature setting was changed several times since the electric heating elements periodically maintained different temperature levels.

During any period at a given temperature level, the electric heating elements cycled on to heat the house then off for a gradual cooling cycle about once or twice each hour. This system maintained house temperatures within a very close tolerance except at times when the elements were off for a period as evidenced on January 1 or the electric element thermostat setting was changed as is evidenced several times during the test weeks shown.

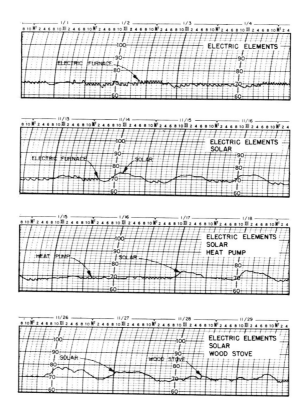

Fig. 5 House temperature patterns using selected
heating system combinations

In the second graph of Fig. 5, the primary heating system was the solar collector and the electric heating elements supplemented the solar system at night and on cloudy days. On November 13, the increase in house temperature is evidence that the solar system was operating.   The house temperatures increased progressively above the 70° F setting from about 10:00 a.m. until 4:00 p.m. then the house temperature decreased gradually for 3 to 4 hours.   The gradual decline in temperature is an indication that the heat was being returned from storage.   On November 14, the pattern is similar, however, the inside temperature increased more rapidly and peaked at a higher level. The temperature decline began about 2:00 p.m. and no supplemental energy was required in the house until about 2:00 a.m. on November 15.

The third graph in Fig. 5, shows the house temperature when the primary system was solar and the heat pump provided supplemental energy.   Under this mode of operation, temperatures were maintained within a close range on November 15 and 16 when skys were cloudy and the house was on the supplemental heat pump 100% of the time.   On November 17 and 18 when solar energy was available, house temperatures began a rapid increase going above the 70° F thermostatic setting beginning at about 10:00 a.m. and peaking about 3:00 p.m.   Then temperatures declined gradually for about 8 to 12 hours until house temperatures dropped to the thermostatic setting. Then the heat pump took over and maintained house temperatures until excess solar energy was again available similar to that shown in the graph.

In the fourth graph of Fig. 5 three of the heating systems were again operating. The solar system was the primary energy source.   The electric heating elements served as a supplemental system and intermittent wood stove operation allowed the family to reduce its electric energy cost.   Again, house temperatures increased above the thermostatic setting when solar energy was available but instead of a quick decline in temperature, after the sun dropped low in the sky, there was generally a second peak in the temperature or the temperature remained at a high level for several hours after the sun had set. This second peak or continued high temperature often lasted until after midnight. The electric heating elements were generally needed for only a few hours during the early morning. On sunny days, the solar system maintained uniform house temperatures through the day. In the evening (6:00 p.m.) the owner generally built a wood fire that kept house temperatures above thermostat setting throughout the evening. The continued high temperature or a second peak was prevalent on most days throughout the operation of Mode IV.

Table 3 shows the percent of the energy supplied by each of the several heat sources, including heat from the lights and other household electrical use.   In some instances the total energy caused the house temperature to exceed 70°F, thus indicating that solar heat went to storage. This condition was prevalent during periods when the solar system was collecting energy and/or when the wood stove was being operated. Such high house temperatures were due to the availability of uncontrolled excess energy.

DISCUSSION OF RESULTS

When the house energy requirements for sunny days were compared to house energy requirements for cloudy days, the rate at which energy was used by the house was found to be greater on cloudy days thus indicating that there had been a passive solar contribution on sunny days (Table 1). For such a research house in which it is planned to determine the contribution of an energy related innovation, it is important to calibrate the house´s energy requirement before the innovation and differentiate between cloudy and sunny day energy requirements as in the test house.

Mode I tests point the way for better analysis of energy related innovations and the need for control of the house solar energy supply during calibration.

During Mode II (solar with electric resistance backup) with 868 degree days and 60% cloud cover, the house energy requirement was 0.36 kWh/dd less than it was during a similar period with 964 degree days and a 60% cloud cover found during Mode I operation when there was no active solar collection. Since both cloud cover and degree days of heating were similar, it could be assumed that the total energy supplied by the solar system during this 868 degree day, 4-week period, is equal to the 0.36 kWh/dd difference times the number of degree days or 312 kWh which is equal to $12.49 if electric costs are assumed to be $0.04/kWh.

During Mode III (solar with heat pump backup) with 686 degree days and 62% cloud cover, the house used 0.57 kWh less energy per degree day than it did during the 964 degree day and 60% cloud cover period of Mode I operation. This is 0.21 kWh/dd greater savings than was attributed to the solar system in Mode II. Cloud cover for the two periods was similar, therefore, the amount of passive gain was similar during the two test periods. Outside temperatures were about 20% warmer than during Mode I operation. Therefore, it can be rationalized that of the 0.57 kWh/dd, 0.21 kWh/dd savings or 377 kWh savings or at $0.04/kWh, this amounts to $5.76 for the 4-week period that can be attributed to the more efficient operation of the heat pump compared to the electrical resistance heating. It also assumes that the solar system continued to supply 0.36 kWh/dd or 247 kWh or at $0.04/kWh, $9.88 of the house energy needs. One would expect the heat pump contribution to be more significant but several factors are involved. First, 47% (Table 2) of the house heating energy came from lights and other household energy uses. In addition, 18% of the heat pump energy use was consumed by the resistance heating elements. Thus, only 38% of the house energy was supplied by the heat pump compressor. Because so much of the house energy needs came from the double duty energy sources and from the electric resistance heating elements, the heat pump did not contribute appreciably to the house heat. If the house had not been earth embanked and it had not been so well insulated, then more energy would have been required and the contribution of the heat pump would have been greater.

Taking a closer look at the heat pump operation in Mode III, Table 1, 1116 kWh of the energy to heat the house was supplied by sources other than the heat pump. From Mode I calculations where the house required 3.31 kWh/degree day when the electric heating elements were used and the 1116 kWh was adequate to supply the heat for 337 of the 686 degree days. This leaves the heat for 349 degree days to be supplied by the heat pump compressor. Since the compressor used 764 kWh, the heat pump used 764/349 or 2.19 kWh/dd for its portion of the heating load. Comparing the rate of energy use with electrical resistance heating where the COP was 1 in Mode II to the rate of energy use if the heat pump portion of the load in Mode III, the COP 2.95/2.19 is a COP of 1.35. From the manufacturer's literature, the COP rating for this heat pump was found to be 1.65.

The wood stove was used intermittently. For long periods during the day, the only backup heating was electric resistance. Even with such long periods without wood backup, the average energy use dropped to 1.88 kWh/dd or a reduction of 1.43 kWh/dd due to the combined value of solar and the wood stove. For the 812 degree day period, this amounts to 1161 kWh and at $0.04/kWh, the contribution was $46.44. Again, assuming this solar contribution to be 0.36 x 812 = 292 kWh at $0.04/kWh or $11.68, this leaves the wood contribution at (1.43 - 0.36) x 812 = 869 kWh at $0.04/kWh or $34.75. Placing the value of wood at $0.66/100 kWh and 100% efficiency, $5.73 of wood was needed. In fact, about 4/5 of a chord of wood was burned during the 4-week period, therefore, at a cost of $40.00 for wood, there would be a negative saving of $5.25. It must be noted that the house was over heated most of the time when the wood stove was operating, therefore, a more conservative use of the wood stove would have resulted in a greater over all savings.

In order for a better analysis of energy related innovations to be performed, the need for control of the house solar energy supply during calibration is necessary. Further tests will be conducted using tents or other shading devices to control the sun's energy during calibration. In like manner, methods must be developed to calibrate the speed and direction of infiltration losses in test houses.

REFERENCES

1. Zornig, H. F., L. C. Godbey and T. E. Bond. A Low Cost Solar Attic Heating System for Houses. Housing Educators Journal 2:17-28. 1977.

2. Newman, Jerry O. and Luther C. Godbey. Standard Modular Foundation Panels for Houses of All Shapes. USDA Technical Bulletin 1541. September 1976.

CHAPTER III

URBAN PLANNING FOR UNDERGROUND SPACE USE

SESSION DEVELOPER: David Mosena

THE ROLE OF THE REAL ESTATE DEVELOPER
IN THE FUTURE OF THE UNDERGROUND INDUSTRY

Forrest R. Browne, President
Great Midwest Corporation
Kansas City, Missouri

I am a real estate developer. My theme is that significant participation of real estate developers will be required if the underground industry is to experience meaningful growth. I shall outline today the prerequisites for this growth of our underground space industry. But before doing so, I would like to define a "real estate developer" and to share with you the perspective of a real estate developer, with particular emphasis on the profit motive and the development process.

It seems to me that a real estate developer is an entrepreneur, a risk taker, who looks very objectively at the elements of every proposed real estate project from idea through completed development. He is motivated primarily by the profit potential of a real estate development. When the developer is convinced that a proposed real estate project is feasible, he becomes the quarterback in the complex development process which is necessary to bring the idea to fruition.

Our company is in both the mining business and the real estate development business. We have mined 20 million tons of limestone and have created 450 acres of underground area. We have designed, constructed, and leased about three million square feet of industrial buildings in this underground, here in Kansas City, Missouri. While I am fascinated and impressed by many of the technical presentations at this conference, I am often reminded that an idea may be no better than its potential for action and implementation. And, it is in this area of implementation that the real estate developer can make a vital contribution to the future of our industry.

Let me describe, from the perspective of a real estate developer, the five elements in the development process which are required in either underground real estate development or the more traditional, surface land, real estate development.

The first element in the development process is typically the location; that is, the land identification and control. The second element in the process is defining the product, the physical building which is to be constructed. Thus, this element in the development process includes the design, the engineering, and the construction phases. The third element in the process is identifying the user for the completed building. In a real estate developer's jargon, this is the marketing function; finding the tenant and negotiating a lease or perhaps finding the buyer and entering into a sales contract.

The fourth element in the development process is the financing of the project. Typically, the first three elements will have been well-defined, as a condition precedent to the financing. That is, the land will have been purchased, optioned or controlled, the building designed, and the user identified before the financing becomes feasible. Finally the fifth element in the process is the property management function. Property management commences after the improvements have been constructed and the tenant or the owner has taken occupancy of the building. Then the property must be maintained, heated, lighted, air conditioned, and the property taxes paid, along with the myriad of activities that accompany ownership or occupancy of commercial, industrial or residential real estate .

In summary, the development process is similar for either surface or underground real estate. It can be described as containing five elements, including the location, the product, the user, the financing, and the property management functions.

If we are to assess realistically the future of our underground industry, I believe that a logical starting point is defining "where we are now". I would like to discuss, from the perspective of a real estate developer, the current "state of the art".

We might look at the first element of the development process which we have been describing. In order to effect substantial growth in underground space utilization, it is imperative to assure the real estate development industry that land is available for underground development; that the geology is right, that the laws, the building codes, the obtaining of building permits and the neighborhood response are acceptable and conducive to successful, profitable development. If such land in metropolitan areas is not available and if such assurances cannot be made, the development will be stymied.

During the last few days in this Conference, there has been much focus on the design and engineering of underground space, the product. It seems to me that tremendous progress has been made in this area in the last few years. The technology for underground development is here and available. The experience of the Great Midwest Corporation in developing three million square feet of industrial space here in Kansas City contributed greatly to our Company's knowledge. We understand the many constraints of geology, engineering and construction in mined-out areas. Similar progress has been made in the design and construction of earth-sheltered buildings for both residential and commercial use. I am optimistic about the progress during the past five years in refining the "product" of underground space. I would hope that there would be creative ways to disseminate this knowledge to the general public and, more particularly, to the real estate development industry.

The third element in the development process is the user, which concerns the marketing of underground space. It seems to me that a masssive education process is going to be required as part of a successful marketing program. There is tremendous prejudice on the part of the potential users of underground space. My experience has been primarily in the industrial and commercial areas, but I think that there are similar prejudices in the residential areas. Our company has mounted in the last four years a large-scale, and I might say very expensive, public relations and marketing program. These programs attempt to get publicity for every user of our underground space. We have constructed and leased underground buildings to companies such as International Harvester, General Foods, Ford Motor, Great Western Sugar and Admiral Overseas Corporation. We have succeeded in getting positive press and electronic media coverage for these and other users of our underground space. We have been featured in the Wall Street Journal, in a front page article about two years ago. Our underground facilities were featured in April of this year on NBC News in a comprehensive report on civil defense. User acceptance of our product has commenced. However, we have a long way to go. If one believes, as I do, that the user is the ultimate judge of the feasibility of our underground product, we need to continue aggressive efforts in this direction of telling the underground story.

The fourth element in our development process is the financing, obtaining the funds with which to construct the underground space. Arranging long-term financing for real estate development in today's marketplace is most difficult. Those of you who have attempted to obtain long-term mortgages on homes know of the scarcity and high cost of such loans.

If traditional long-term real estate financing is difficult, underground real estate financing is very nearly impossible to obtain, except for the most select of projects. Our company has been successful in placing approximately twenty million dollars of long-term financing for our underground buildings in the last four years. We are fortunate to have established borrowing relationships with two major life insurance companies in the United States. Because of our track record in underground development, though the interest rates are high, long-term financing is available to us. My prognosis of the future is that, while such financing will be available, it will be a "tough sell". Financing of underground space will require a mighty creative approach on the part of either real estate developers or users of such space. I believe that in this area of financing, as in the area of marketing of the underground space, a massive public relations and educational program is going to be required. Lenders will have to be convinced of the viability of the underground concept.

The fifth and final element in the development process is the property management function. That is, maintaining the underground facility once it has been conceived, designed, engineered, financed, constructed, and occupied. This is an area where I am perhaps the most optimistic. In the industrial use of underground space, for warehousing, distribution, office and light manufacturing, we have experienced significant cost savings in both construction costs and property management. Our utility bills are a small fraction of what normal utility bills are in this part of the United States for heating and air conditioning. We project for our underground tenants a ninety percent savings in energy for heating and air conditioning. I am satisfied that the actual energy savings approaches ninety-five percent. Similar savings in property management occur because of less physical depreciation and deterioration of the properties. As an example, our six miles of paved, underground roads and our underground parking lots deteriorate at a much slower rate than above-ground roads or parking lots. Underground there is, of course, no freezing and no thawing. There is no rain, no snow, no winds and no significant changes of temperature.

My blueprint, or long-range prognosis, for the future of our industry envisions, first, significant industrial and commercial uses of underground space, followed by significant residential and institutional uses. Industrial uses have the potential to familiarize large numbers of people (including financial institutions) with the concept of underground utilization. Large scale underground residential development will be hampered primarily by building codes and availability of home mortgage financing, both constraints of which are local in nature. Thus, I conclude that underground industrial uses have the greatest potential in the near term.

Many of you have read, as I have, that if the United States is to accomplish its long term economic goals and objectives, there must be in the 1980's, a tremendous reindustrialization of our nation. There is some evidence that our Federal Government is committing itself to a policy which supports massive reindustralization of domestic U. S. industry. From the perspective of a real estate developer with particular experience in the industrial field, I feel that a dynamic future of our underground space industry could result from our going "piggy-back" on the twin national goals of reindustralization and energy conservation. Our underground space industry has an opportunity, and perhaps a responsibility, to assume a major leadership role in accomplishing our nation's twin goals of reindustrialization and energy conservation. To the extent that we expect to do so and to experience dramatic growth in the industrial utilization of underground space in the 1980's, it will be necessary to attract the attention of the very large manufacturing, processing and distribution companies as they make their investment decisions. It is only in this manner that we can hope for significant participation in our nation's progress towards reindustrialization.

Looking into my crystal ball for the 1980's, I see three important prerequisites for growth of our underground space industry. These prerequisites are creative financing, public sector support, and a commitment to a long-term payoff. I would like to discuss briefly each of these areas and to indicate why I feel that progress in these areas will likely be a prerequisite to a dramatic growth of our industry in the 1980's.

The first prerequisite for growth is the area of creative financing. I have previously referred to the availability of short-term and long-term funds as being one of the required elements of real estate development. Given the current and anticipated high cost and scarcity of long-term financing, the successful underground real estate developer will be required to consider creative alternatives. This will include establishing joint ventures, or partnerships, with long-term lenders. Certain life insurance companies and, increasingly, pension funds are beginning to emphasize this form of long-term financing, whereby a percentage of the ownership of the completed real estate project will belong to the lender.

Another area of creative financing which I envision for the underground developer is the pre-sale of entire real estate projects. That is, a sales agreement between the developer and the long-term lender is executed early in the development process, prior to construction. This sales agreement becomes a "takeout", in much the same manner as a traditional long-term mortgage commitment. It allows the developer to obtain the required interim financing from a commercial bank, for example.

Yet another area of creative financing, which may be required during the 1980's for underground development, is the condominimum concept. While there may be some legal obstacles to underground condominimums, which share access roads or parking lots or rail docks, there is no reason why individual underground buildings cannot have separate ownership. This would be similar to individual apartments in a high-rise condominium. While industrial real estate developers have generally avoided condominimum projects, it may be required for the successful underground developer to consider such an alternative in the future.

The second prerequisite for growth of underground space is support from the public sector. The conservative political mood in the United States is, of course, unsympathetic to expanding government financial support of private individuals or private companies. It seems to me that we must be very realistic about this political climate as we map the plans and strategies for underground development in the 1980's. There are, however, important areas of public sector support which may be attainable.

There can be policies of state, local, and federal governments which facilitate the development of underground space. There can be policies to minimize the regulations on such development; to modify existing building codes, to issue building permits on criteria very different from these which regulate traditional surface real estate development. I'm thinking of public sector support, say, to extend post office services to underground buildings.

You may be interested to know that the U. S. Post Office will not deliver mail to the fifty-seven companies which occupy buildings in our 450 acre underground industrial park. They will not deliver mail along our six miles of lighted, paved roadways although hundreds of automobiles, trucks and rail cars enter the underground complex each day. This regulation is an obvious example of government discrimination of our industry, and it must be changed if we're to grow in the future.

Public sector support needs to be improved in the area of fire protection. Creative fire codes for underground buildings need to be developed and enforced. Similar to the response to the construction of high-rise buildings, investment in specialized training and equipment for firemen is required. These are obvious areas of needed public sector support.

The third prerequisite for growth of our underground industry is an emphasis and focus on longer term returns on investment. That is, we must learn something from our colleagues in business and industry in Japan, who take an investment approach that is far longer term than is customary in the United States.

The underground space industry is in its infancy around the world. Those of us who are developers or financial investors in this industry will have to be patient. This need for patience has certainly been the experience of our Company. We have approximately twenty-five million dollars invested in our underground building complex, and only recently are we beginning to see a satisfactory return on investment, after having been in this business for over ten years.

From the perspective of a real estate developer in the underground space business, we have identified three prerequisites to growth of the underground space industry. That is, the need for creative financing, the need for public sector support, and the need to take the "longer view" of the potential financial pay-off of underground development.

I believe that the final need, the delayed return on investment, will be the most difficult to accomplish. It is the tradition in this country to look for immediate gratification, immediate satisfaction and immediate return on investment. This is a philosophical bent which our business lenders and investors have consistantly favored in recent years. I am hopeful that, as a nation, we can re-orient to a longer term perspective in our evaluation of business and investment decisions. If we begin to take the "longer view", I am optmistic that the underground concept will be a prime beneficiary. Then perhaps the future growth of our underground space industry will be directly related to our success in attracting entrepreneurs, specifically real estate developers, into the industry.

KANSAS CITY'S UNDERGROUND ASSETS
AS VIEWED FROM CITY HALL

J. Harold Hamil, Councilman-at-Large,
Chairman, Plans and Zoning Committee,
Kansas City, Missouri

I am honored to have this place on the program of what I look upon as a significant and purposeful conference. Let me say at the outset, though, that I make
no claims to being a technical expert in the field of your central interest.

I have been reminded many times since my first election to the City Council that
about the best I can call myself these days is a politician, and that many people
have a low opinion of anyone thus identified. I would remind you, though, that
my first election to public office was just a few months before I became eligible
for Medicare. Under such conditions, some of my friends insist, I am a politician without a future. Such a politician, they say, is less a menace to society
than one who can look forward to 30 or 40 active years.

This is one of those occasions when I try to erase the politician label--as much
as one can--and emphasize my earlier experience in the field of journalism. A
journalist, as an old friend once put it, is a stand-in for the person who cannot
see or hear for himself. The journalist quotes freely from other sources--some
with attribution, some without, and some, as the cynics say, by fabrication.

The journalist strives for, but may not always achieve, objectivity. Rarely does
he achieve it to the satisfaction of us politicians.

My report to you today will be more that of an onlooker than a participant. Most
certainly I must avoid technical detail. I hope, though, that I can properly emphasize the importance of Kansas City's underground assets and point out some
of the activities, concerns and hopes of those of us who look on and--within the
range of our responsibilities--participate in the development and utilization of
underground space.

Kansas City, as have most major cities, evolved through a succession of discov-
eries and adaptations of natural resources. Early explorers, hunters, trappers
and traders marked the site as the place where one had to choose between turning
sharply northward and staying with the Missouri River or heading up the Kansas
(or Kaw) and staying on a westerly course. In due time many travelers made

401

this the point of departure from the Missouri for overland trips to Oregon, Santa Fe and other points.

Railroad builders, following the water courses, brought their lines from every direction for a meeting near the confluence of the Missouri and the Kaw. Highways followed. In due time the search for building materials brough development of limestone quarries, and these, it turned out, were most easily developed by horizontal mining along the valley floors, in easy reach of railroads and highways.

Finally came the realization that mined-out limestone caverns were easily adaptable to a wide variety of uses. And so we come to consideration of an asset important enough to have been a concern of policy makers and management staff at the City Hall of Kansas City, Missouri, for more than a decade. I speak for Mayor Richard Berkley, City Manager Robert Kipp and my Council colleagues when I say we are pleased to share our experiences and our opinions with all who represent other parts of the country. We are glad you are here.

To a greater extent than the public generally recognizes, I would say, any conscientious holder of office in municipal government is sensitive to the needs and problems of economic development. Mayors and councils of Kansas City for many years have put a lot of effort into the encouragement of business and industry. They have teamed with the Chamber of Commerce and other private groups to promote the city's assets and advantages among outside firms looking for new locations or firms already in our community with need to find additional space through relocation or expansion.

In the last three or four years our concern has been heightened by the constant reminders of the extent business and industry are shifting from cities of the northern and eastern sections of our country to the so-called southern Sun Belt.

There are several ways to look at Kansas City's place in this southward migration. I am reminded of the recent advertising campaign of one of our banks. It points out that there are five banks larger than it is and 107 smaller. It proudly spots itself as the smallest of the big banks and the largest of the small ones --"just the right size."

We haven't formalized Kansas City's claim for recognition in quite that manner, but we do hear a lot of talk about our being on the northern fringe of the Sun Belt and the southern fringe of the snow belt.

If I were organizing a campaign to sell Kansas City as an industrial and commercial center, I might consider paraphrasing the bank's claim and referring to our city as, "Just the Right Place."

I see all sorts of possibilities of doing that while at the same time chipping away at some of the arguments for locations in the Sun Belt or in more northerly climes than ours.

I would, of course, put proper emphasis on Kansas City's central location, its central relationship to a network of highways and railroads. I would call attention to our airport and to the availability of virgin land for all sorts of industrial development.

But there would be, in addition to these more or less trite and routine claims, a special pitch in behalf of the vast potential for development of our underground spaces, the peculiar nature of these spaces, their potential for adaptation to a great variety of uses and, especially, the economies of space development underground.

Not the least of our underground assets is a considerable body of knowledge that is available to anyone interested in exploring our caves--if I may engage in a little bad punning. High on Kansas City's list of underground assets are the results of more than a dozen years of studies and surveys and the practical experience of businesses that have put underground space to a variety of uses.

In preparation for this talk I went to the City Development Department and armed myself with copies of reports and studies on file there. Thumbing through this material, I couldn't help concluding that in any list of Kansas City's underground assets there must be room for the name of Dr. Truman Stauffer, Sr., of the geoscience department of the University of Missouri, Kansas City. Back in 1971 he produced what he called a "Guidebook to the Occupance and Use of Underground Space in the Greater Kansas City Area." Five years later he came out with an update under the title, "Underground Space: Inventory and Prospect in Greater Kansas City."

These documents and writings in which Dr. Stauffer shared authorship with others are must reading for anyone wanting to draw on Kansas City's experiences in the development of underground spaces. The City of Kansas City is deeply indebted to Dr. Stauffer and to the University which has permitted him to make extensive studies of ways to utilize our mined-out limestone caves.

Along with the University and Dr. Stauffer, the City Development Department of the City of Kansas City and the Underground Developers Association have contributed to better understanding of what our underground assets are and what they can mean to the community as their development proceeds.

I might mention that Joseph Vitt, former director of the Development Department, Ben Kjelshus, project manager of the underground space program of that department, and Don Woodard, a former department staff member, now vice-president of Great Midwest Corporation and general chairman of this conference, have all contributed to our growing collection of underground literature.

During the last ten years there has been a tremendous surge of interest in foreign-trade zones. Marshall V. Miller, a Kansas City attorney, has been a leader in this movement as president of the National Association of Foreign Trade Zones. Miller and others have given Kansas City standing near the top among metropolitan areas with designated foreign-trade zones. Two of the zones here are in underground space--one in Kansas City, Kansas, and one in Kansas City, Missouri. Storage of merchandise for uncertain or prolonged periods is a requisite of many who utilize foreign-trade-zone privileges. The constancy of conditions in underground warehouses has a special appeal to traders in many lines of goods of foreign origin.

As chairman of the Plans and Zoning Committee of the City Council I get involved quite early in the location of new industries. To those wanting open spaces we usually have something to offer. To those wanting to be near the airport we are especially blessed with tracts already zoned for industry. We have above-ground industrial parks with utilities in place.

Within the past year and a half we worked out arrangements with a national company to locate a distribution center in one of our new industrial parks. This firm, wanting to build more than a million square feet of warehouse space under one roof, required a number of things that could not have been provided in underground areas.

I couldn't help speculating, though, on what that company might have done if it could have utilized an underground location. It could have reduced air-conditioning and heating costs to a minimum. It could have omitted the cost of those many acres of roof. It could have reduced by millions of dollars the cost of construction and--over a few years--the cost of heating and cooling. It could have realized other economies, but circumstances were not right for it to do so.

There are all sorts of commercial activities that can utilize underground space. And Kansas City can offer more underground sites than can any urban center in the United States. We in city government feel a powerful obligation to make it as easy as possible for business and industry to take advantage of the underground potential.

Within the big question of what government can and should do in this regard are many smaller questions. How do we adapt our zoning regulations to the special needs and problems of underground development? Do we take for granted that space on the surface is zoned the same as convertible space beneath it?

Do we zone all underground space for industrial use, or do we restrict certain areas for more limited use--for offices, as an example, or some limited commercial use?

What about underground residential development?

Are we making proper provisions for public safety?

Some of these questions have received serious consideration. Some have not surfaced to the extent they demand definite answers. They are coming, however, and we are preparing to deal with them.

From the standpoint of the total community there is still the problem of selling underground development to the extent many of us feel it is saleable.

There is no question but that there are certain image problems that have to be dealt with in the promotion of underground space. Deep in the human psyche is a feeling that there is something wrong with moving into a hole in a hillside. Our ancestors of ancient times came out of such holes in what we look upon as sort of a landmark decision. It takes a little imagination to see the advantages of going back.

We talk a lot in city government circles about the need for city involvement in dolling up the approaches and entrances to underground developments. Most of them are quite drab and far short of reflecting the efficiency and sophistication one finds after getting inside.

Initial concepts of the uninformed are influenced by the words and expressions of everyday description and experience. One person says "cave." Another says "abandoned mines." The next recalls damp air, odors, dark and spooky passage-ways. These initial reactions aren't always easily overcome.

A few weeks ago I attended a social function in one of our underground develop-ments. There were about a thousand persons in attendance. It was a rainy night, but all our cars were parked in the underground space. Nobody got wet. Nobody heard the thunder or saw the lightning. Many were in one of our underground spaces for the first time. Comments were interesting. The note running through most of them was surprise. One businessman of my acquaintance was among the first-timers. "This," he said, "has to be the wave of the future."

He talked about the economies of heating and cooling. He was sold by this one experience. He had seen the light.

That brings me to a story:

    A church member objected to some church improvements. The minister said he wanted to start with the purchase of a new chandelier. When the pastor asked why the member objected, he replied: "Well, first no one can spell it; second no one can play it, if we do get it, and third, what we really need around here is more light."

We in Kansas City should be throwing a stronger light on our underground assets. This conference, in my opinion, helps us do that. Your being here helps us ap-preciate what we have, and we hope that our experience helps you realize what underground space can mean to your respective communities. Whether you are dealing with the wave of the future might be debatable. There is no question in my mind but that you are dealing with a wave of the future.

USE OF UNDERGROUND SPACE IN CHONGQING, CHINA

By Zhu Keshan and Xu Sishu
Chongqing Institute of Architecture and Engineering
Chongqing, China

I. GENERAL SITUATIONS

The "mountainous city" of Chongqing is situated at the confluence of the Jialing river and the Chang (Yangtzs) river. Some 2000 km upstream from Shanghai. The steep peninsula rises 30-50 meters above the rivers and along them is dotted with 19 small narrow plateaus between gullies and ridges. (map 1)

In addition to difficult topography, the subtropical climate adds further strain -- not only a prolonged hot summer with 38 days in the average year over 35°C during the day and over 25°C at night, there are also 6 8.8 foggy days in winter. The city has been congested with the ever increasing population in recent years, so that the building density in the metropolitan area exceeds 85%, while the streets occupy only 8.7% of the residential area. New housing projects are going on with an average annual rate over one million sq. m. More public facilities and recrea- tion spaces are needed. Surface preservation is a major challenge. Different ways have been tried. Several high-rise housing projects have been built and use of underground space has been explored. The latter has appeared especially promising.

Use of underground space in most of our large cities has gained popularity in recent years by combining civil defense shelters with various peacetime uses. Ad- vantages of underground thermal stability has soon been recognized. People welcome underground teahouses and restaurants,especially during the summer season. Under- ground food storage demonstrates both economical and technical superiority. Venti- lation utilizing natural draft from underground galleries provides an easy way to condition room temperatures during the hot summer days. A second advantage is pollution control. Garages, noxious chemicals, oils and other inflammables are much safer underground. But the main advantage is surface preservation particularly because of the limited space. Currently, more than 20% of the shelters have been put into various uses in Chongqing alone. A brief description of several examples will illustrate the main ideas:

II. USE OF SUBSURFACE FOR SURFACE PRESERVATION

Sichuan Ship Repair Yard could not make any expansion on the narrow beach in front and hills behind. Since 1965, they have tunnelled through the hills to get 7275 sq. m. for general and precision manufacturing plants, a 400-KW underground power station, an oxygen station, a 1200-ton watertank and a small auditorium. The river beach has been transformed by laying excavated rocks into a high bank with a flat surface over 30,000 sq. m. where shipyard, workshops, warehouses, and offices are erected. Both surface and subsurface communication routes come into being. The productive capacity has been increased ten-fold with a moderate investment. The underground plants have been in operation more than 10 years. By natural and mechanical ventilation the relative humidity is maintained at 65-75% with an almost constant temperature of 21oC. No corrosion or damping of electrical devices has occurred.

An Underground Rest Center for the Chongqing Bus Terminal is located in the downtown center. They couldn't find even a single sq. m. to provide a rest place for drivers, conductors or other workers in the hottest summer days. Since 1969, they have started to build an underground rest center by themselves. Right now a 1260 sq. m. space has been completed for rest rooms, teahouses, conference rooms, and another 2,000 sq. m. subspace is ready.

Chongqing Watch and Clock Industrial Co. is surrounded by the rich fields of the People's Commune. The company couldn't afford to compensate the high cost for expansion. Since 1967, they have been able to tunnel more than 8,000 sq. m. underground for warehouses, watertanks and living facilities (dining room, kitchen, restrooms, conference rooms). Other working facilities such as an oxygen station and a power-station will be housed in lined caverns. An underground cinema theater, a hospital and food storage cellars are in the planning stage.

III. LOW COST UNDERGROUND STORAGE UTILIZING THERMAL STABILITY AND TIGHTNESS OF THE EARTH

China has a long tradition of storing cereals, grains, vegetables, sweet potatoes and other foodstuffs underground to maintain the quality of the stored products for use during the severe winters in the northern provinces, where the natural 15-16°C of the underground space may prevail with a relative humidity controlled at 60-70%. Since the founding of the People's Republic of China in 1949, various kinds of underground storage have been built all over the country. Chongqing, as the communication hub of Sichuan province, the "land of abundance", is destined because of its geographical location to be the largest products collecting and distributing center in the Southwest of China. Its strategic location and geologic conditions have been of particular benefit to the experimental development of underground storage for warm climate regions with high humidities.

Chaoyang Underground Granary is an experimental facility utilizing an existing air-raid shelter. The natural air temperature in the tunnel is about 18°C with a relative humidity of 80-100%. For air-tightness and damp-proofing, two variants were selected, both lined with stone masonry. But one chamber is waterproofed outside of the lining by asphaltic membrane and sealed by a small door to form an air-tight chamber, while the other is sealed inside by an air-tight huge thin bag of plastic film. In this way, the chamber temperature could be maintained between 16°-20°C with a relative humidity around 66%.

A comparative study was made with the same lot of husked rice and identical packing and quantity stored both in the underground chambers and a nearby surface granary over a period of more than nine months. It was observed that though there is little difference between the two variants of underground granaries, the latter are decidedly superior to the surface ones. The main findings may be summarized as follows:

Thermal stability -- when the open air temperature was 30°C, temperature in the surface granaries followed up immediately to 29°C while temperature in the underground chamber remained practically stable around 20°C. The moisture content in rice -- surface minimum, 12.7%, maximum, 14%, rice grain colored slightly. Underground -- minimum, 12.6%, maximum, 13.9%, no change in color. Activity of hydrogen peroxide enzyme of the underground grain was 36% greater than that of the surface one. The reduced carbohydrate of the surface grain was 22% more than that of the underground grain. These results were equivalent to say that in the surface granaries microbes and insects were more active and the rice grains themselves could breathe more freely, both conduced to the deterioration of the quality of stored rice.

It was also observed that pests developed some 2 to 3 months earlier in the surface granaries and had to be controlled by pesticide twice, while in the underground granaries control needed only vacuumizing with an air pump during the same period in summer. It may be concluded that underground granaries are safe even in hot summer, with practically no bacteria growing, no selfheating, no coloring, slow quality deterioration, no need to use pesticide for pest control, thus avoiding pollution, and good preservation of nutrition value, taste and flavor of the grain. In addition, it is cheaper and easier to manage an underground granary. It was estimated the maintenance fee could be cut down by more than 25%.

The Dabengiao Underground Refrigerated Storage illustrates energy-saving features in prominence. It has been in operation since 1975, with a storage capacity of 500 tons, and a refrigerating capacity of 12 tons per day. The investment is about 26% less than a comparable surface plant. Furthermore, heat flow measurements have shown that since February, 1976 temperature in the storage tunnels has been continuously lowered, and the longer the operation time, the less the heat flow needed. Thus load on refrigeration and refrigerating time is reduced. The tunnel temperature could still be maintained stable when 40-50% of the refrigerators were shut down and the operating time could be cut down by 55% if no ice-making or freezing were being done, thus reducing consumption of electricity, oil, ammonia and the cost of production. Generally speaking, when all the refrigerators were put into full operation during the first 8 months, the consumption of electricity, oil and ammonia and the cost of production was relatively high. Since then, even under full load conditions, if all the machines were shut down for 24 hours, without ice-making or freezing, the tunnel temperature would rise less than $1^{o}C$. This illustrates well the energy-saving features as well as the reliability in case of an accident. To take advantage of the topographical conditions and respond to people's demands, there are already ten similar refrigerated underground storage facilities in the different districts of Chongqing with a total storage capacity over 10,000 tons.

The Xiejiawan Underground Vegetable Storage has a capacity of 200 tons to provide the market in off seasons. It is another instance of converting air raid shelters for peacetime use. The natural tunnel temperature is $17^{o}-21^{o}C$ with 90% relative humidity, favorable for cabbages especially. Twenty-one species of vegetables such as potatoes, onions, tomatoes, strip beans, cauliflower, rutabagas, etc., totaling 155 tons, were stored 1-3 months underground, with 90% in good condition.

The Jiulongpo Underground Fruit Storage is a similar example, but with mechanical ventilation. Comparative study has shown that for oranges, 75% were preserved over a period of 43 days for surface storage while 92% were preserved over a period of 56 days with underground storages. Even longer periods of preservation are under experimental investigation.

For economy and efficiency, a strict operation program has generally been adopted: keep the tunnels tight during humid seasons and keep the tunnel doors open during dry seasons. Mechanical dryers are operating only in case of necessity. On receipt or delivery of large quantities of goods during wet seasons, it is advisable to schedule placement in the midnight to dawn period. Use isolation of storage areas, with portieres, air curtains or other automatic sealing doors at entrances.

IV. UTILIZATION OF NATURAL VENTILATION OR UNDERGROUND GALLERIES FOR COOLING

Chongqing is also famous as one of the "hot furnaces" in China due to her long, weary hot summer from mid-May to September. Modern air-conditioning is not only expensive to initiate and operate but consumes a substantial quantity of energy. Our air-raid shelters provide a convenient means to cool cinemas and theaters, commercial buildings, factories and other public buildings by utilizing natural ventilation. The high efficiency and low cost of the method earn such popularity that even the Shancheng Wide-Screen Cinema has substituted the technique for its original air-conditioning units.

The May First Cinema has been the first public building to cool its auditorium, projection room, etc., since 1973. Formerly, the room temperature in summer often reached 36°C and the auditorium was full of unbearable sweating smell even with 20 ventilator fans of fairly large capacity. With cooling by underground ventilation the auditorium temperature is at least 4°C lower than the outdoor air temperature, so one feels fresh and cool. Measurements have shown that the higher the outdoor air temperature, the larger the temperature difference between inside and outside.

At the Chongqing Cigarette Factory tobacco for smoking has to be processed by steaming, boiling and roasting. The air temperature in tobacco and cigarette wrapping workshops would usually be some 5-6°C higher than the outdoor air temperature and might even reach 40-50°C in the hottest summer days, full of dust and steam, before cooling by underground ventilation started to be utilized two years ago. The arrangement is simply to set up 2-40 kw suction fans at the gallery entrance to draw up 540 cubic meters of cool air per hour into the workshops through a ventilation duct about 300m in length regulated by a series of vent-doors. Not only is the air temperature in the workshops 3-5°C lower than the open air temperature, the dust environment has also been improved by the humid cool air.

Drawing up underground cooling air and spraying a water jet provides an efficient and economic way to control humidity and temperature to a satisfactory degree in knitting and weaving mills in various textile factories and other hot-working mills.

The problem of heat transfer in tunnels has been investigated and it has been shown that only a short distance near the entrance is disturbed by natural ventilation according to a logarithmic relation. It has also been observed that the cooling efficiency is maintained after five years of operation. Hygienic standards can be met by adequate regulating devices such as filters and vents.

V. VARIOUS USES AND PROSPECTS

Underground space provides a new field for development in Chongqing. There have been built various underground workshops, warehouses, classroom laboratories, meeting rooms, teahouses, dining rooms and other utility rooms which enable better use of surface installations in addition to surface preservation.

Underground warehousing other than food storage has also been attracting attention. There are various kinds of underground storage facilities for general materials, oils, chemicals, dynamites, inflammable and toxic materials, etc. The isolation feature of the earth for safety and pollution control has been recognized so as to reduce the safe distance between storages as much as 50%.

The underground environment is also favorable for biological and agricultural studies. Experimental investigation has been conducted on planting mushrooms, fermenting yeast, bacteria and enzymes, raising earthworms, etc., underground. For the best advantage of the city and the people, a coordinated effort to provide the optimum arrangement of underground installations, and to improve surface installations, is urgently needed to adopt local topographical and geological conditions. Construction costs could be reduced as the excavated rocks might be used as linings or aggregates, and the excavated earth could be used to fill up gullies if well planned in advance. A subway system for passenger transit and freight transport seems particularly appealing though much has to be done in the way of economic evaluation and scientific research. An energy-efficient underground will surely answer the call "to provide further space for the people" and "to keep spring blossom underground." In line with China's economic readjustment, the development of underground space in Chongqing will open up a new vista.

THE PSYCHOLOGICAL AND PHYSIOLOGICAL ECOLOGY OF INDOOR
ENVIRONMENTS

George Rand, Ph.D.

U.C.L.A. School of Architecture and Urban Planning
Los Angeles, California 90024

ABSTRACT

Presentation of emerging field of study called "indoor air pollution" as it affects
underground building. Literature is reviewed and stories of the negative impacts
of modern buildings on occupant health are summarized.

KEYWORDS

Health; indoor air pollution; architecture; buildings; environmental psychology.

A WARNING FROM THE FIELD

"The place stinks, man, it stinks," goes a comment by an Oakland High student reported in the San Francisco Chronicle (March 14, 1981) about a newly opened $9.5
million building. Studies of the move to the modernly fitted high school building
found a significantly higher frequency of such symptoms as itching, dryness of nose
or throat, frequent sneezing or stuffy nose, eye irritation, sinus trouble, and
more than one headache a week in students and staff. No causes for these prevalent
symptoms has yet been found.

Social workers interviewed in a newly opened San Francisco office building for
social service employees complained of similar low-level symptoms. For four of
the workers effects of moving into the building were so extreme they had to obtain
medical releases from work until they could be transferred to older office buildings.
A survey was made of medical histories and symptoms in the new facility which
sampled 250 of the building's 800 employees. A majority complained of newly acquired symptoms they attributed to the move six months prior to the new, tightly
sealed, energy efficient high-rise. The problem was traced in part to cleaning
and maintenance fluids (formaldehyde, trichlorethylene) building up and recircula-
ting organic chemicals in the indoor atmosphere. Also, air-handling systems were
exchanging air at the prescribed rates but not reaching desk top levels. Air was
functionally "stalled" in many instances.

A number of cases have been reported of upper respiratory infections and eye irritation traceable to use of fertilizers and pesticides for treatment of office
plants. Landscapists use fertilizers and pesticides unconsciously. They assume

413

air handling systems will whisk away these chemical products when in fact they re-circulate them. In one instance an epidemic of facial redness and other symptoms was traced to these chemical contaminants. There are many instances traceable to the use of paints, laquers, cleaning solutions in the same unconscious manner.

A large number of reports exist concerning environmentally-induced symptoms due to incredible oversights such as: sulphuric acid given off and circulated in the air by audio-visual copying machines; cleaning fluids and solvents stored improperly near air-intake ducts; buildups of carbon monoxide due to intake of auto emissions from cars, trucks, or other equipment into work spaces. In one case on the UCLA campus, a building had to be closed temporarily after symptoms of drowsiness and nausea were reported by occupants. The cause was traced to air intakes at the street level sucking in exhaust of trucks delivering concrete to a neighboring building site.

In West Germany a brand new building has been shut down for a year or more to try to "bake" out chemical toxins that exist in wall panels, floor surfaces and soft vinyl-plastic fittings by turning the heat to its maximum level.

Hundreds of such reports have begun to form a crescendo of concern about what is now called "indoor air pollution." Air fresheners, solvents, adhesives in building products, cleaning fluids, fire-retardant chemicals to prevent aging of paints and finishes, all threaten to turn the insides of tightly sealed, energy-efficient buildings into virtual gas-chambers. It is important to address the design of these indoor spaces -- the composition and qualitative characteristics of their atmos-phere -- with some ecological insight. Like regulating the food we eat -- reducing intake of fats and sugars -- there is need for adoption of a form of advice about our "environmental dietary needs. What are the long-term health and productivity impacts of working and living in artificially-constituted environments? Are there cases of people -- like those addicted to sugar -- who are adapted to the negative effects of impoverished, even toxic, indoor environments?

By its nature, architecture is an imprecise activity; it remains largely a "craft" industry and requires wide physical tolerances. Systems-building efforts in the U.S. housing industry attempted to apply space-science, NASA-inspired methods to the building of school facilities and mass-housing. Constructors ended up making gerry-rigged modifications to factory-built steel trusses and other elements when they actually got them down to the building site. Buildings simply refuse to "clip together" as designed. Steel responds to air temperature and moisture and despite its elegant abstract proportions, beams are never totally straight. In parallel fashion, automobile production inspired building of factory-assembled mobile home units made of lightweight plastic panels. Later we discover that formaldehyde used as a bonding agent leaches out of these wall and floor panels at high temperatures and can produce suffocating and toxic atmospheres in the tightly sealed enclosure of a mobile home unit. Several cases of death due to formaldehyde poisoning have been reported in sensitive older residents.

Obviously any activity in life brings with it certain risks and rewards. The ques-tion is whether a different kind of ecologically sensitive architect could produce settings that are healthful as well as economically sound and functionally efficient. These warnings to the field to suggest caution about naively accepting the assump-tions we make about HVAC systems and building materials. For the most part, archi-tects design with what is available as the "shelf stock" of the building industry. It is especially important to be self-conscious about the health-qualities of in-door environments that are totally or largely dependent on artificial controls. Poorly functioning or unhealthfully operated mechanical systems, toxic fumes from furnishings and fittings and maintenance fluids can nullify the benefits of archi-tecturically competent buildings.

MEASURING HEALTH IMPACTS

Part of the problem in the field is that people do not respond uniformly to poor environmental conditions. Some users are highly allergic, others are not. Many, such as a secretary interviewed at the University of California, cover up allergies for fear of being labeled as malingerers. They feel there is nothing to be done and find themselves part of a growing group of people in our society whose immune systems have been injured or overtaxed, leading to extreme and often systemic allergic reactions.

In designing offices, schools and other buildings to uniform standards, we imagined some "average person" for whom it is an ideally-suited setting. A uniform environment creates functionally unequal places for people with different physiologies. Typically, these needs for individual adjustment to environment are treated as insignificant. Colds, arthritis, allergies, eye irritations, etc., have enjoyed widespread acceptance as low-level occupational diseases. As long as symptoms remain within limits treatable by self-prescription, they are discounted as expected occupational hazards. In no small measure, part of the reason for this discounting process is the fact that most likely victims in homes and offices are women, who are in any case suspected of hypochondriasis. Health records are not kept diligently on low-level environment-related illnesses. One reason for this is that the nature of illnesses is not sufficiently severe to allow toxicological studies. In short, no one keels over dead from these effects due to isolatable toxins. Air quality and quantity may be poor and yet not traceable to a single source. On the other hand, there is good reason to suspect psychogenic causes in some instances of contagion for which traces of chemical toxins cannot be found.

There is some danger that environmental concerns can become the subject of malingering of employees. Since we run to the coffee machine and imbibe symptom-masking drugs before thinking to blame it on the handsomely engineered environment, it becomes very difficult to separate fact from fiction. The whole syndrome, however, deserves a close examination. On the side of studies of the environment, the results are a good deal more clear. Recent evidence suggests that the concentrations of some pollutants in buildings can exceed those levels commonly occurring in the outdoor environment. Chemical and biological contaminants released into indoor environments are unavoidable products of human activity. Typical indoor contaminants include gaseous and particulate pollutants from indoor combustion (heating, cigarette smoking), toxic chemicals, odors from cleaning fluids, micro-organisms from humans, odor-masking chemicals which are themselves pollutants, and a wide assortment of chemicals released from indoor construction materials and furnishings (asbestos, formaldehyde, vinyl chloride).

These contaminants in excessive concentrations may impair the health, safety or comfort of occupants. The introduction of outdoor air by infiltration, by opening doors and windows or by ventilation with fan and duct systems of varying complexity was the traditional way in which the health of occupants was protected. New engineering controls of indoor air quality use controlled flows of air to reduce levels of air contaminants by diluting them in fresh outside air, or via recirculation systems that incorporate filtering and chemical cleaning devices.

Researcher Craig Hollowell and his colleagues at Lawrence Berkeley Laboratories have been examining the mix of indoor air pollutants in energy efficient buildings. Once indoor air is heated or cooled, exhausting it for ventilation purposes represents a major energy loss. Minimal ventilation is used as a means of conservation. As less fresh air is introduced into building, the quality of the indoor air decreases.

Recent research in residential buildings has shown that air exchange rates less than one air change per hour may allow the concentrations of certain contaminants (formaldehyde, nitrogen dioxide, carbon monoxide and radon) to reach levels at which there is a health risk to occupants. In institutional and commercial buildings, CO and particulates from tobacco smoking, emanation of formaldehyde and other organics from certain building materials could provide a substantial risk at low ventilation rates. Since there are no metering devices in offices and schools, we must rely largely on hearsay accounts of health effects of poorly vented environments.

Because of the low level nature of these health effects, a very special set of studies is needed. Since most people are in office and commercial buildings some of the time, exposure history of people to toxins cannot be traced simply. This has stymied epidemiologists because they cannot trace illnesses to a specific substance or toxin which can then be eliminated or reduced.

Probably the only way to attack this problem is for companies with many employees or health insurers with large numbers of clients to provide health histories for analysis in relationship to large number of different work environments -- large and small offices, new and old, automated and conventional data processing. Likewise, there is a series of minor (and major) illnesses which could be correlated with office settings or at least investigated to determine the environmental component in their creation: low-grade pulmonary, digestive, opthalic, dermatological diseases.

IMPLICATIONS FOR A CONSCIOUS ARCHITECTURE

It may be a matter of time before we recognize the connection between modern buildings and the process of desensitization that has gripped our culture in this country. By the time we recultivate sensuous concerns -- joy in the sights, sounds and odors of nature -- we may find that a large portion of our culture has lost these sensibilities. Simple visual forms and sharp color contrasts have substituted for the organic complexity of natural settings. Designed settings depend to a great extent on control of perception, reduction of diversity in the work place to a set of rationally defined elements. The rest of the organism may suffer in these overly rationalized settings in which sexual appetites, olfactory pleasures and other biological needs are made to fit work schedule requirements and limited in expression to the electronic equivalent of a Skinner box.

Instead of controlled environmental conditions -- using room deodorants to mask odors and fluorescents to mask movements of the sun and weather -- there is a need to build settings which exploit these environmental effects positively. The enclosed office building was designed to increase productivity by eliminating concern with external environmental conditions providing uniformly "perfect" work settings 365 days a year. Simple irritants like dust, and the blowing of papers when the door was opened, were eliminated. The new interest in solar design, attention to building envelopes with possible storage uses for "coolth" or warmth, use of atriums, all suggest revival of concern with a more ecological approach to design that needs desperately to be assured a place in thinking about design under the ground. They make us conscious that the notion of environmental control has radically reduced environmental diversity. Air is demoisturized and tempered to create uniform conditions. Wind and weather are totally controlled, eliminating arousing eddy currents that produce sensations on the surface of the skin. This leads to a less dynamic sense of the environment. Sounds are acoustically "trimmed" so that high-frequency and low-frequency waves are eliminated by soft absorbent surfaces and lack of resonant materials. Light waves are abbreviated by flourescents, eliminating the red-violet end of the spectrum. In all these ways our environmental "diet" is reduced, abstracted, controlled and reconstituted just like processed foods. Quadrophonic stereos and other devices compete to provide complex sounds that compensate for concrete shells -- a benefit that was previously guaranteed by wood joists and hardwood floors.

TECHNOLOGICAL TRANCE

In underground building it is particularly important to be conscious about the
nature of "habitation" -- lived experience -- suggested by the built environment.
The potential benefits of earth cover can extend beyond energy savings to a radically new way of thinking about materials and buildings. Future technology is likely to make the most ubiquitous materials available as a source of energy -- hydrogen
to run machinery through hydrolysis of water, silica (sand) for high speed laser-
optic transmission of messages and communication. In a similar manner materials
for buildings can be based on materials that are ubiquitously available. This connection to the earth extends deep into the unconscious mind of architecture; the
field has always felt reverence and devotion for transforming readily available
materials into radically beautiful and profound structures. The great arches of
Gothic cathedrals, patterns that grace the walls of Islamic Mosques all were created with bricks and tiles arranged by subtle plays of the geometry of mortar layers, brick layers and painted surfaces to create astonishing diversity with a small
number of simple materials.

The modern movement in architecture introduced Cartesian rationality into architectural form. The "grid" allowed it to span great distances. Massing and layering of subtle building elements were subordinated to gratuitous spans using an
abstract Euclidean framework. The concept that a building was constructed from the
groupd up, brick by brick -- tectonic design of a building -- gave way to an abstract conception of atectonic design. Buildings were conceived by a mental act,
a series of lines, axes, coordinates in an abstract space. Ornament was virtually
banned from architecture as were any outward signs of tectonic design unless, as in
the case of Mies Van der Rohe, the welded joints of steel members revealed the
abstract inner order of the constructional system.

Underlying this conception is a key idea. From bricks and their arrangements architects (and Masons) learned to use simple geometries to create diverse forms. With
the discovery of principles of reinforced concrete came the idea of three dimensional space ("the box") raised on large columns to carry the weight of the floors,
to allow freedom in the puncturing of exterior walls to reveal the function of
internal spaces. The key idea -- the box -- led architects to believe in the notion of environmental plasticity -- an underlying fantasy of the universal super-
element that could be molded into any form and reshaped at will. The invention
of petroleum-based plastics added great gusto to this belief system.

This set of beliefs has produced a built environment paralleled by a modern belief system concerning the plasticity of human development and the arbitrary nature
of human activity. There has emerged a paradoxical loosening of the connection to
the natural sources of architecture in human labor and the earthly environment. The
culture has become gripped in a "technological trance"

An example will help here. Abstraction and rationalization of work began as far
back as the turn of the century when Frederick Taylor introduced the notion of
scientific management. With the shift to abstractly-conceived post-industrial work
(corporate, banking financial transactions as opposed to manufacturing) dominating
the current scene the elements of labor have become still more ethereal. Most work
does not result in anything actually happening in a wordly sense. So it seems
fitting that the setting of this work, increasingly performed on computer terminals,
be in abstractly-conceived office buildings. Recently, in part due to the energy
crisis, TVA, the State of California, the State of Illinois and others have begun
to experiment with radical new office building types that offer atriums, daylighting, trombe walls to store coolth, all on a very large building scale never before

tested with modern building materials.

From these experiments we expect to see a dawning consciousness, especially on the part of workers, that something has been missing in the earthly environment of this largely symbolic work. There is a missing sense of change in the quality of light from morning until evening, of seasonal changes in air temperature and quality and weather shifts. A responsive environment will have in it huge fans, shades and pumps which will create dynamic events -- things moving and changing -- in response to outside environmental conditions. Workers will be forced to respond to these events, or at least contend with them in a new way.

Imagine a large office building on its side (a large version of Frank Lloyd Wright's Marin County Civic Center) which is half buried in an earthen "parkland." On the dark side there are dimly-lit spaces for computer consoles. The other side is deeply penetrated by daylight and is structured for face-to-face interaction and small group meetings. The notion of the environment of work that grew out of the image of wind blowing papers across the office and distractable workers idly looking out the window at the landscape no longer addresses the needs of current workers. The benefits of a real landscape are only partly replaced by its simulacra, e.g., planted offices, offices fitted with fine art.

The underlying issue has to do with the "technological trance." Industrial-technological categories have shaped the way we think about environments. Despite the fact we spend eight to ten hours per day in work settings we demand little of them, that they accommodate our personal needs, allow us to "personalize" them or at the most radical extreme that we participate in the design and management of the environment.

Underground architecture should invoke the richness of a new landscape, new in-tuitions about the environment unparalleled in history, except perhaps by the 18th Century English landscapists. It should not fall prey to casual assumptions by state-of-the-art materials science and mechanical engineering, but demand an equal creative contribution from these parallel professions.

APPENDICES

I.   Symptoms of Indoor Air Pollution

- headaches
- upper respiratory infection--frequent colds and sore throat, chronic coughing
- skin redness, rashes
- eye irritation
- lethargy
- dizziness
- stress
- mental lapses

II.   Long Term Impacts (largely unknown due to difficulty in performing systematic epidemiological studies in the area since almost every person in the society is exposed to these effects to some degree).

- augmentation of chronic ailments such as asthma, allergies, heart and lung diseases in children and elderly
- increased risk of cancer, especially in combination with other chemicals, pollutants, cigarette smoke

III. Potential Sources and Settings (any enclosed building with modern air handling systems).

   A. Offices and Schools

- plasticizers in furniture used to keep them pliable
- fungicides, fire retardant chemicals and static-resistant treatments in carpets
- sealing and bonding compounds in office partitions and wall surfaces
- cleaning and maintenance fluids used for carpets, drapes, surfaces
- dry and wet process copying machines
- typing correction fluids, solvents in cement, sprays and other materials used in office work

   B. Homes

- improperly vented, incompletely combusted products of working and heating -- kitchen ranges, fireplaces, heaters
- chemical products in rugs, walls, paints, furniture, clothing
- solvents in hair sprays, deodorants, oven cleaners, pesticides, laundry aids, floor and furniture polishes and air fresheners

IV. Indoor Air-Pollution Hazards

- Formaldehyde--offgasing from urea-formaldehyde insulation used in walls, partitions, cupboards, furniture; permanent press clothing, rugs and drapes made from synthetic fibers; major bonding ingredient of particle board.
- Nitrogen dioxide--byproduct of combustion from gas ranges and heaters that are improperly vented.
- Carbon monoxide and hydrocarbons--(benzeprone). Source in tobacco smoke, wood and coal burning stoves, auto emissions drawn into buildings and recirculated.
- Radon--radioactive gaseous element from radon in soil and concrete. Found in concrete block, bricks and other building materials.

V. Other Hazards

- Airborne infection, e.g., Legionnaire's Disease, embedded in insulating materials in and around ductwork.
- Narrow (abbreviated) light spectrum from flourescent lamps.
- Poorly engineered video display tubes. Low intensity, long-term electromagnetic radiation due to microwave transmitters.

VI. Symptoms and Sources

- Colds, fatigue, bacterial infection, respiratory problems from poor air quality and ventilation, "scaling in" and recycling toxic fumes, vapors, dusts, bacteria. Temperature and humidity variation.
- Eye, nose, throat irritation, coughing, headaches, drowsiness from toxic substances (organic compounds) in the atmosphere; ozone from photocopiers; methanol from duplicators; solvents in correction vluids and erasing compounds.
- Eyestrain, short-term loss of visual clarity; headaches, neck and back pain; dizziness, nausea due to poorly designed equipment, work stations and video displays.

REFERENCES

Berk, J.V. and others (July 1979), The effects of energy efficient ventilation rates on indoor air quality in a California high school. Lawrence Berkeley Laboratory Report, LBL 9174.

Budiansky, S (September 1980). Indoor air pollution. ES&T Outlook, 14, 1023-1027.

Conoley, G (February 1980). Living may be hazardous to your health. American Way.

GPO (September 24, 1980). Indoor air pollution: an emerging health problem.

Hollowell, C.D., R.J. Budnitz, and G.W. Traynor (1977). Combustion generated indoor air pollution. Proceedings of Fourth International Clear Air Congress, Tokyo, 6 84.7, Japanese Union of Air Pollution Prevention Associations.

Houck, C. (December 4, 1979). The case against artificial light. New York Magazine.

Jonnes, J. (June 9, 1980). Living with microwaves. New York Magazine.

Krueger, A., and S. Sigel (July 1978). Ions in the air. Human Nature.

Lyman, F. (December 1980). Behind closed doors: energy conservation can be hazardous to your health. Environmental Action.

Rand, G. (October 1979). Caution: the office environment may be hazardous to your health. AIA Journal.

Repace, J., and A.H. Lowrey (May 1980). Indoor air pollution, tobacco smoke, and public health. Science.

Repace, J. and others (June 22, 1980). Total human exposure to air pollution. Presented, 73rd Annual Meeting of Air Pollution Control Association, Montreal, Canada.

Rothchild, J. (June 1978). They blight up your life. Mother Jones.

Silver, F. (Spring 1979). Buildings affect your health. The Human Ecologist 1.

Yaglou, C.D., and W.N. Witheridge (1936). Ventilation requirements. Transactions of the American Society of Heating and Ventilation Engineers. 42, 133-163.

PERSPECTIVES OF PLANNED TWO-TIERED USE OF SPACE IN KANSAS CITY, NORWAY AND SWEDEN

Truman Stauffer, Sr., Director, Center for Underground Space Studies
Department of Geosciences, University of Missouri-Kansas City

ABSTRACT

The subsurface is increasingly being used as space adjustment for otherwise congested surface functions. Planning the use of two tiers of space differs in the
Kansas City approach from the Scandinavian approach. These two perspectives are
treated in this paper with emphasis on a 3rd phase of planned development.

KEYWORDS

Planning subsurface, two-tier space use, underground use, vestiges of mining,
national policy, lithothermal energy, Phase 3 Subsurface Development.

Kansas City is the world capital of underground space use of mines converted to
commercial purposes. The subsurface dimension of our city is increasingly being
used as space adjustment for otherwise congested surface functions. This can
especially be valuable in urban areas where highly competitive bidders for use of
the land make vertical expansion a possible attempt to create space. High-rise
apartments and skyscrapers have made multiple use of the atmosphere, a common part
of our urban landscape. Now, inversely, the subsurface is providing space for
warehousing, factories, and offices, allowing retention of the surface function
for aesthetically valued open green space, urban uses, and as a continued tax
base. The conversion of mined areas into components of a viable urban plan has
helped ameliorate the longstanding problem of urban-quarry conflict. The results
of this venture into underground usage are economic savings, conservation of space,
and reduction in energy needs when compared to similar surface functions.

Kansas City is a laboratory and model in the sense of its natural equipment of
stone, physiography, geography, and creative intuitive men. The existence, character, dimensions, and accessibility of a massive, thick, level, marketable
limestone creates ideal conditions for horizontal mining and subsequent use of the
mined space. The dissected topography, exposing numerous bluffs, with competent
beddings of limestones as an overburden required room and pillar mining on a hori-

421

zontal plane to avoid the expense of their removal and, as a bonus, they provided a stable overburden to protect the rooms created below.

The various flat plained valleys of the Kansas City area have provided the incentive for the location of industry where transportation lines were easily constructed. The combination of industry in the valleys and mined out space in adjacent bluffs provided the laboratory and the crucible in which the occupancy and use of underground space or the secondary use of mined out space was produced. When industry needed space, it could most economically move into the vacant rooms of the mine.

The limestone industry in the entire midwestern United States, is now being perceived as a part of the entire spectrum of urban land use which does not stop with the initial phase of mining but continues to a secondary use of the space acquired by mining plus the surface space as well. Kansas City in this way served as a model in the development of two tier use of its urban space. This is in effect creating space in an ever increasingly congested world.

All mining in this area is now done with anticipation of space rental at some later time. For this reason, blasting methods are used which carefully preserve the integrity of the ceiling, and pillars are spaced to provide a grid pattern so that rail or truck lines may be easily placed along straight routes.

Where limestone mining was once the primary use and the space left by mining only a byproduct, new uses for this space have given a second and continuing role for mined areas. These secondary uses are greatly exemplified in the Kansas City area. Low cost of underground space has enabled the warehousing economy of Kansas City to expand into its role of national leadership. Major secondary uses are warehousing, factories, and offices in that order. One-seventh of Kansas City's warehousing is now underground. As a major railroad hub, Kansas City is an ideal place to store food and other items being shipped across the United States. Intervals of storage in the underground warehouses are now a part of the Kansas City economy and storage of food is now a capital resource. Eighty freight cars, each capable of holding 100,000 pounds of food, can be accommodated at one time on Inland Center's two underground rail spurs and many jobs are created by the receiving, handling, and redistribution of goods (Figs. 1,2).

The geographic location of Kansas City in the great "food basket" of the United States and midway between the western area of the United States, which produces some 50 percent of the nation's processed and frozen foods, and the eastern region, which buys two-thirds of the production, is ideally suited for a storage-in-transit point. The first underground freezer storage room was developed in 1953 and Inland Center leads as the world's largest refrigerated warehouse handling 8 million pounds daily. Over a pound of food for each person in the United States can be stored in this facility at a given time. The Department of Agriculture reports that Kansas City has 34,473,000 cubic feet capacity for frozen food storage about one-tenth of the total such capacity of the entire nation, most of which is underground (Fig. 3).

Amber Brunson was the first to quarry rock as a secondary process with the thought of obtaining the underground space for a factory being his primary objective. Occupation of the underground factory began in 1960 and his facilities have since been an object of interest, both nationally and internationally. Brunson led in pre-planned mining and pillar arrangement which he aligned to serve the purpose of secondary occupance. Tunneled into one of the numerous limestone bluffs that characterize the terrain around Kansas City, the Brunson Instrument Company manufactured surveying and optical instruments which were used on the moon. Personnel in the 140,000 square feet factory 77 feet below ground

PHOTO BY *Wilborn & Associates* KANSAS CITY, MO.

Fig. 1. Truck dock entrance to Inland Center showing rock bluff in which warehouse is developed.

PHOTO BY *Wilborn & Associates* KANSAS CITY, MO

Fig. 2. Rail service inside Inland Center.

PHOTO BY *Wilborn & Associates* KANSAS CITY, MO.

Fig. 3. Food storage in Inland Center.

has ranged as high as 435 among whom are technicians in optical tooling. Precision settings are made at any hour in this vibration free environment whereas
only the low traffic hours of 2:00 to 4:00 a.m. could be used in a former surface
location. The economy of construction, building his plant at a third of a comparable surface cost and creating a vibration free environment were his reasons
for locating underground (Fig. 4.).

Fig. 4. Inside the Brunson Instrument Company.

One operation, Downtown Industrial Park, has an extensive two-tier development with 44 acres of industrial park on the surface and a choice of either an elevator or a ramp entrance to an additional 28 acres of offices, industry and warehousing over a hundred feet below in a former limestone mine. Additional mall shopping space is created about 150 feet below at a cost slightly over half of a comparable surface structure. Heating, air conditioning, maintenance, and security can be provided at a savings which may reach 60-70% less than a similar surface location. At a time when national attention is focused on energy conservation and any energy savings of 6-7% gains public attention it would seem that underground utilization which approaches an energy savings of 60-70% merits being in the national spotlight. This is achieved through use of lithothermal energy. The amount of energy conserved varies with the type of function one compares. An underground factory or a general warehouse requiring only worker comfort would need less than 10% of comparable surface heating to raise the constant natural subsurface temperature of 45-54° to 60-65°. In fact, the author was comfortable in an unheated underground warehouse when the outside temperature was 2°F and at another time when the outside temperature was 107°F. However, one cannot assume this economy for a special function such as refrigerated and frozen food space. A general figure of 50% reduction in use of energy through use of the subsurface for refrigerated space is an ideal which is still subject to being partially offset by added energy needs such as added daytime lighting and electric forklifts and carts in order to control atmospheric pollution. However, capital outlay for extra equipment to handle temperature extremes is not required since the temperature range in an underground site is less than 6 degrees over the whole year (Fig. 5).

Fig. 5. Offices representative of where 88 tenants occupy the subsurface at Downtown Industrial Park.

Security is more easily attained in a subsurface site due to less entrances to guard or control. During the riots and burnings of the late sixties in Kansas City valuable goods, supplies and technical equipment were hastily stored in the vacant subsurface areas for protection. Additional civil defense uses of the underground are evident in the storage of a 400 bed hospital and survival supplies in mined out space.

The easily controlled humidity plus the ease of security has made the underground facilities ideal for film and records storage. Company standby records for use in case of flood, storm, or fire as well as bank and university microfilm are increasingly being stored in underground vaults.

Contrary to what may be expected, the underground rooms are easily kept dry with a minimum amount of dehumidification. Outside surface air does not rush into the subsurface so that air once dried is easily and economically kept at whatever amount of humidity one desires. Metal machinery and factory equipment does not rust and metal parts may be stored without damage. Allis Chalmers Farm Equipment Company makes use of a large underground storage area for a parts warehouse (Fig. 6.).

Fig. 6. Allis Chalmers' assembly line being installed in the underground at Inner Space Corporation.

Subterropolis, a development of Great Midwest Corporation, is most diversified in its use of mined space. It has 20 tenants in warehousing and storage, 8 tenants in manufacturing, and 7 tenants in records storage, offices, and maintenance. Out of a total of 20 million square feet of mined space (17 million with deduction for pillars) 5,700,000 square feet has been developed into leasable space and common areas which include roads, rails, and docks. Another 1.2 million could be made ready immediately. Two recently acquired tenants are General Foods Corporation leasing 239,000 square feet, and Admiral Overseas Corporation, a Foreign Trade Zone facility for assembly of TV sets, leasing 110,000 square feet (Fig. 7.).

Fig. 7. Offices of the Admiral Overseas Corporation.

Not all underground developments are large. Bannister & Holmes Business Park has 96,000 square feet in use with 8 tenants. Leavenworth Underground is just putting 400,000 square feet on the market.

The United States government owns and operates a 1.5 million square feet facility in Atchison for the storage of production equipment for the Defense Department. At another site in the same city, Atchison Underground Facility is developing 60,000 square feet for warehousing.

Midcontinent Underground Storage operates a 600,000 square feet storage area with 85% of this in freezer storage. Kroger, Fleming, Morton, Banquet and Marriott are among the tenants.

The examples cited and others in the Greater Kansas City area have added 20,000,000 square feet to the commercial development in the area by use of a second tier of space. The economic importance of such a feat is registered in an increased tax base, added employees, capital gains, low cost lease space for gaining and retaining companies within the city, and the retention of surface space for other uses.

Planning for the integration of this development into an overall comprehensive plan for optimal land use in the city has been mostly nil. One basic reason for the lack of such a plan is the nature of this development which begins with a quest for rock. Nature has placed rock in unyielding locations. These ventures all begin as quarries and as such had to situate themselves where the rock could be found. Planning had no voice in their location. Development of these sites has been after-the-fact. Planning has been relegated to the position of taking space already mined at fixed sites and converting it to a stable, usable, and marketable facility. In this, the individual planners for these individual operations have done a commendable job. Some of our underground developments have achieved the planners' dreams of efficiency, design, and aesthetics.

Attempts are now being made to develop building codes, and zoning law which will assure common commitments to safety, construction techniques, and orderly maintenance of health standards, fire protection, and legal descriptions of property. Mr. Ben Kjelshus, City Development, Kansas City, Missouri, has been most diligent in developing these standards.

Quite the opposite is true in the major underground developments in Sweden and Norway. Comprehensive planning preceded any action. The development of underground space was not predicated on there being an economically successful mining operation. When the need for any function arose planning could address the questions of where it should be located and whether it should be a surface or a subsurface development. This freedom for comprehensive planning of underground development has not existed in the Kansas City region where a commitment of location is already made to a mine site based totally on a successful mining operation. Underground development in Kansas City shows vestiges of mining in its entrances, use of bluffs and siting.

By preplanning there are several factors which may be included. The Swedish people as well as being conscious of their need for civil defense also have an admirable concern for maintaining beauty in the natural environment. The combination of these two motivations resulted in an underground garage where Stockholm's garbage trucks are brought, cleaned and stored until their next run. The underground space obtained is also calculated as to its civil defense potential for shelter and storage of supplies. The surface is a beautiful glacially rounded granite hill with shrubs and flowers slightly out of downtown Stockholm, into which the Sellberg garage for storage of up to 226 medium size lorries is carved. The trucks are steam washed on entry and stored. A repair shop for any needed service is also maintained.

The city of Stockholm and its suburbs form the most populous region of Sweden and handling its waste efficiently has understandably presented a tremendous challenge to planners.

The Kappala union, composed of several suburbs, have combined their needs for waste treatment into one overall plan which has resulted in pumping the sewage through tunnels from the suburbs to the underground facilities at Lidingo, another suburb of Stockholm, where after treatment it goes into an arm of the Baltic Sea. One would scarcely believe that beneath this beautiful surface shown in Fig. 8 and in this quite obdurate granitic rock briefly exposed in Fig. 9 there is a complete sewage treatment system as depicted in Fig. 10.

Fig. 8.   Surface area over underground sewage treatment plant.

The citizens of Lidingo would not have consented to being the recipient of the waste from the other areas had it not been for the capability of locating the treatment process underground.

The nature of the rock overhead can be seen in Fig. 11 which shows a walk through one of the many steps in the purification process and also gives some indication of the size of these chambers. Silt that is filtered out in the process is sold as commercial fertilizer to build the soil and aid in productivity.

The Kappala underground sewage works comprises 37.5 miles of tunnel built in granite rock and was conventionally built with drill and blast methods. It serves 10 suburban communities with an estimated capacity for 540,000 persons. The Kappala Union was granted permission to build tunnels under private property by the Court of Water Law which avoided the necessity of obtaining costly surface property rights. It was decided to place the plant in rock because of environmental requirements and land space shortage. The plant site can be expanded to serve up to a million people. The total cost for tunnels and sewers, pumping stations and treatment plant was 180,000,000 Swedish crowns.

Planners in Giorvik, Norway excavated a swimming pool and recreation center conveniently located in short walking distance from the people and into a hill of syenite rock. The entrance is very well designed and gives no appearance of being a rock quarry. Inside the proper temperatures are easily maintained for year round use.

Fig. 9. Rock exposure revealing type of rock in which the
treatment plant is placed.

These Scandinavian examples show what can be done where planning is exercised
prior to site selection and prior to rock removal. Control of all facets of
construction are guided to a desired end use.

There are two categories of planning subsurface development which are easily defined. One, is the planning by the individual developer which has been done well
in the Kansas City region, the other is the comprehensive plan which includes
regulating agencies and in this the United States must concede to others. Urban
areas are lacking the precedent and example of national commitment to planning
the use of the subsurface. There needs to be strong effort to develop common
factors in subsurface into national policy which includes geographic, demographic,
economic, and geologic input. With the proven capacity of underground space to be
converted into a major resource and the established potential for two tier development, the need for planning is evident.

There are 3 phases to underground development as exemplified in the Kansas City
area. These are:

Phase 1.   An era of mining with no thought or planning for the second use of the
           mined space left after the rock was removed. The result of this era
           is millions of square feet of waste land with great amounts of sub-
           sidence. Most of this can never be used without leveling.

Phase 2. This is our current era when we are attempting to harmonize mining and
the secondary use of the mined space. The result is commendable but
many vestiges of mining mar the aesthetics and control the limits
within which planning secondary development can occur.

Phase 3. This is an era yet to come wherein the use of the underground space
stands on its own merit. Mining, where used, is servant to and not
master of the subsurface site. Energy saved by use of the subsurface
may well be the threshold over which this phase will enter. Planning
will be sovereign in this phase and many aesthetic, functional, and
structural deficiencies now existing will be amended. We will no
longer weaken bluffs by perforating them with multiple openings. We
will tunnel back far enough before opening rooms to allow for proper
weight distribution of the overburden. We will account for shale expansion before investment of large sums of capital to be followed by
huge expenditures to correct floor heave. In general, we will incorporate the knowledge learned through Phases 1 and 2 with unrestricted
use of new studies.

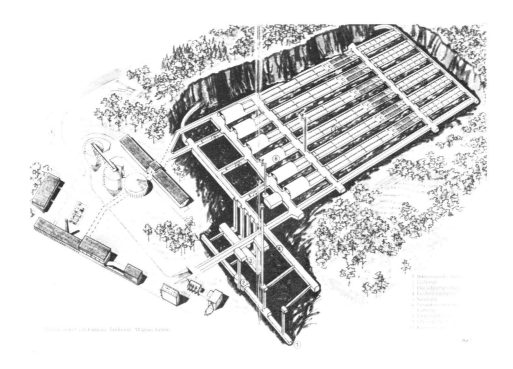

Fig. 10. Cut-a-way diagram of Kappala underground sewage treatment plant.

432

Fig. 11. Observation walk through a processing tunnel.

REFERENCES

Energy Cost Savings in Kansas City's Underground Space Development, Technical
    Bulletin 4, City Development Department, Kansas City, Missouri, 1979.
Stauffer, Truman, Sr., "Energy Use Effectiveness and Costs Between Surface and
    Subsurface Facilities," Subsurface Space, Vol. 2, Ed. Bergman, Proceedings of
    Rockstore 80, June 23-27, 1980.
Proceedings of the Symposium on the Development and Utilization of Underground
    Space, Eds. Truman Stauffer, Jerry Vineyard, Department of Geosciences,
    University of Missouri, March 5-7, 1975.
Underground Utilization: A Reference Manual of Selected Works, 8 volumes, ed.
    Truman Stauffer, Sr., Published by Department of Geosciences, University of
    Missouri-Kansas City, 1978. (Available through Center for Underground Space
    Studies, Room 201, G-P Building, UMKC, 64110.)

PUBLIC-PRIVATE COOPERATION IN DEVELOPING AN
UNDERGROUND PEDESTRIAN SYSTEM

Donald Reis

Program Manager, Central Business District Division
Department of Urban Planning, City Hall, 1500 Marilla
Dallas, Texas, 75201

ABSTRACT
Issues related to the development of a Central Business District underground pedestrian system from the public sector point-of-view are highlighted, with focus on six issues: Need, Design, Community Commitment, Financing, Operations, and Planning Implications.

KEYWORDS

Pedestrian system; design standards; central business district; underground walkway; public sector role; public finance; retailing; streetscape; capital investments programming; building security; public relations.

INTRODUCTION

Dallas, Texas enjoys a robust economy. In recent years, the City has been among the top three in the country in terms of housing starts, office development, and corporate headquarter locations. Presently, only New York and Chicago rank ahead of Dallas in the latter category.

While the results of the economic boom and current low unemployment (currently 5%) can be seen throughout the city, perhaps the most dramatic evidence is the changing skyline of the Central Business District, the city's largest employment area. During the last decade, more than nine million square feet of office space was built downtown, half of it between 1979 - 1981; another ten million square feet will be completed before 1985. A conservative estimate of CBD employment suggests an increase from the present level of 130,000 to nearly 180,000 by 1990 and 200,000 by the year 2000.

The rapid expansion of the CBD has created significant economic benefit for the City but at the same time has also generated congestion problems on streets and sidewalks. Sidewalks are narrow; many are only one-half the width required to accommodate present peak volumes. The downtown also lacks alleys which forces trucks to use traffic lanes for parking. Deliveries must be made

434

across sidewalks through main building entrances or with sidewalk elevators that virtually force pedestrians into the street. Pedestrians congest street intersections which slow traffic. With the exception of two or three routes, all transit lines either pass through or terminate in the downtown which further adds to the congestion.

Congestion is not the only downtown problem faced by pedestrians. High wind velocities can occur year round and moderate winds are an almost daily occurrence. In addition, summer temperatures soar above 100 degrees. Both conditions create an uncomfortable pedestrian environment.

The community's response to the problems of climate and congestion has been the development of a climate-controlled pedestrian system, both above and below grade, which presently provides nineteen pedestrianway connections. The system links 14 million square feet of office space, 750,000 square feet of retail space, 1,300 hotel rooms, and 10,000 parking spaces. By 1985, new development linked to the system will nearly double present space totals.

How did the system come about? People in Dallas pride themselves on community spirit, a "can do" attitude based solidly in the free-enterprise system. The Wall Street Journal, among others, labeled Dallas "a developer's town," a city of tremendous opportunity and few constraints. While the label is basically correct, it should be clearly noted that the heart of Dallas' lifestyle is a long-standing, successful, and well utilized system of public and private cooperation which has resulted in projects such as the pedestrian system. Other successful projects include ThanksGiving Square, a major downtown park and chapel dedicated to international thanksgiving, which is tied to the underground pedestrian system and built over a city-owned underground truck terminal, and the Reunion complex, a $200 million mixed-use project that to date has resulted in a 1,000 room hotel, a 50 story observation tower topped with a revolving restaurant, and a city-owned and operated 20,000 seat arena. Other joint public-private ventures are currently under way.

PEDESTRIAN FACILITIES PLANNING

The need for a pedestrian system to reduce street congestion and pedestrian/vehicle conflicts has been formally recognized for at least twenty years. The 1961 Dallas Central District Master Plan noted that "pedestrian movement in the CBD is the most neglected form of transportation." The report clearly identified the conflict inherent in multiple use of the street right-of-way by automobiles, buses, trucks, and pedestrians and recommended that "pedestrianways be part of a defined circulation system."

Four years later, DeLeuw, Cather and Company produced a transportation plan for the CBD. Among the proposed capital improvements was a series of pedestrianways that would extend throughout downtown, both above and below grade. Some general design standards were also included.

In September 1969 a third CBD plan prepared by Ponte-Travers Associates was presented to the Dallas City Council. The report proposed the functional reorganization of the Central District into what was termed a "Multi-level City Concept", a vertical separation of pedestrians, vehicular circulation, goods delivery, and other transportation functions. The consultants recommended six integrated facility systems to serve the Central Business District: a pedestrian network of shopping malls and plazas, a green space network of parks linked by landscaped boulevards, street and traffic adjustments, a parking plan, a mass transit plan, and a freight delivery plan for underground trucking compartments.

The Report's pedestrian facilities system component proposed the development of a sheltered network of shopping malls, plazas, courts, and passageways throughout a 150 acre core area and the creation of a pedestrian environment linking all major buildings totally separated from street level traffic and outside weather. The grade separated system of pedestrianways would reduce vehicle/pedestrian conflicts, relieve traffic congestion, and improve pedestrian circulation. Projected large scale development/redevelopment would be the nuclei from which the pedestrian system would spread as new structures were completed. The system's cost would be borne primarily by private enterprise with only 10% of the projected system involving public participation.

The City Council accepted the Report and referred the proposals to the City Plan Commission for study and detailed recommendations concerning each of the system components. During the period between 1971 and 1975, four of the component plans were produced and adopted as guides to development.

Among the systems plans adopted by the City Council was the Pedestrian Facilities Plan. This city-staff-prepared document was adopted by City Council Ordinance in April 1975 after extensive review and analysis of the Ponte-Travers proposals, informational briefings, and public hearings. The review process utilized the assistance of a special Central District Committee composed of representatives of downtown business interest groups, City Plan Commission members and key city staff.

The plan endorsed the concept of a "Multi-level System" for pedestrian circulation throughout the high density center of the CBD and recommended surface level pedestrian facility improvements in the form of building setbacks and sidewalk widenings. It also recommended pedestrianways connecting major employment centers, retail, hotel, entertainment and convention facilities with peripheral garages and core area transit. The plan also advocated the designation of a pedestrian precinct, an area within the center of the CBD where densities and pedestrian activities were the highest and improved facilities and amenities for pedestrians could be emphasized. The pedestrian system implementation recommendations also encouraged developers to make necessary provision for designated facilities with the understanding that the city would share the cost of facilities over and under public rights-of-way.

Late in October 1979 consultant Vincent Ponte prepared an update to the Pedestrian Systems Plan. The update listed project improvements made between 1975 and 1979, evaluated new opportunities, called for the incorporation of the pedestrian facilities plan into the City's Zoning Ordinance, and encouraged the formulation and adoption of design standards. Ponte's recommendations have been reviewed by city staff as well as the local business community and developed as specific proposals to be submitted to the City Plan Commission and City Council.

UNDERGROUND SYSTEM

Planning for the pedestrianway system did not distinguish between above and below grade connections by giving one priority over the other; both are needed to create a true workable system. Nevertheless, the earliest sections of the system were underground, and today, most of the system is below grade.

The first significant segment of the underground system was developed during the 1965-1975 decade as a shopping promenade. From that anchor, the system has expanded in 4700 feet of walkways connecting eight office buildings, six parking garages, two banks, and a number of shops and restaurants. The system makes six connections beneath city streets, the last three involving city participation in cost sharing.

A second section began developing in 1977 with the completion of ThanksGiving Square. The project, outlined earlier, not only offered an underground truck terminal and central CBD open space but a hub for below grade pedestrian facilities connections as well. Two of the connections are in place, and three other major links will be in use in the next three years as new high rise office structures are completed.

Presently, there is a major gap between the two major underground sections due to cost and engineering problems that have been encountered in making the connection. The importance of the connection is well recognized, and a feasibility study for developing the link will soon be completed. Tying the two individual sections together is a city priority.

LESSONS LEARNED - A TWENTY YEAR EXPERIENCE

Looking back over the history of the Dallas CBD pedestrian system, it's interesting to note that following formal recognition of the congestion problem in 1961, four years elapsed before any formal systems planning took place and construction began. Another four years passed before pedestrian facility planning was fully integrated into the CBD planning process as part of the circulation system. From that point, a dozen years have been devoted to further system planning and construction. That experience has provided insight into the political, fiscal, design, operations, and planning factors that may be encountered in building an underground pedestrian system.

Whether an underground pedestrian system is entirely a public venture or a shared responsibility between the public and private sector, as is the case in Dallas, the network represents a major capital improvement which requires detailed planning and design and the participation of local government and the business community. Everyone has a vested interest in the system in terms of its design, construction, operations, maintenance and economic consequences. Making the system work takes formal public-private cooperation and a clear understanding of the issues and goals that form the planning foundation. Experience in Dallas demonstrates that six items must be addressed and fully explored: answering the question - is there a true need for the system?, designing the system, building commitment, financing methods to be used, operational responsibility, and implications of system development.

IS THERE A NEED?

Since a pedestrian system represents a substantial capital improvement with sizeable fiscal implications, the need for a system should be identified and documented. What are the problems to be addressed? Are there alternatives that could provide a solution at less cost and time? Is there agreement on what the basic issues are and how they should be addressed? Would the system add to the economic viability and marketability of the area it serves? Do costs exceed benefits?

Documented need provides a common reference point upon which to build a system concept and community consensus. It also provides the opportunity to examine motivation. Is the system the "best" solution or is it being pursued because it's the trend in CBD development and appears to work elsewhere? Too often, programs for improving the aesthetic and economic character of downtown are built around the "quick-fix", the "new" technique, or the emulation of suburban shopping centers.

In Dallas the sidewalks are narrow and congested. The climate can be uncomfortable, and pedestrians must compete with street loading and sidewalk elevators. There is clear need and no viable options. If the climate is mild however, subsurface conditions limiting, or if overall CBD design would be diminished by burying part of the circulation system, then an underground pedestrian system may well not fit local need. Other factors that may influence the decision whether to build an underground system include: costs, opportunity to serve neighboring development, ease and convenience of vertical movement between levels, location of transit connections, and utility placement.

DESIGNING A SYSTEM

Once the need for the system has been determined and documented, attention should turn to system design. Between 1961 and 1969 several different concepts were proposed for the Dallas CBD which hindered expansion of the system; the 1969 Ponte-Travers Plan fixed the concept of the multi-level system and pedestrian precinct. It is essential that a clear statement of design be prepared outlining such items as: the ultimate size of the system, locations of horizontal and vertical connection points that tie the system together, means for integrating the system into the physical fabric of downtown, major design features, timing and phasing of the project, and the relationship of the pedestrian system to other CBD planning. A design framework is essential. Attempts to create an underground pedestrian system on an incremental basis without an overall plan will almost surely fail.

In addition to plans outlining where the system will be built, it is equally important to outline design standards which will guide actual architecture, engineering, and construction. The mere fact of a space being below grade, frequently without visual contact with surface landmarks, may tend to make it oppressive to some people. It is important, therefore, to establish design standards for ceiling heights, corridor widths, use of color and lighting, floor covering, graphics, lighting, security, emergency access, maximum noise levels, directional signing, identification of street level access points, and barrier-free criteria. The goal is to create a pleasant, psychologically inviting environment.

One of the best means for creating a stimulating environment is the provision of retail space along the pedestrian system. Retailing not only provides a return to the developer who includes his property in the system but adds pedestrian interest as well, with a panorama of variety, color, merchandise, and activity.

One other important feature of system design and properly established standards is that both provide reference points for new isolated development. Some development may occur in areas where no immediate opportunity exists to tie into the pedestrian system. Knowing where the system will be built and to what standards permits a developer to plan for future connection and to use the space where the link will be made for some profitable interim use. Such conditions can do much to encourage support for the system from the private development sector.

BUILDING COMMITMENT

Once need has been documented and a system designed, attention must turn to building support for its implementation. Since it's unlikely that the physical framework for a pedestrian system and attendant design standards will be formulated without assistance from the private sector, as well as some public review, the foundation for securing a commitment will be in place. But building a pedestrian system is not a short term project, nor inexpensive, nor of total benefit during its initial phases. A system generates large capital costs which often must compete against other projects for scarce resources. It may generate other public costs for increased maintenance and security, and it may require the private sector to provide space for connections that may or may not permit interim use. The real benefit of a pedestrian system emerges only when its sections are interconnected; after twelve years, the Dallas system is just now reaching that point. Staying with a project that is incremental in its makeup for that long and supporting it year after year until it reaches its full potential only occurs when a full commitment is made at the start. That support is achieved more readily if the need for the system is well established and a plan has been developed as a focal point for future planning.

Achieving a commitment also requires enthusiastic leadership capable of pulling together diverse groups into a coalition of support. Garnering that support often requires taking risks, stimulating creative solutions to issues, balancing public and private interests, and turning limited perspectives into longer term perceptions.

Often, existing organizations may either provide the leadership or form the core for establishing a coalition representative of the general community. In Dallas the Central Business District Association, an organization representing a broad range of downtown business interests, has provided direct leadership for pedestrian facility planning. The Association provides funding for facility planning, organizes subcommittees to review projects, works as an intermediary between the public and private sectors in solving problems, initiates numerous public relations activities, and works directly with city staff and management in melding shared interests into workable projects. The leadership exercised by the Central Business District Association, in conjunction with the City, is an excellent example of the kind of public-private cooperation that is positive, workable, and fosters commitment.

A well conceived and executed public relations program is also essential in securing the commitment and the resources needed to build the system. The program should be more than a cursory publicity package as is too often the case; rather, it requires a strategy for placing the message of need and benefits for a pedestrian system before the many publics who will build and use the system. It is vital that the local governmental jurisdiction be committed to building the system and in a manner that is economically feasible from the private sector point-of-view. In addition, merchants will demand to know the likely impact of below grade retailing on their street level businesses, and system users need to know where the system is, the buildings it interconnects, and that it will be safe and easy to use. A public relations program must carry those messages as well as others if serious commitment is to be achieved. The old cliche, "out-of-sight - out of mind" applies to an underground pedestrian system. In spite of more than a decade of planning, programming, and public relations, many downtown workers in Dallas do not know that a below grade system exists. Overcoming that problem requires a professional public relations response.

One other means for achieving commitment is to find a developer willing to start the system in a significant way. In Dallas a 1965 project known as Main Place, now one of the two major underground pedestrian segments, became the real impetus for building an underground network. The success of the project required a commitment from both the developer and the City, which was made, and which directly let to other investments in the system. Construction provides solid evidence of intent and visibility which can hasten collective agreement as well as committed programming and planning.

FINANCING THE SYSTEM

Given the capital intensive nature of a pedestrian system, financing is a constant issue as new increments are added, costs continue to increase, and resources become increasingly scarce and must be shared among competing interests. Building a stable financial base can take as many forms as there are systems; however, most are funded with bonds, federal funds matched by local shares, economic development corporations or similar entities, by the private sector, or combinations of the previous. Equally important to finding a reliable source of funding is the matter of cost sharing between the public and private sectors. Utility relocation costs, responsibility for system components under public property or rights-of-way, and sale or trade of subsurface rights are just a few of the cost-sharing issues that may require negotiation and settlement. Since 90% or more of a pedestrian system falls within private property as part of new project development or building renovation, the matter might appear to be easily solved; however, public costs can still be substantial.

For example, the City of Dallas uses bonds to finance the public share of pedestrian facility connections using a formula where the city pays one-third of the cost and developers or existing property owners on both sides of the street, one-third each. During the 1965-1980 time period the City spent nearly eight million dollars from bond sales for facility connections; a significant portion was used to support underground links. The use of bonds represents a tangible City commitment to pedestrian comfort and safety as well as an improved circulation system in the downtown. It is also a clear indication to the private sector that the city will assist in making connections between project sites and will not require physical improvements to accommodate the system if links cannot be made across public property.

440

Whatever the financing program used, unexpected costs will assuredly occur, and the financial structure must be flexible enough to permit some alteration to basic agreements. It is important, therefore, to establish guidelines for such circumstances and to state in the beginning the conditions that must exist before a deviation from standard practices will be considered and other financing arrangements used. In Dallas the city may deviate from its one-third share policy from time-to-time but only in rare instances and only where the benefits to the public are sufficient to warrant the change. The trust and credibility that bind public-private projects together can quickly evaporate if the ground rules, particularly those involving financial matters, change too often and imply inconsistency. On the other hand, absolute adherence to standard practice can create rigidity that may preclude private sector involvement. How that balance is achieved must be a local decision related to specific conditions.

OPERATIONAL ISSUES

As the pedestrian system develops questions regarding operations will arise, particularly for sectors built with both public and private funds. Questions related to liability, general control of the system (i.e. who controls loitering), and general security must be addressed. With the investment of public funds comes the issue of operating schedules; is the system open 24 hours or only during periods that connecting buildings are open for business? Maintenance responsibility must be delineated, means for heating and air conditioning established, access for the handicapped assured, and plans for emergency services developed. Most operational issues are readily handled through formal agreements and procedures and, with some built-in flexibility, will provide effective control of the system. Also, operational issues should be explored early on in the planning process and policies set inasmuch as they may affect construction of the system.

For example, security is an increasing concern not only within the system itself, but particularly for connecting buildings. More and more building owners are tightening internal security, mainly through the use of a lobby security center. Often elevator operations are designed to prevent passage through the lobby level to floors either above or below the security checkpoint without a transfer from one elevator bay to another. Such circumstances maintain security but complicate the movement of people from the subgrade pedestrian system to building floors above the lobby.

A second security issue with design implications is the system itself. Tunnel sections should be designed without blind spots, access to lighting panels, or dead-end connection points that will provide future linkages. If possible, the system should be equipped with closed circuit television monitoring or an alarm system of some type. To be fully effective, the pedestrian system must have well marked openings to the street, some of which may be unsecured. The potential problems for an underground system are obvious. The safety and comfort of pedestrians will be key factors in the success or failure of a system; therefore, security and emergency services should be part of the design of the system and its operation and involve public safety departments.

One issue that may arise, although it is not truly an operational problem, is the developer who builds a project on a site designed as part of the network and refuses to connect his building into the system, or where appropriate, provide for a future link. Experience in Dallas suggests that in the first case security is most often the issue, while in the second, it may be the loss of permanent leasable space. The solution to such dilemmas is not universal; much depends on local conditions and applicable legal options. While some jurisdictions may be able to mandate connections through zoning, that option is not always available, as is the case in Dallas.

Another option is the use of variance approvals. Many large projects require variances of one type or another which provide the opportunity for negotiation and trade. Not only do variance discussions offer possible solutions to the problem, they also permit the reason for non-participation to be stated and reviewed, often leading to agreements to make connections.

Open and direct discussions provide a third opportunity which permits an exploration of the issues and available options for reaching a mutually acceptable solution. Confusion, lack of knowledge, and preconceived ideas often are at the root of problems dealing with project connections and can be harmoniously dispatched by effective communications.

While the number of problems dealing with connections seems to decrease as the system expands, one vital missing link can heavily affect the entire system and preclude achievement of its full potential. Procedures for likely connection problems should be established as part of the overall operational policies and practices, and appropriate legal remedies should be in place ready for use.

PLANNING IMPLICATIONS

Once a pedestrian system is designed, it is important to assess the implications that construction might have on the planning and development of the CBD as a whole. A major consideration is the consequence for street activities; Will the system be so effective that the vitality of the street is lost? Will retailing patterns change, and if so, are the new arrangements beneficial to the downtown as a whole? Can the system be integrated with transit stops, public plazas, and governmental buildings, or will public facilities remain isolated from offices and retailing?

While all the likely implications from the Dallas system are still not known, the quality of streetscape is considered as important as the pedestrian system itself, and design work has been completed to assure the street does not lose its vitality. For all its problems, downtown Dallas is walkable; six months or more of the year, the climate is conducive to pedestrian activities. Both the pedestrian and streetscape systems have been planned together to assure that one does not develop at the expense of the other.

Another implication involves the relationship between the construction sequence of the pedestrian system and downtown development patterns. It is possible, given the long term implementation period for system development, to create preferential sites for new buildings through the construction of connecting links. During the earliest stages of development, the impact of the system is minimal; however, as connections are made and the various links become a well established network, development patterns may become skewed from those desired, sidewalk and street capacities altered from the expected norm, and demand created for improvements out of line with established capital improvements programming.

The positive side of the matter is that the timing and sequencing of connections may provide a useful development control in appropriate situations. Obviously, the likelihood of development patterns being affected will vary greatly with local circumstances and intentions, as well as other real estate market forces. Still, a pedestrian system is a large-scale capital investment with the capability to affect the real estate market place, and the implications it has for CBD planning must be carefully analyzed.

OVERHEAD VERSUS UNDERGROUND

Perhaps the most frequently asked question concerning the Dallas pedestrian system involves local preference for a given type of connection; Is one better than the other? The answer is straight forward; both have effective uses.

Overhead pedestrian facilities generally are less costly than underground connections and appear to offer a more secure and psychologically satisfying environment to pedestrians. An overhead system permits exposure to light and air, as well as frequent orientation to the street. Such connections may encourage second level retailing opportunities.

Underground facilities, on the other hand, often provide greater retailing opportunities for a developer, resulting in more total leasable space in a building. Underground areas also offer real possibilities for creating "people spaces" that may not be appropriate at second levels in the structure. They also work well in relation to subway systems, as part of large-scale projects, and in situations where towers are set back but subgrade construction extends to the property line. They may also be preferred where overhead connections have a negative aesthetic impact on the street environment. The Dallas system is now about 70% below grade; completion of the system for the CBD Core will, however, result in a fifty-fifty split between above and below grade connections.

A pedestrian system is a major downtown design element, a large-scale capital improvement, a development tool, and a real estate marketing device. How it is designed, the implications it has for downtown development, the support it receives, its method of financing, and how it will be operated all depend on meeting a local need and creating development potential. That need and potential may require a split system, as in Dallas, an overhead system of the type found in Minneapolis, or an underground system similar to Houston. The "best" approach is the system that clearly and effectively meets community needs and benefits the downtown of which it is part.

REFERENCES

DeLeuw, Cather and Company, (1965). Long Range Transportation Plan for the Central Business District, 78-81.
Department of Urban Planning, City of Dallas, (1961). Dallas Central District Master Plan, 140, 167-174.
Department of Urban Planning, City of Dallas, (1975). Pedestrian Facilities - Dallas Central Business District, 7-10.
Myrick, Newman, Dahlberg and Partners, Inc., (1980). Dallas Central Business District Streetscape Guidelines, 7-11.
Ponte-Travers Assoc., (1969). Dallas Central Business District Plan, 22-64.
Ponte, V., (1979). A Report on a Sheltered Pedestrian System in the Business Center, 1-8.

UNDERGROUND BUILDING:
A NEW FORM OF ARCHITECTURAL DESIGN
FOR THE TOTAL BUILDING MARKET

G.R. Scafe

President, Terra-Dome Corporation
Independence, Missouri

ABSTRACT

Underground construction offers a practical, viable, and necessary approach to solving the problems of the building industry. Mass production techniques, paired with superior design concepts, can make underground building competitive in costs to conventional, aboveground building, creating high quality homes and commercial space. Builders' profit margins can remain high and the homeowner's quality of life can be improved.

KEYWORDS

Mass-production; commercial applications; subdivisions; multi-unit housing; design concepts; systems; modular systems.

We live in a time in which people are searching for more efficient, economical, and effective methods of building. Due to the energy crisis, many of us are turning to earth sheltered, earth contact, or underground housing. This new market, developing over about the last three years, has been largely based on individual homes built on single parcels of land. Often, building has been done by the home-owners themselves. This trend marks an experimental stage in the course towards a viable solution to the problem of making efficient, affordable housing available to the total population. But I believe that in order for underground and earth sheltered housing to make a major contribution to contemporary lifestyles, the building market, as a whole, must be affected by a new form of architectural de-sign. I hope to illustrate in this paper how underground building can achieve this goal.

Two criteria definitive of good building techniques must be used in order to affect the total building market: one, that the structures can be built quickly, and two, that they are affordable to the homeowner. Mass production is a key that I see in solving our housing problems. Comfortable, energy-saving, beautiful, and livable homes, both single-family and multi-unit, are then easily possible, as well as cost-efficient, stylish, and versatile commercial space. Good design, teamed with mass production, will create living and working environments that are enjoyable and conducive to happy, fulfilling lives that are marketable and profitable.

We're on the road to overcoming the sterotype of underground living as dark, damp, dead, and depressing. These negative feelings can easily be switched to the positive side. Attractive exterior design and landscaping is the first step in changing these opinions. After viewing the exterior and opening the door to an underground home, I like to see the average reaction as, "Wow! If this is underground, I want it!"

Unique design properties of underground building enable us to be more creative in planning a home's functions than conventional, aboveground housing. For example, the controlled climate available in underground homes presents many possibilities for recycling systems, plants, indoor gardens, and allergy-relief. The earth cover makes landscaping one of the most important aspects of exterior design. Exciting lighting effects can be achieved easily, and acoustical properties make stereo systems sound like live music! Interior floor plans and exterior designs are limitless.

Underground building represents a step forward in ultimate living style. The underground homes Terra-Dome is now building illustrate this. Not only are energy-efficiency and recycling systems easily incorporated, but cost, resale value, style, and comfort are strong selling points. Responsible design allows us to give our homeowners the benefits that will add to the quality of their lives. I want to make sure our homes provide feelings of security and warmth. Our underground homes are just that--not houses, but <u>homes</u>. This atmosphere is conducive to better relationships between husbands and wives and children. A motto at Terra-Dome is: "We have better families because of the homes we live in."

A home is often the biggest investment a family will make. I want to be assured that the resale value of our homes and buildings will be higher a year--or two hundred years--from their purchase dates. Underground homes will be much better investments than conventional, aboveground housing. Higher quality materials and workmanship, along with superior design concepts, allow this to happen. Energy-efficiency will take a back seat in comparison to design advantages. Another motto at Terra-Dome is: "Energy is a by-product of design."

To achieve this large-scale concept of underground housing, we must be competitive. This necessitates mass production. Mass production requires <u>systems</u>: procedures that are executed over and over again, cutting labor costs and saving time. The custom market will always exist, but not everyone can afford to hire an architect or engineer. Systems eliminate the need, in most cases, for these expenses.

Terra-Dome structures are built with a modular system, each module measuring 24 feet square and having its own domed roof. The dome is designed to add spaciousness and provide a weight load bearing capacity that is 20 times greater than that of a flat roof. While many underground homes can only support a maximum of 24 inches of earth, Terra-Dome roofs are engineered to hold eight feet plus a bulldozer. This additional earth cover allows for more of the earth's moderating effect, creating a 60 to 90 percent reduction in energy costs. The modules can be arranged in any number or configuration. Each module is connected to the next by an open arch 16 feet wide. This arch can be framed in, left open as a passageway, or filled with windows or sliding glass doors.

Each module is constructed of concrete and reinforced with steel. The footings are poured first and then the aluminum side forms and special fiberglass dome forms are set in place. The concrete is poured in one continuous, monolithic pour from the footings to the top of the dome, leaving no cold joints at the roof line. The concrete is a 3500 pound mix, poured with a low slump and vibrated as it's poured. An admixture is added for waterproofing. The walls of the modules are 10 inches thick. The domed roofs are 14 inches where they join the walls; 5 inches at the apex.

Because of the mechanized dome form and pouring system, the structure can be built quite rapidly, reducing costs. Approximate time for construction of one module after the footings are in is two and a half to three days. The modules can be built on a variety of terrains and soil types.

ELEVATIONAL
(Rectangularly and geometrically projected)

BERM (Rectangularly degreed into flatland/slope)

ATRIUM (Rectangularly shaped with an open patio)

Waterproofing, either a liquid polyurethane elastomeric membrane, a bentonite product, or cementitious material, is applied to the structure after pouring. Urethane foam or rigid styrofoam boards are used for insulation to one foot below frostline and on the exposed exterior. An exterior drainage system carries water off the roof to ensure a waterproof structure.

The largest segment of the building market is planned multi-unit developments. These developments, of very high density, are often poorly designed and of low quality workmanship. A mass production system designed for underground building will allow the same level of density and the same high profit margins for builders, without the accompanying problems. A park-like environment can be accomplished easily with landscaping, playgrounds, and gardens--on the roofs of the homes. Everyone likes the idea of living in a park! Both small and large homes can be built without affecting adjacent property values, because interior square footage can be disguised by earth sheltering. This naturally promotes a larger buyer market.

This concept calls for not merely nice landscape design, but total design that places homes <u>within</u> a park. The homes are laid out so that entrances contain a maximum of privacy and are out of the view of other entrances. Cars and parking are hidden by earth berming and landscaping. All homes are interconnected to a greenbelt that promotes the natural interaction of people with bicycle paths, jogging paths, playgrounds, and common recreational areas. Because of these park-like features, layout of the road system, lots, and entrance exposures must be much different than that of typical subdivisions. For example, a road may be placed five or ten feet below grade level to obtain maximum earth cover, passive solar effects, and landscape design.

Apartment complexes are also economical when built with underground systems. A 20 unit apartment complex is in the planning stage at Terra-Dome. The modular system is easily incorporated into variously sized apartments. A 225 foot skylight covers the atrium which contains an indoor swimming pool and a driving range. A playground and partially covered parking area fit nicely into the landscaped terrain.

The cross sections and floor plans pictured below illustrate various layouts that can be accomplished with a modular system in a multi-unit concept. Two levels can be used to overlook an atrium or create more space deeper into the earth. Separate apartment units can be stacked on top of each other, or two-story apartments can be formed. An infinite number of designs are possible. Common walls cut construction costs. Concrete construction and earth covering make the units relatively soundproof. An apartment development done in this manner can be geared to high fashion living or low income housing, with energy savings for all and significant profit margins for the builders.

448

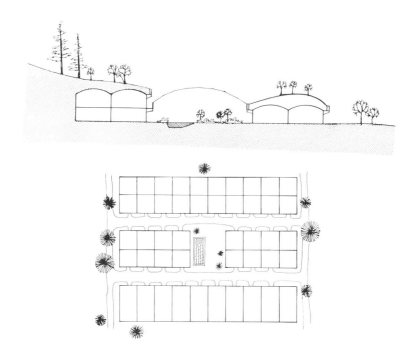

Similarly, larger subdivisions and even planned city developments can be economical, profitable, and beautifully livable. Community centers, shops, restaurants, theaters, grocery stores, racquetball clubs, and schools are only a few services that can be built and contained in a single development. When a community is self-contained in this way, the bond between the people living there deepens and encourages healthier living. In areas of sudden high populations, such as "boom towns", an underground city of this kind would eliminate the ugly trailer parks that are wasteful and depressing.

Because of the variability of the underground modular system, commercial space can be economical and versatile. Since load bearing walls occur only every 24 feet, interior partition walls can be removed later if the function of the building changes. A home could later become a restaurant, a storage facility, a barn, a retail space, an office, or even another home.

An office building in the planning stage at Terra-Dome contains 36,000 square feet. Weight load bearing capacity is great enough, due to the domed roofs, that parking can be located on the top of the structure. Glass elevators carry people down into the atrium-entry way. Again, the modular system allows fast erection, cuts labor costs, and provides a reasonably-priced structure that offers high profit margins for the builder.

In Lexington, Nebraska, a 3300 square foot underground medical clinic is nearing completion, built with this system. The owner realized that the advantages of underground made it a natural choice: energy-efficiency, fire-resistance, low exterior maintenance, and economy.

Underground building, through the use of good design and mass production, can contribute greatly to our standard of living by:
    -improving the social-psychological development of our homeowners
    -providing homes that are affordable to homeowners and profitable to builders
    -providing energy efficiency
    -building homes that are beautiful, comfortable, safe, warm, and dry
    -allowing high-density, well-landscaped subdivisions that enhance the lives of
     their inhabitants
    -allowing for low-cost, appealing commercial developments

In order to positively affect the lives of our neighbors, we must offer them the option of underground living and working. This means that the building market must become active in underground single-family, multi-unit, and commercial buildings. As we see mass production and responsible, superior design begin to work together, more of us will be able to reap these advantages. A new form of architectural design will take shape across the building industry, composed not of surface building, or even earth-sheltered building, but underground building. I am looking forward, in the years to come, to watching these contributions develop. We should all be proud that we are involved in such a great enterprise.

For further information on Terra-Dome systems and adapting underground building to multi-family and commercial applications, call or write our office. Printed materials and slides are also available.

Terra-Dome Corporation
14 Oak Hill Cluster
Independence, Missouri 64050
(816) 229-6000

CHAPTER IV
DEEP MINED SPACE

SESSION DEVELOPER: J. Gavin Warnock

TUNNELS OF TOMORROW

Edward Cross, Secretary-Treasurer
Tunnel Workers Union Local 147, New York, N.Y.

Who Will Build Them?

The art and practice of <u>tunneling</u> is not new but rather dates back to very ancient times. Back as far as the Babylonians in recorded history. Men dug the early tunnels in a search for minerals. But later tunnels were dug for water and its delivery and still later for transportation. In Egypt, water was stored in underground chambers and moved in underground conduits.

The early Romans, in a book by Sextus Julius Froninus the Chief Engineer of the Rome Water Supply, told of remarkable achievements in constructing tunnels for the delivery of water - the book tells of nine aqueducts, ranging in length from 13½ miles to 56 miles. The Appian Aqueduct, almost completely tunnel, was 16 miles long. There were no tunnel unions, and as even today no schools for training tunnel workers. Yet even in those days of long ago, Cicero wrote that contracting for Public Works is one of the best ways to get rich honestly.

Then as now, rock tunneling was a grueling and dangerous job. In most cases slaves were used by Romans. They built fires against the face of the rock, then hurled buckets of <u>water</u> against the heated rock which cracked into small pieces. But water was a necessary part of life as we in the Northeast are learning today. No life can exist without water and tunnels provided the best conduit.

Tunneling for transportation has a more recent origin. The Milpas Tunnel in England is regarded as the first exclusively for transportation. It ran into the mountains to a coal mine and was part of a canal that took the coal to Manchester, an industrial center. Coal was shoveled directly from the mine into canal boats. The project was so successful that before long the owner, Duke of Bridgewater, opened a Grand Trunk Canal 130 miles in length which included 5 tunnels. The first American tunnel recorded was the Auburn Tunnel, 1818-1821, 450 feet long, in the Schuykill coal area. This was followed by the first American Railroad Tunnel, the Allegheny Portage near Johnstown, Pa., 1831-1833.

These tunnels were built by adventurous men daring enough to believe in the practicality of tunnels and confident in their ability to construct and operate them profitably. We must give due credit to those individuals who were willing to risk their money to finance tunnel construction as tunneling is a risky business where a contractor can go broke on a single bad job. It also took engineers with

454

imagination and daring. Then, as today, there is not much teaching of tunnel construction in our engineering colleges or trade schools. But let's not forget
those individuals who actually did the work, both slaves under the Babylonians and
Romans who had no choice and the men in later years who chose tunneling. They
were usually miners who may have been lured by high wages, the challenge of a new
frontier or perhaps economic adversity, to try a new line of work where they could
get a job even without previous experience in mines. I refer to projects such as
the Gauley Bridge disaster where Union Carbide Co. contracted Rinehart and Dennis
Co., heavy construction firm from Charlottsville, Va., for a tunnel in West
Virginia and because of its isolated location, recruited workers from the Northeast, Southeast and even the Midwest and left a trail of death and misery behind
them.

Today many workers are finding their jobs repetitious, boring. They seek another
trade that offers a challenge or one that calls for imagination and innovation.
Those people are potential tunnel workers. Or people - I say people and not men
because now we have women in the tunnel building work force - who have lost their
jobs or foresee the possible loss of jobs to automation and mechanization. They
too will contribute to a new type of tunnel worker for tomorrow. Tunnel work,
aside from the wages which made the work originally attractive, offers the challenge many people seek in the effort of earning a livelihood. Seldom are two
shifts exactly alike. Most tunnels, because of expensive machinery and equipment,
used 2 or 3 shifts going around the clock. Each day presents new problems and
challenges and today's tunnel worker is generally known as a "can do" worker.
They learn to be innovative, to keep the job moving even on a temporary basis
until permanent repairs or replacements can be made.

Yet aside from the coal mines and the metal and non-metal mining industries -
talc, salt, etc. - there is no training ground for tunnel workers as such. The
trade was absorbed by the Laborers' International Union, formerly the International Hod Carriers. Because they are classified as Laborers, there are few
apprentice training requirements. True, the Laborers' International Union has
established training centers. Also training members in the use of machines and
special equipment is now a part of the Heavy Industry Construction's efforts.
The day of the simple laboring task has long passed.

Today's member of a Laborers' Local probably has a high school background - indeed some have college backgrounds - and is competent to operate the complicated
and expensive equipment found on road, tunnel and dam work. In my Local, Tunnel
Workers Local 147, we have had members who represent many professional groups -
lawyers, judges, doctors, insurance executives and executives of construction contracting companies. As energy development booms in the 1980s and 1990s, interest
in underground work also will develop.

As time passes, the equipment needed or at least used in tunnel construction will
become mechanized almost to the point of being automated. But true automation
will not come in our time, nor is it likely in our children's time. For many
years to come there will be the need for the worker in the tunnel. Thus far not
even the most sophisticated equipment can turn the eye from the shaft to tunnel
and go in far enough to set up the Tunnel Boring Machine. That takes workers and
we will continue to recruit workers from the coal mines, where today the lure of
pay may be even more attractive than in tunneling, and from the Metal and Non-
Metal mines. Many of our members came from lead, zinc and copper mines. The
economic uncertainty in those mining fields, with prices for ore fluctuating, has
driven many workers to seek out tunnel work. Workers will also be recruited from
the big cities where the flight of industry has left hundreds and thousands unemployed and with little hope of finding manufacturing jobs. The making and importing of clothes, shoes and electrical products by other countries has thrown, and

will continue to throw, many men and women into situations where to provide for their families they will need t seek new areas of opportunity. There will be such opportunity in the tunnel construction fields.

Tunnels will be needed for a wide variety of activities. Some old, some new. No city of substantial size could exist without tunnels that provide for water, sewage, utilities and for many, transportation. Underground transportation is still the best answer for the rapid movement of people and products.

In this country we can expect new frontiers in tunnel construction. This conference takes up the issue of Underground Space. Because of space limitation, energy costs and safety, more and more underground projects will be planned and completed. Underground storage and underground power plants are coming into their own here although they are nothing new in Europe. And as the war threats escalate and little wars break out around the world, we will look into the question of shelters for civilians and perhaps even factories going underground. The proposed Program will use thousands of workers, many for underground work that will last for years. Two large American companies have started to use underground facilities in their expansion plans, Mutual of Omaha has a three level underground addition in Omaha, Nebraska. It is reported to have cost less to build than a comparable building above ground. American Cyanamid in Wayne, N.J., has constructed an addition to an older underground building recessed into the side of a hill.

It is reported that millions of people in China live and work underground. They are warmer in winter and cooler in summer by virtue of the earth insulation. Here also, workers were recruited from available sources, many of whom had little or no previous construction experience. However, before one gets the idea that anyone can get a shovel and start digging a tunnel, let me remind you that every industry has certain inherent hazards. Potentials for disaster in underground construction are many and cannot be ignored. And, as in other fields, the experienced worker will be more productive than the inexperienced. The worker who learns the skills needed under proper supervision will most likely stay on the job longer and become a better and more productive worker.

A steady, experienced worker will also be a safe worker. Keep in mind, one reason for the more acceptable integration of tunnel workers compared to other trades is the recognition of the underground hazards that face all those underground. In New York, we learned to become brothers under the skin long before it became fashionable or the law of the land. The mistake a worker makes alongside you at the face or tunnel heading or a mistake back the line can wipe out an entire crew and jeopardize the whole project. In tunneling as in most industries, but maybe more so, Safety Pays Big Dividends. The workers realize that the pay, no matter how big, is of no value if you do not live to spend it or if you end up in a wheel chair. The contractor also knows that what he may save in safety training or equipment could cause heavy insurance losses or higher compensation premiums on the next tunnel job. Also a safe work place will keep the labor turnover low. This means more skilled and experienced workers are available to complete the project and to train the new workers who replace those who leave. We have all seen construction jobs where workers leave after a week or two or even after one day. I have gone over records of employment where many workers left after only a half shift and never returned to collect the pay due them.

Tunnels and how they are built have long posed a mystery to the average citizen. Even today the sophisticated New Yorker will ask "how can you build a tunnel through the water?" Few realize the underwater tunnels are actually under the river or harbor bed. They tell, in N.Y., that shortly after the Civil War, a man known only as Crazy Luke hung around Wall Street and would wait outside clubs and restaurants to badger bankers and industrialists about financing a tunnel under the Hudson River between New York and New Jersey, a mile long tunnel that trains would go through. In the end, the story goes, they put Crazy Luke away in an asylum. Yet less than 40 years later in 1908 a group of bankers and financiers made the trip from 14th Street and 6th Avenue in Manhattan under the Hudson to Hoboken on the Jersey side. We assume Crazy Luke rolled over in his grave.

Tunnel construction is a fascinating field and many find it attractive because it, by its nature, signifies progress and achievement. But many people lured by what appears to be big money and the glamour of underground work are disillusioned by the reality of being in the tunnel. It is hard, dirty and dangerous work. But as bad as it may first seem, it is also fascinating to some. They still show a couple of movies on late night television about building subaqueous tunnels. It might be the Holland tunnel where in one, Victor McLaglen virtually digs with his bare hands at times and emerges covered with mud and a later film with Fred MacMurray and Claudette Colbert doing a cleaner but equally heroic job of constructing a tunnel under the Hudson. They were compressed air tunnels. Actually the tunnels under the Hudson were a model of manning for their day and will continue for years to come.

The Tunnel Workers Union took in as new members a limited number of men and introduced them into the tunnel in small numbers in the regular work crews. In that way the older hands could supervise and train the inexperienced worker and in time the new worker acquired the skills needed to be an efficient member of the crew or gang as they were called. This same practice of infiltrating a few new members among a larger group of older members was carried out successfully in the construction of the Queens Midtown Tunnel, Lincoln and Brooklyn Battery Tunnels.

When San Francisco decided to build their subway system, the unions there sought the help of locals in other cities where there had been substantial tunnel work and where skilled tunnel workers were available. They came from Seattle, Chicago, Boston and Philadelphia as well as Local 147 in New York City. Local 147 at one time had 200 of its members working on the BART System and another 70 members working and training inexperienced workers on the Los Angeles Water Tunnel out of Bakersfield, California.

As tunnel work increases there will be greater emphasis on training and development of safety standards. Employers must sooner or later embrace training and safety programs and fund them. As mentioned before, we in Local 147 over the years put a few new members into gangs of predominantly older members and we were able to maintain an enviable safety record at least as far as fatalities went. Local 147 set up their own training classes and the Kiewit Company, contractor on one project, was most cooperative and this resulted in improved safety and a resultant increase in productivity. Eventually the Union included in its Labor Management Agreement provisions for a Safety Miner at each construction shaft site. The only duty of that Safety Miner was to police safety on the job. Those selected as Safety Miners were required to have 10 years' tunnel experience. In addition we sent them to the Bureau of Mines Training Facility in Pittsburgh, Pa., and with help of Jim Bennett and Jim Greer taught them underground safety. Most of our accidents come from falls from drill Jumbos or from falling rock out of the roof or rib of the tunnel, again on men working on or near the drill Jumbo. There were two haulage fatalities and two fatalities involving explosives. In both instances the follow-up accident was repetitive and never should have happened. All resulted in temporary shut down of the projects and a natural slow down when the men went back to work. It is difficult to maintain a constant work schedule for a day and at times a week after a fatality or serious accident. It remains on the workers' minds and one wonders if they will be next. For this reason as well as respect for human life Safety Pays for everyone.

The worker of today and tomorrow, be they man or woman, will demand better working conditions and better pay and will have an eye on pension possibilities. The employer who treats his workers with respect and accords them the dignity all people want will be the successful contractor of tomorrow. Workers want to know what the dangers of the job are, what the employer is doing to minimize the chances of injury, what health hazards exist on the project and what the employee can do to help reduce the hazards. Not only that mandated by the law or rule, but what will be done voluntarily by the employer because it is the right thing to do. I am sure we all know of employers whose record of concern for their employees was so extensive that the employees developed a loyalty to that employer greater than the workers' loyalty to his union or even to authority. Look after your employees, show them concern, give them dignity and your employees will respond. They are human beings just like you but with less affluence.

Don't get caught up in the macho image so many contractors, and I admit, unions and their members have developed on the question of women tunnel workers. Women are part of the world of tunnel building. While it may seem a new and intrusive situation, such is not true. Women have been in mines for years. It might be recalled by some of the more social minded that one of the arguments for enactment of child labor laws was the employment of boys and girls from ages 8 to 11 years to push ore and coal cars in tunnels. This in the mid 1800s. Today women work in tunnel construction in many parts of this country and other parts of the world. They may work by government mandate or because of a shortage of workers in general. But they are a part of the work force and will continue to be. We have seen women workers in mines who were capable and efficient. In fact in the Climax Molybeendum mine in Henderson, Colorado, they not only utilize women workers but they have an entire drill crew composed of women and they have a relief foreman, a woman, who takes over any crew of men or women if the regular foreman is out. Women not only produce, but according to managers, take better care of equipment. Women, like men, seek the means to a better life for themselves and families. Many support families. Some find the work challenging and interesting. Like the men in tunnels, they can look around at the close of the day's work and say "I helped do this." I doubt any find it glamorous and seek work for that reason. They find hard work, dirty work underdifficult conditions and many hazards, and in some places hostility from men co-workers. But they persevere. The tunnels of tomorrow will need women as well as men. As mentioned, they are a part of the Industry.

Tunnels in the years ahead will not be all done by the miracle methods of automation and mechanization. True, more tunnels will use tunnel boring machines (TBMs). More TBMs will be virtually automated where as in some pipe jacking jobs virtually all work is done mechanically. The day will come when the recognized tunnel worker will be a worker who mans a console, push buttons and levers that will control machines that will do the work. But there are and will always be those tunnels that require physical workers. Tunnels in ground not consistent for any great length, shallow tunnels in populated areas not suitable for open cut work but not long enough to justify the cost of TBMs and tunnels that make frequent turns will all need workers. In some instances the human worker will out-produce the machine. Many professionals will scoff at that statement but a look at the records of some recent projects shows that machines, even forgetting down-time for repairs, have not always kept pace with human workers.

On the Red Hook Sewer Project in New York, the Digger Machine was removed and the worker with shovels and air spader replaced it. Because of boulders in the bottom, the use of the digger became time consuming and impractical. Machines no matter how reliable cannot at present be programmed to cover all possibilities one may find on a tunnel job. There will still be a need for miners and drill runners on tunnels small and short in length. Some of today's tunnel workers will learn new skills or upgrade existing ones. But it is most likely that the government, State if not Federal, will one day require some training be given to applicants for tunnel work. It is done in metal and non-metal mining under the orientation process so that new workers have some idea of what to expect underground and what will be expected of them when they report to a regular work gang. Tunnel work will always face uncertainties. No one can accurately predict ground ahead of the face or heading or say with certainty what the ground will be like between the test borings. And the greater distance between borings the more likely the changes in ground. As long as such conditions prevail, the ideas of surmounting difficulties with machines alone and no workers in the tunnel will remain remote.

While mechanization and automation will reduce the number of workers on a particular project, the demand for tunnel workers will grow with the increase in the number of underground projects that will be developed. There may actually be a scarcity of tunnel workers for a time and a renewed interest in training. Because of the nation's defense requirements, tunnel workers will be able to pick and choose their jobs. Many will have to adopt the nomadic habits of today's tunnel worker and go to the jobs wherever they may be, but for those workers, it will pay handsomely.

And the truly skilled miners, drill runners and blasters may one day find themselves labeled an endangered species and be sought after by contractors and owners to do tunnels not adaptable to machines. And they too will quote Cicero, "Contracting for Public Works is one of the best ways for getting rich honestly."

CONTRACTUAL ISSUES

David G. Hammond

Vice President
Daniel, Mann, Johnson, & Mendenhall
201 N. Charles Street
Baltimore, Maryland 21201

ABSTRACT

As our underground engineering works become larger, more expensive, and more complex, the final cost of completion and efficient prosecution becomes less likely. Many of the difficulties stem from the adversary relationships established by past and current contractual practices for which some remedies will be discussed in this paper.

KEYWORDS

Forms of contract; adversary contractual relations; claims handling policies; contract changes; differing site conditions; disputes resolution; escalation considerations; underground geotechnics.

INTRODUCTION

Slowly, much too slowly, there is a dawning realization that contracting practices and policies are very significant factors in the cost and the feasibility of construction projects. This is particularly true of underground construction projects where the traditional contracting practices and policies establish adversary relationships which result in delays in completion of projects and therefore in higher end costs to not only the participants, that is contractors, engineers and owners, but especially to the taxpayers who ultimately pay the final cost.

It is important to note that the owner is usually not an individual, but a collection of taxpayers called upon to foot the bill for the final cost of any project. You will note the emphasis here on public projects, although the same considerations apply, to a lesser extent, to private construction contracts. Private contracts are different only in the opportunity that they afford to the "owner" to exercise judgement and discretion early on as to what is the stockholders' interests as to whether to go underground and under what contractual arrangements.

While private owners are in a better position than representatives of public agencies to assess and act on the merits of enlightened construction practices than are public staff members, there is not a very full recognition among private owners as to the possible merits of taking the cash and letting the credit go.

459

MANAGEMENT RELATION TO CONTRACTUAL ARRANGEMENTS

Of particular significance to underground construction is the relation to contracts of the way that the project is organized and the management philosophy and organizational structure which is established for conducting the project. Not the least important ingredient, of course, is how the people who will participate at various levels of the project carry out, or are allowed or required to carry out, the responsibilities assigned to them.

While design contracts come after planning and construction, and procurement contracts come late in the game, contractual arrangements for all of these contracts must be considered and established from the very beginning and woven into the planning, organizing and establishment of the management philosophy and structure. In all of the considerations, the end functional purpose of the project, whether it be underground transportation, water supply, sewage treatment, or whatever, together with the time considerations and related costs must be constantly borne in mind.

Much recognition has been given recently to the importance of considering such things as clear establishment of the roles of the owner, the engineer, and the contractor in underground projects. Related to this is recognition of the necessity for making equitable allocation of risks, liabilities, responsibilities and authority and incorporating them clearly in contractual arrangements.

BETTER MANAGEMENT OF UNDERGROUND CONSTRUCTION

The National Academy of Engineering, in 1978, completed a study intended to lead to better management of underground construction projects. In the conduct of this study, the considerations which I have just mentioned are very much in evidence. The study is of underground construction, being sponsored by the U.S. National Committee on Tunneling Technology. However, the committee has noted that most of its conclusions and recommendations would be applicable to any large or super project whether underground or not. This study stems from an earlier study conducted by the USNC/TT, more directly related to my topic. The USNC/TT issued a report in 1974 entitled "Better Contracting for Underground Projects".

FORMS OF CONTRACTS

In addition to the traditional lump sum or fixed price contract, there are a number of other approaches that have been used on underground projects with varying degrees of success.

Let's examine some of those. First, of course, is the one that I mentioned, the fixed price or lump sum contract, which has the merit, theoretically at least, of establishing the firm funding requirement for accomplishment of that particular piece of work. It traditionally has been required by political agencies and most private bodies. Generally, but not always, these are based on sufficient time to have a fully engineered design and specific set of contract requirements. Also, they generally have placed most, if not all, of the responsibility on the contractor for performing the work at his bid price, regardless of what eventuality may occur during the course of carrying out the work under the contract.

Another form is the so-called cost-reimbursement contract, which has several variations, such as inclusion of incentive-penalty provisions, and payment provisions which may either be cost plus a fixed fee, or cost plus a percentage of the cost (not allowable on Federal contracts). This type of contract, of course, increases the risk to the owner that the final cost may vary materially from the initial contract price. On the other hand, it gives the owner greater flexibility in having necessary work performed expeditiously and economically under the contract that is different from what was originally specified either because of design developments or changed local public, political, or physical conditions.

A third type, so-called fast-track, is employed when time is considered paramount, and it is desirable to start construction before there is time to complete a fully engineered design. This, of course, adds considerable uncertainty to what the end cost will be. This approach is most applicable in projects where the completion of construction will lead to operation of a facility which will be revenue producing several years earlier than might be accomplished by the use of the traditional fully engineered fixed-price approach. This may justify considerably higher construction costs, or at least the initial uncertainty of what the final construction cost will be.

PUBLIC INTEREST

Generally, public projects cannot clearly show this justification, or at least the procedures are not allowable under laws or regulations of public bodies. The public is usually presumed to feel that it is better protected with fixed price contracts. Few political leaders seem willing to make a proper assessment of what contractual arrangements would really serve the public interest and then lead public opinion in that direction. In private projects, however, management can make this kind of decision based on its own trade-off analyses.

SURVEY ON CONTRACTING PRACTICES IN THE UNITED STATES

It has been over five years since the Committee on Better Contracting Practices for Underground Construction of the U.S. National Committee on Tunneling Technology (USNC/TT) issued its report. This report included seventeen specific recommendations for improvements in contracting practices. It is appropriate now to assess the extent to which these recommendations have been considered and acted upon by the various agencies involved in construction contracting.

Most studies of problem areas conclude that there is a communications problem. I'm afraid that we will find that this is the case with the recommendations for better contracting practices. However, in some cases, they have been largely adopted, and while all of the experience has not yet been gained, it is possible to assess the validity and desirability of some of the specific recommendations.

I will discuss one project from personal knowledge in which thirteen of the seventeen recommendations have been adopted and incorporated into construction contract documents and in contract administration -- that is the Baltimore Region Rapid Transit System.

Let's run down the major issues:

Changed Conditions or Differing Site Conditions

We have, with the blessing of our enlightened client, the Mass Transit Administration (acting on the recommendations of its enlightened consultant), incorporated into all of our contract documents (not just those for underground) the Federal Government Construction Contract Differing Site Conditions Clause.

The propriety and effectiveness of this clause is tied to the topic we will discuss later, especially for underground contracts, and that is the degree of information, factual or conjectural, which is given to the contractor and whether or not the owner disclaims responsibility for it.

In implementation of this clause, it is essential to keep records as to what conditions are actually encountered in comparison to conditions which were either predicted in contract documents, or which a prudent contractor could have been expected to infer as being conditions in which he might have to work. In other words, it is important to establish the parameters and the procedures that will apply to the application of the differing site conditions clause.

While all the returns are not yet in on the Baltimore project as to whether this use has been advantegeous to the owner, preliminary results would indicate that if nothing else, it considerably reduced the amount of contingency included in bids received. To date, five or six claims have been received varying from relatively small amounts to $2 million to $3 million claims. Basically, they allege actual conditions different from those the contractor had a right to expect.

Generally, they have to do with rock at different or higher elevations in tunnels than was shown in contract documents. There is always a problem in addressing this kind of claim as to not only are the site conditions different, but how different are they and what effect, if any, did it have on the contractor's operations and therefore on his costs. To date, no claims have gone through to litigation, and it is not presently known what the settlements will be.

The State of Maryland has established a Contract Appeals Board in the Department of Transportation. The claims are in various stages of being heard by the Mass Transit Administrator, or if denied by him, on appeal to the Contract Board of Appeals.

Subsurface Information

As mentioned previously, the degree of information which is given to bidders is relevant to the differing site conditions clause. In our case, we give complete information to bidders, not only boring logs and opportunities to examine cores, provide geological reports and water data, but more importantly, bidders are provided with a design summary report. This is the designer's report of how he viewed the ground conditions, his design assumptions and basis: i.e., how he expected the ground to behave.

In keeping with this philosophy, disclaimer or exculpatory provisions are not utilized in our construction contracts. We do, however, make a distinction between data obtained early and with less sophisticated techniques or equipment, and later, more complete data. Such data are so labelled.

Variations in Quantities Clauses

On some other matters, such as variations in quantities clauses, we include a standard provision in our contract -- where estimated quantities are included in the bid and contract documents -- that variations greater than, plus or minus 25%, will be handled by an equitable adjustment.

Time extensions, if appropriate, are also allowed under the variations in quantities clause. To date, we have had little experience with variations greater than this tolerance. While we have had no disputes or disagreements in this area, the methods or procedures for handling these are spelled out in the contract documents.

Exculpatory Clauses

As for exculpatory clauses, in general, we do not use them. For site delays, it depends on the nature of the delay. In case of right-of-way delays, which are common, the Federal Suspension of Work clause is utilized.

There is one area somewhat peculiar to Maryland which relates to the availability of facilities constructed by others to which the contractor must interface. Here, in accordance with Maryland law, an exculpatory clause is utilized which gives him a time extension for the affected items, but reimbursement of costs is made only when the delay is directly due to the owner's negligence.

Change Orders

Our change order procedures are similar to Federal provisions. Basically, the change must be within the scope of the contract. Oral orders must be confirmed in writing. Provisions are made for equitable adjustment for time and cost effects of the change. The contractor had 30 days to assert a claim for adjustment under the contract.

As is not uncommon, our experience has not been completely satisfactory in being able to resolve change orders in a timely fashion. This is partly due to the inability to readily resolve differences of opinion as to equitable adjustment and sometimes due to the changes occurring while the change order negotiation is taking place.

We use the Change Notice procedure to affect a direction for a change. Payment is based, however, on change orders which, as I indicated, can be delayed. In some cases, and we consider it undesirable, work will be done on a force account basis. Where this is so, progress payments are made in a routine fashion.

Change orders, unless on a force account basis, of course, have to be assessed as to what their effect was on what the contractor would otherwise have been doing, what his actual reasonable allowable costs are, and what mark-ups should be provided for. No mark-up formula is provided in our contracts, except for force account work.

Where there are long drawn out differences regarding change order adjustments, they may eventually reach the dispute or claim stage. In this case, they are then handled in accordance with the disputes and claims procedures.

Retained Percentage

Moving to retained percentage, we do retain, but in a lesser amount than the standard 10%. Our retainage is 5% for the first 50% of the work, including change orders, at which point, if the work has been performed satisfactorily, and reasonably on schedule, no further retainage is held. However, the previous retainage is not returned at that time to the contractor.

By completion or partial completion of the contract, the then $2\frac{1}{2}\%$ retainage is further reduced to 2%. The contract calls for this to be paid 30 days after the formal completion and acceptance of the contract.

Insurances

Insurance coverage is becoming an increasingly important factor in construction contracts. The limits we specify are generally $5 million each for Workmen's Compensation, for General Liability, and in some cases, as much as a $20 million umbrella requirement. We specify a $1,000 deductible for each occurrence. Our reasoning on this is to give the contractor some incentive to be reasonable in the conduct of his work, even if it costs him a few dollars extra, instead of relying on being covered by insurance, theoretically, at no cost to him.

On our project, Wrap-up insurance is provided for the owner and to all construction contractors and to the General Consultant. Wrap-up insurance covers Workmen's Compensation, General Liability, and All-risk construction. Automobile and Aircraft/Watercraft insurance are not included in Wrap-up and must be provided by the contractor.

Disputed Work

As I mentioned earlier, machinery is provided in the contract for settlement of disputes. Disputes, as you know, come in various sizes and degrees of importance. We try to have as many potential disputes as possible resolved and kept out of the dispute category by the field construction management agency. Where this is not possible, or it is taking too long, the matter is moved up to the construction manager and owner level.

The dispute machinery is established in that any disagreement that the contractor has with a settlement by the resident engineer or the construction manager may be appealed to the owner. The head of the Mass Transit Administration holds a hearing and has the authority to make a "final" decision.

If the contractor is still not satisfied with that resolution, which as you might expect generally does support the actions of the construction manager, he can, within 30 days, appeal to the Maryland Board of Contract Appeals. This is a board created specifically to handle MDOT contract claims and disputes.

To date, only four or five disputes or claims have been processed as far as the Board of Contract Appeals. They generally are in the category of alleged differing site conditions. Even here, as I mentioned earlier, the dispute is not so much whether the conditions are different or not, but as to the nature and degree of the difference, and therefore, the amount of additional compensation to which the contractor is entitled.

Escalation Clauses

In the matter of escalation clauses, it depends, in our case, on the contract type. To date, our client has not been willing to use escalation clauses for construction contracts. We have recently again recommended it because of experience in bidding in this climate of large uncertainty as to what the future escalation rates are going to be. We have included, however, escalation provision in certain procurement contracts, notably for procurement of running rail and for our vehicles.

For vehicles, for example, a formula is included in the contract, specifying the percentages for labor and materials, and limiting the applicability of the escalation clause.

While future events may prove this is not completely equitable to one side or the other, it is the only practical way in which to use an escalation clause -- that is, to make it by formula depending on and tied to specific indices. That brings it down to a mathematical entitlement, regardless of whether that is exactly the way events turned out or not.

Liquidated Damages

For liquidated damages, we do specify these. As you know, the theory of liquidated damages is that actual damages would be difficult, if not possible, to ascertain even at a later date. Liquidated damages are specified as a contractually agreed amount in lieu of trying to determine actual damages. This is a very difficult area in which to specify what are reasonable amounts.

If, for instance, you specified liquidated damages that were likely to come close to your real damages, the amount would be so high as to scare off all bidders, or to require all bidders to put such a large amount of contingency in the bid that the low bid would be unaffordable.

We try to tailor our liquidated damages to straddle this gap and make them large enough so that it is not attractive to contractors to incur these damages; but not so large as to either scare them off or to scare them into putting a large allowance in their bids for possible collection of liquidated damages.

SUMMARY

We include these provisions in all construction and in all large procurement contracts. We have used them for small procurement contracts, but are presently re-evaluating whether that is a reasonable thing to do from our own standpoint.

We believe that the owner, as well as the contractors on our Baltimore project, will have benefitted at the end from the use of these revised and improved contracting practices.

I keep hoping and pushing to see a broader acceptance and use of them to demonstrate their value and to achieve general improvement in contracting practices in the underground construction industry.

RISK AND INSURANCE

C.A. Muller

Clarkson Puckle Insurance Broking Group, Ibex House,
Minories, London EC3N 1HJ

ABSTRACT

The author presents an update report on the insurance aspects of projects underground.   Stress is laid on avoiding and minimising hazards and their importance
to those concerned with the steps necessary to develop an effective insurance
programme.

A simple classification of the principal types of operations currently carried out
in underground space is provided with the author's technical consultants' input of
factors to be considered in assessing risks.

Stress is laid on the adaptability of insurers' to new types of commercial developments and an example of reduction in the price of insurance is given.

The research project being undertaken by the author's company into designing a
'Comprehensive Underground Space Insurance policy' CUSIP, is shown in flow-chart
form.   With the attendance at Think Deep coming from many parts of the world, a
summary of the insurance markets in various categories of countries is provided.

The paper reaches the conclusion that adequate insurance can be made available at
reasonable cost providing that risk management techniques are followed from an
early stage in the planning of a new development.

KEYWORDS

CUSIP; RIMS; risk management; underground space.

INTRODUCTION

The general theme derived from the above title is to continue the research undertaken since the 1977 and 1980 Rockstore Conferences in Stockholm and to update
those especially interested in the commercial development of underground space and
in particular the premium cost comparisons between above ground and underground.

Whilst we would readily accept that insurance represents only a small part of the capital expenditure on construction and the cost of maintaining an effective underground space operation, the price and availability of adequate insurance can be an important factor in any viability study.

It is right that the title of this paper emphasises the word 'risk' since this is the first and most important aspect which should be addressed by the purchasers of insurance.   Indeed the change in title in many corporations from "insurance manager" to "risk manager" is reflected in the growth both in reputation and size of their professional bodies throughout the world known in the USA as RIMS. After all, any major investment of funds i.e. premiums should be preceded by an objective study of the potential hazards and how they can be avoided or minimised before an approach is made through the normal insurance market channels to establish the cost of buying insurance protection, purpose-designed to meet the specifications.   Where a corporation has been in business for many years statistics will be in the hands of the risk manager indicating previous loss experience.   However, this may well not be available in this instance;   thus there is the exceptional importance of operating sophisticated risk management programmes which, though their form may vary, are all designed to reduce instances which would give rise to insurance losses.   Since these techniques are well-established it is not necessary to elaborate in this paper their particular application to underground space operations.

It is, however, worth noting that as expressed by de Saventhem (1977), an insurance buyer will need to locate those insurers who are familiar with under- ground operations and who recognise the attractive underwriting features so as to obtain the best terms.   Even within the same insurer, opinions may vary as to the hazards involved and it is therefore necessary to find the underwriter who has the greatest confidence in the successful operation of an underground project. We remain confident that the insurance costs of such a successful operation can be reduced to an acceptable cost level and the following section of this paper indicates some of the aspects which, during our investigation, and for which this paper forms an interim report, are determining factors.

As international insurance brokers based in London, England, we with our USA correspondents, H & W Underwriters (Agency) Inc. of Kansas City, engaged the services of the Schirmer Engineering Corporation of Illinois.   This Corporation has had much experience in providing technical input to insurers such as Lloyds of London.

Investigation has been made of a number of different types of underground space principally in the commercial field.   Later we deal with different categories; we believe currently we are approaching those operations where the cost and availability of insurance may be a decisive factor in their viability.

Our review of a limited number of examples of the use of underground space reveals several main features and in this respect it should be emphasised that we are examining the main exposures of loss being fire and consequential loss (use of occupancy) therefrom.   Most insureds will require additional coverage such as extended perils but the main hazard is clearly fire and losses related to it. Many of the aspects revealed take on an added importance due to the necessity of fighting the fire from the inside rather than the exterior as is the case with most above-ground structures.

The evaluation of the fire department takes on added importance inasmuch as:

a) Fire department personnel may refuse to enter the underground space

b) Access for fire fighting might be impossible or difficult due to smoke and heat.

Establishment of a well-trained, properly equipped, and frequently drilled private fire brigade would be a necessity. It should be noted that the U.S. Occupational Safety and Health Authority (OSHA) now requires private fire brigades to be properly trained and equipped with proper protective gear. For underground space this could include:

a) Portable breathing apparatus

b) Protective clothing

c) Training in operation of smoke control equipment

d) Training in operation of emergency power supply (if installed)

e) Training in use and operation of all fire fighting and protection equipment, i.e.

    Sprinkler systems
    Alarm systems
    Fire pumps
    First aid fire equipment and hand hose

f) Cooperation and communication with local fire departments

The roof supports in the underground space should be non-combustible.

Smoke handling and control is of prime importance, especially in a space where perishables or similar highly damageable commodities are stored. It is also important in relation to fire suppression activities and should include:

a) Postitive pressure in tenant storage space - during normal operations

b) Negative pressure (exhaust) in common use areas - during normal operations

c) In a fire situation, the fire area should be programmed to be converted to maximum exhaust with other areas converted to maximum supply.

Provisions of an emergency power supply would be considered important to maintain lighting, air supply - circulation, and operate smoke removal equipment in the event of loss of main power supply.

A properly installed sprinkler system with adequate water supplies properly maintained should be installed in all areas containing combustible storage or processes.

An emergency notification or fire alarm system should be provided and properly maintained to alert the private fire brigade and the local fire department.

Provisions should be made to drain sprinkler and/or water from fire hose streams to a safe remote location.

a) In earth sheltered space above grade level, this could be proper drainage to outside atmosphere.

b) In earth sheltered space below grade level, this could be proper drainage to remote areas of the mine.

In any case, it should be ascertained that values in the underground space would not be subject to flooding.

The proper fire separations between storage areas or hazardous operations should be provided by means of fire walls or adequate open space to limit amount of exposure to agreed upon limits.

Special attention should be paid to storage of flammable liquids. This hazardous commodity should be stored and handled according to NFPA Standard No. 30, "Flammable and Combustible Liquid Code," as they are stored in above ground buildings.

Storage of hazardous gases should be excluded. Items such as liquified petroleum gas for mobile vehicles used in space should be stored in an approved location outside of the underground storage space.

Hazardous operations should require a special review and analysis to consider if special protection that would be required will be available in the underground space.

Establishment of an on-going loss control programme would be a necessity. Frequent loss control inspections should be made. Management should establish a written loss control policy which would be signed by the senior officer and understood by management and employees alike. This would include compliance with loss control recommendations on a priority basis.

It was noted that mining operations continue in some underground space, and in others, mining operations have terminated. No special problems seem to be present when mining operations continue in areas of the mine distant from the space used for storage.

Underground storage space need not be solely interpreted as mines or caverns. This space could also include tunnels previously used as subways or highways.

To enable insurers to quote premium figures for the cost of fire insurance, one of the standard underwriting types of report can be adapted to the particular needs of underground space. At this point it may well be asked how one would differentiate the hazards involved in different types of operations (e.g. manu-

facturing processes) in similar physical circumstances underground.   We consider
that static factors such as outlined above should together with conventional loss
control measures be the basic factor = 100% in establishing a premium rating
system.   It does not appear that sufficient actuarial information is available
to calculate premium based on previous loss experience.   Nevertheless, insurers
would be expected to offer some insureds "good experience" discount off the 100%
base rate.   Equally other loadings or discounts plus or minus would be applied.
We are confident that leading insurers with whom we have discussed this rating
concept would be prepared to indicate an attractive base rate and most likely this
would be in the best conditions no more costly than above ground.   With this base
rate we would then examine the various types of occupational hazards which would
attract percentage reductions or additional premiums calculated on the above base
rate.   The main categories could be as follows:-

   a) University - schools and similar non-commercial
      operations;   offices could be also included.
      In this regard we would exclude earth-sheltered
      private houses which are already the subject of
      our separate investigation (Muller and Taylor,
      1980).

   b) Industrial uses.   As with above ground, this
      would be divided into numerous sections depending
      on the operations.   We would not expect very
      hazardous work to be undertaken underground
      because of the danger to life: insurers would,
      in the circumstances, quote high premiums and
      demand high safety standards both in technical
      and 'housekeeping' methods.

   c) Storage of dry goods or commodities such as occurs
      in converted mines in the vicinity of Kansas City.

   d) Storage of oil and similar products and gases under
      pressure.   Many of these in the USA will be under
      government control and/or form part of major inter-
      national corporations' activities and whose
      insurance needs are not of the conventional type.[1]

   e) Nuclear waste depositories.   Because of the
      special insurance situation in all countries whereby
      insurance is controlled by nuclear pools there is
      no possibility of devising a special insurance
      programme in the conventional markets.

Our researches continue and it is hoped that at future meetings we will be able to
make further reports on how the insurance industry has developed its thinking.
It is worth mentioning that all new commercial developments find the insurance
market responding, perhaps hesitantly at first, but finally on a competitive
basis by offering coverages at highly attractive terms.   An example was when jet
aircraft came into commercial use and the high hull values presented capacity
problems the international market quoted premiums in the range of 4% on the value.

---

[1]Final report by International Research & Technology Corpn of McClean, Virginia
 'Cavity Degradation Risk - Insurance Assessment' on underground energy storage
 by compressed energy storage (CAES) and underground pumped hydro (UPH) system is
 an important contribution in a specialised area.

472

Good experience combined with technical advances has enabled airlines of proven
reliability to obtain coverage at 1% i.e. a saving of 75%.   There can be few
'commodities' purchased which have reduced in cost by this extent over the last
ten years or so.

Delegates to Think Deep may find of interest the following flow chart indicating
our approach on behalf of clients to the planning of our comprehensive underground space insurance programme, CUSIP.

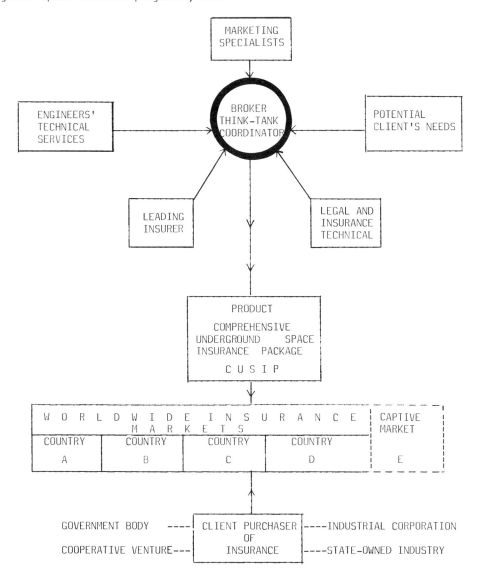

Fig. 1 . C U S I P DEVELOPMENT PROGRAMME

Fig. 1.   CUSIP DEVELOPMENT PROGRAMME

Country A   National Insurance Company e.g. many third world countries
Country B   National Reinsurance   e.g. Brazil
Country C   Free Insurance market   e.g. United Kingdom
Country D   Regulated Insurance market e.g. USA

Country A   Clients requiring CUSIP insurance package have only one source and
             competitive alternatives are not available.

Country B   Whilst clients can have a choice of insurers the local companies will
             normally require reinsurance through one channel and the terms will
             therefore be fixed by reinsurers.   Some negotiation may well be
             possible.

Country C   Providing that the amount of insurance is not so large as to require
             a pooling of insurers' capacity competitive quotations for CUSIP will
             be obtainable.

Country D   Notwithstanding the regulations, in USA competitive pressures exist
             between domestic insurers themselves.   Traditionally Lloyds of
             London has played an innovative and often price reducing role.
             Insurance markets in the USA tend to operate on a highly cyclical
             basis with interest rate on funds being a significant factor.
             Currently insurance premiums are one of the few 'commodities' which
             can be purchased at cheaper cost than, say, ten years' ago.

CAPTIVE   Many major industrial groups do not feel the need to buy insurance for
MARKET E   their smaller loss exposures in the conventional market.   Favourable
             loss experience enhanced by stringent safety measures can often make
             the ownership of a 'captive' insurance company a tax effective
             proposition.   Providing that full investigation is made into the
             legal and tax implications such a company can be established in
             e.g. Bermuda.   Coverage such as CUSIP could be insured by such a
             company providing, as with other categories of insurance, reinsurance
             protection against catastrophies is purchased.

In conclusion, we welcome contributions of whatever nature which may help the
development of our CUSIP programme and in particular news of individual risk
managers' experience on behalf of their corporations of purchasing insurance
throughout the world.

Being enthusiasts ourselves in all forms of Underground Space the challenge is
one we are happy to meet.

REFERENCES

Muller, C.A., and R.A. Taylor (1980). No cause for apprehension about costs
    insuring earth-sheltered homes. Underground Space Vol 5, Pergamon Press, London.
    pp. 28-30.
Muller, C.A., E.M. de Saventhem and R.M. Aickin (1980). Environmental and other
    insurance aspects of underground storage projects. Subsurface Space. Pergamon
    Press, Oxford and New York pp. 83-86.
de Saventhem, E.M. (1977).   Insuring risks underground - some general considera-
    tions.   Underground Space Vol 2. No. 1. Pergamon Press, London. pp 19-25.

THE QUALITATIVE EFFECTS ON PETROLEUM RESULTING FROM
PROLONGED STORAGE IN DEEP UNDERGROUND CAVITIES

Harry N. Giles* and Peter Niederhoff**

*Strategic Petroleum Reserve Office, U. S. Dept. of Energy
Washington, DC 20585, U. S. A.
**Kavernen Bau- und Betriebs-GmbH, Rathenaustr. 13/14
D-3000 Hannover 1, West Germany

ABSTRACT

A sampling and analysis program was undertaken to assess the qualitative effects
on petroleum resulting from prolonged storage in underground cavities. Two salt
dome solution cavities and a converted potash mine containing crude oil, and a
salt dome solution cavity containing distillate fuel oil were studied. A series
of samples was collected from each of the cavities and each sample was subsequently
analyzed comprehensively. Overall, the analyses did not indicate any deleterious
qualities which would require special handling at a refinery for the bulk of the
crude oil. At the oil/brine interface in one of the solution cavities a thin, relatively dense and viscous sludge layer was detected. This layer is thought to have
formed through gravity settling of impurities originally dispersed in the overlying
crude oil. Although not sampled, there is evidence of the existence of a similar
layer in the other crude-oil-containing solution cavity studied. It is unknown
whether a sludge layer exists in the lowermost level of the potash mine. The distillate fuel oil samples exhibited some slight differences in quality but, overall,
the product conformed to DIN standard specifications. These studies demonstrate
conclusively that crude oil and distillate fuel oil can be stored in large underground cavities for prolonged periods without undergoing deleterious changes in
quality.

KEYWORDS

Petroleum; crude oil; distillate fuel oil; underground storage; long-term storage;
strategic storage; quality; stability.

INTRODUCTION

In the United States, the Energy Policy and Conservation Act of 1975 established
the requirement for a Strategic Petroleum Reserve (SPR) of up to one billion (10^9)
barrels (1.6 x 10^8 m^3) of oil. The Federal Energy Administration, now a part of the
Department of Energy (DOE), was tasked with implementing the SPR. The majority of
the oil comprising the SPR will be stored in solution-mined cavities in salt, and in
conventionally dry-mined cavities in salt and other rocks. This oil could remain in
quiescent storage for as long as twenty years.

It has been a generally held view that crude oil and certain petroleum products will not undergo deleterious changes in quality when stored for prolonged periods in mines or salt dome solution cavities. Until recently, however, there had been little evidence to support this hypothesis. In implementing the SPR program, DOE initiated studies on the qualitative effects on crude oil and distillate fuel oil resulting from prolonged storage in deep underground cavities. These studies were managed by Kavernen Bau- und Betriebs-GmbH, turnkey contractors for several of the petroleum underground-storage projects in West Germany where a stockpiling program has been underway since 1971.

For these studies, three of the West German petroleum reserve sites were selected. Near Wilhelmshaven, 33 cavities leached in the Etzel salt dome are used for the storage of crude oil and refined petroleum products by the sites' owner, Industrieverwaltungsgesellschaft mbH (IVG). Mobil Oil AG uses six solution-mined cavities in the Lesum salt dome, near Bremen, for the storage of crude oil and distillate fuel oil. The disused Wilhelmine-Carlsglück potash mine at Hulsen, in the Verden/Aller District, has been converted and is now used by Wintershall AG for crude oil storage.

SAMPLING AT THE ETZEL AND LESUM SITES

General

To obtain oil samples truly representative of the depths from which they were to be collected, it was necessary to preclude any mixing caused by downward passage of the sampling device. To accomplish this, a special sampling device was constructed (Fig. 1). This device was designed to be lowered on a wireline to the assigned depth where an electric motor would rotate a cantilevered arm 90° to a horizontal

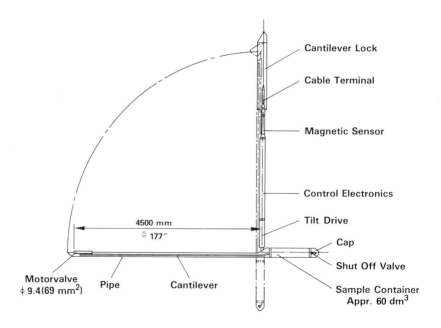

Fig. 1. Sampling device used at Lesum and Etzel.

position. An electrically-actuated valve would then open at the tip of the arm, 4.5 m from the centerline of the device. Up to 6.0 l of oil would then flow into an evacuated sample chamber. The valve would then close, the arm rotate upward to a vertical position, and the entire device drawn to the surface.

Once the device was on the surface, a nitrogen line was coupled to the sample inlet valve, the sample drain valve connected to a series of nitrogen-purged tinplated cans, and the entire sample displaced using nitrogen pressure. Once the device was empty of oil, it was cleaned with a suitable solvent and then evacuated with a vacuum pump. Sampling was then repeated at the next lower depth.

Etzel

In one of the 33 cavities, K 117, 460,647 m^3 of four different Middle East crude oil streams and a tank farm crude oil composite were in storage as of January 1977 (Table 1). No oil was pumped into or out of the cavity for the nearly two years which preceded sampling operations in August 1978. Based on records of the incremental volumes of crude oil stored and knowledge of the geometry of the cavern derived from sonar caliper surveys, it was possible to construct a profile of the cavity and specify theoretical depths to the boundaries between the different types of crude oil (Fig. 2). Using this profile, 22 depths were selected for the collection of samples. These depths were selected so that at least two samples from each of the four different crude oil streams and the oil/brine interface region would be obtained.

TABLE 1 Oil Stored in Etzel Cavity K 117

Date Stored (month/year)	Quantity Stored (m^3)	Cumulative Quantity (m^3)	Type Oil
7/74 – 8/75	422	422	Blanket oil (distillate)
9/75 – 12/75	18,829	19,251	Arabian Light crude oil
1/76	14,224	33,475	Iranian Light crude oil
1/76	61,083	94,558	Basrah crude oil
1/76 – 4/76	11,391	105,949	Crude oil composite
4/76 – 5/76	44,454	150,403	Arabian Light crude oil
6/76 – 7/76	38,328	188,731	Iranian Heavy crude oil
7/76 – 8/76	25,338	214,069	Arabian Light crude oil
9/76	93,662	307,731	Iranian Heavy crude oil
10/76 – 11/76	12,389	320,120	Crude oil composite
12/76	40,826	360,946	Iranian Heavy crude oil
12/76	54,871	415,817	Crude oil composite
1/77	44,830	460,647	Arabian Light crude oil

Prior to sampling, it was necessary to first depressurize the wellhead, which resulted in 2,531 m^3 of crude oil being produced at the surface. Throughout the subsequent sampling operations crude oil was continuously produced at the surface due to creep closure of the cavity, with an additional 385 m^3 being produced.

Following depressurization of the wellhead, the 2 3/8 in. freshwater dilution string was pulled. Then a Baker bridge plug was run and set at a depth of 1,390.2 m in the 8 5/8 in. brine string (Fig. 3). This was done so that as the brine string was pulled the brine contained therein would not cascade into the oil, causing subsequent analytical results for the oil's water and salt content to be erroneously high.

Following these precautions, sampling of the oil commenced in the manner discussed earlier. Near the oil/brine interface a dense, viscous sludge layer was encountered which could not be sampled with the cantilevered-arm device. To obtain samples of

478

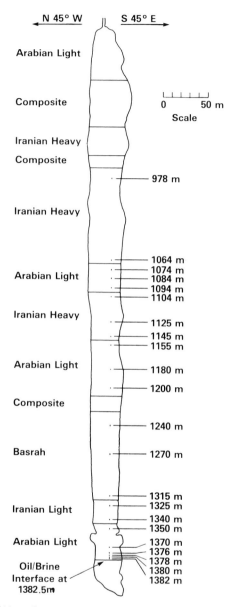

Fig. 2. Profile of Etzel cavity K 117 showing theoretical
layering of crude oil and sampling depths.

this layer, a conventional oil well sampling device was used and two samples were
obtained. Altogether 22 samples were obtained from Etzel cavity K 117.

Lesum

At Lesum, over 250,000 m^3 of Arabian Light crude oil had been stored in cavity L 103

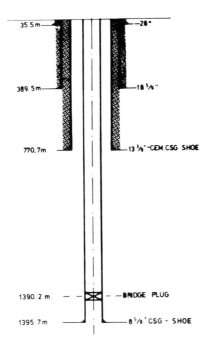

Fig. 3. Casing and brine string completion of
Etzel cavity K 117.

for nearly six years at the time of its sampling in September 1978. Relatively
small increments of crude oil were pumped into or out of the cavity during the in-
tervening years, but over 90 percent of the crude oil was effectively in undisturbed storage during this time. Sampling depths for L 103 were evenly spaced, except
in the oil/brine interface region, because the cavity contained only a single ge-
neric crude oil (Fig. 4). Sampling operations for this cavity were conducted in
the same manner as for Etzel cavity K 117, except that less oil was produced at the
surface during depressurization and sampling because of the comparatively smaller
size of the cavity and its shallower depth. A Baker bridge plug was run and set at
a depth of 752.0 m (Fig. 5), following which the 7 in. brine string was pulled in
the same manner as at Etzel. A gamma log was then run to determine the oil/brine
interface depth. Ten crude oil samples were subsequently obtained from L 103. Although the existence of a dense, viscous sludge layer was indicated by the presence
of a 1 m long grease-like coating on the brine string immediately above the oil/
brine interface depth (Fig. 6), no sample of this layer was obtained.

Between October 1975 and February 1977, 197,000 m^3 of distillate fuel oil were pumped into cavity L 104 at Lesum. During the next 11 months only an additional 27,000
m^3 of fuel oil were pumped into the cavity so that, at the time of sampling in September 1978, 90 percent of the fuel oil had been in essentially undisturbed storage
for 19 months. Again depressurization and sampling were conducted as previously
done at Etzel and in Lesum cavity L 103 (Fig. 7). Only a small volume of fuel oil
was produced at the surface during depressurization. and none was produced during
sampling operations. Altogether, 10 fuel oil samples were collected from various
depths within L 104 (Fig. 8). No sludge layer was evident at the oil/brine interface, and such a layer was not expected to be present because of the inherent quality characteristics of the oil.

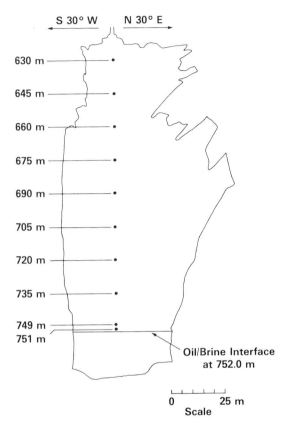

Fig. 4.   Profile of Lesum cavity L 103
showing sampling depths.

SAMPLING AT WILHELMINE-CARLSGLÜCK

General

The Wilhelmine-Carlsglück mine storage facility is inherently different from Lesum
and Etzel. First, it is a conventionally dry-mined, principally horizontal series
of chambers; in contrast to a single, vertical solution-mined cavity. Second, the
mine is essentially dry, as it was developed by dry-mining methods and has no groundwater inflow. Third, the stored petroleum is at virtually atmospheric pressure and
there is a gaseous phase above the crude oil. This latter difference precluded the
use of an electrically-actuated sampling device, because the mixture of gases is
potentially explosive and a short circuit in the control cable could ignite the mixture. Therefore, a mechanical device normally used for sampling of petroleum reservoir gases or liquids was modified for sampling within the mine. This device incorporates a clockwork mechanism to open and close two valves at preset times. Rather
than being evacuated, as was the sampling device used at Etzel and Lesum, the 1.0 ℓ
sample chamber is filled with mercury. In operation, the sampler is lowered by
wireline to the predetermined depth and, at the preset time, upper and lower valves

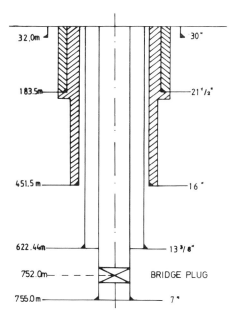

Fig. 5.   Casing and brine string completion
of Lesum cavity L 103.

Fig. 6.   Grease-like coating on brine tubing from the oil/brine
interface level in Lesum cavity L 103.

open.   The mercury drains out of the sample chamber into a receptacle and oil or
gas flows into the chamber through the upper valve.   The valves close at a preset
time and the sampler is then drawn to the surface .   All sampling operations were
performed through a lubricator installed at the shaft head on a 18 5/8 in. casing
string (Fig.9), which was designed to be used as a pump conduit through the shaft's

bulkhead.

Carlsglück shaft

The disused Wilhelmine-Carlsglück potash mine was converted into a crude oil storage reservoir by Wintershall AG between 1971 and 1973. Filling of the lower level of the mine with 347,935 m^3 of a heavy Syrian crude oil took place between August 1973 and September 1974. The crude oil remained totally undisturbed until sampling commenced in December 1979. Initially, three samples of the gaseous phase were obtained at 200 m intervals using the sampling device described above. At the surface, the gas samples were collected in evacuated gas traps for storage and shipment to the laboratory. Next, 6.0 l of sample were collected from four different levels within the crude oil. Because the sampling device had a maximum capacity of only 1.0 l, it was necessary to repeat sampling runs to the same level five time to obtain 6.0 l. As at Etzel and Lesum, once the sampling device was on the surface, a nitrogen line was coupled to the sample inlet valve, the sample drain connected to a nitrogen-purged tinplated can, and the entire sample displaced using nitrogen pressure. The chamber was not cleaned between sampling runs to the same level, but only prior to sampling at a lower level.

It had been planned to collect samples from two deeper levels in the shaft, with the lowermost being in the shaft's sump where it was believed that a sludge layer, similar to the ones encountered at Etzel and Lesum, would be present. When lowering the sampler to initiate collection of the fifth sample from about 602 m, however, an obstruction at about 556 m prevented further lowering. Because of the danger of a spark causing an explosion, sampling operations were terminated. Apparently, a portion of the steel mesh lining the shaft had come free since completion of sampling

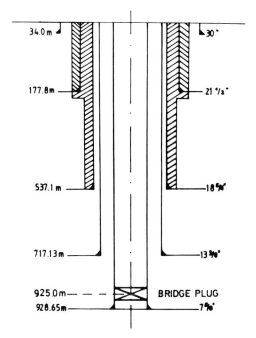

Fig. 7.   Casing and brine string completion
of Lesum cavity L 104.

at the fourth level (598 m) and hung out into the shaft, preventing further down-ward passage of the sampler.

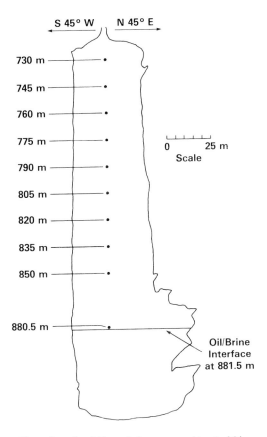

Fig. 8.   Profile of Lesum cavity L 104
showing sampling depths.

ANALYTICAL PROCEDURES

Crude Oil

Deutsches Institut für Normung (DIN, 1976) and American Society for Testing and Materials (ASTM, 1977) standard methods; together with other conventional, widely used analytical procedures were used in characterizing the crude oil samples (Table 2).   Optical density (OD) color of the samples was determined spectrophotometri-cally by measuring the absorbance at 410 nm of a 0.5 percent solution of the oil in benzene.   The trace metals vanadium, nicker, copper, and iron were determined by atomic absorption spectrophotometry.   Nitrogen was determined using an automated elemental analyzer; and neutralization number was determined by potentiometric ti-tration.   All of the conventional methods were standardized so that data for dif-ferent samples would be directly comparable.

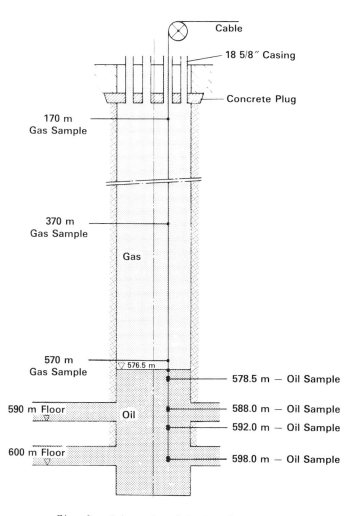

Fig. 9. Schematic of Carlsglück shaft
showing sampling depths.

Distillate Fuel Oil

The distillate fuel oil samples were analyzed using, principally, DIN standard methods (Table 3). Because of their inherent transparency, it was not necessary to dissolve the fuel oil samples in benzene before determining OD color.

TABLE 2 Analytical Procedures for Crude Oil Samples

Characteristic	Analytical Method*
Water content	51 582
Sediment content	51 789
Density	51 757 (D 941 & D 1481)
Viscosity	53 015
Pour point	51 597 (D 97)
Vapor pressure	51 754
Total sulfur	51 768 (D 1551)
Hydrogen sulfide and mercaptans	51 796 (D 1323)
Total nitrogen	Elemental analyzer
Salt content	51 576
OD color	Spectrophotometric absorbance
Neutralization number	Potentiometric titration
Trace metals	Atomic absorption spectrophotometry
Distillation yields	51 751 (D 86) and D 1660

*Methods without a D-prefix are DIN standards. Methods with a D-prefix are ASTM standards. ASTM standards in parentheses are the equivalent of the DIN standard.

TABLE 3 Analytical Procedures for Distillate Fuel Oil Samples

Characteristic	Analytical Method*
Flame point	51 758
Pour point	51 597 (D 97)
Water content	51 582
Carbon residue	51 551 (D 189)
Ash content	51 575 (D 482)
Distillation	51 751 (D 86)
Viscosity	53 015
Density	51 757 (D 941)
Copper strip corrosion	51 759 (D 130)
Total sulfur	51 768 (D 1551)
Salt content	51 576
OD color	Spectrophotometric absorbance

*DIN standard methods, except as noted. ASTM equivalent methods are shown in parentheses.

RESULTS AND DISCUSSION

Etzel Cavity K 117

Results of the analyses of the 22 Etzel samples are summarized in Table 4. Down through sample 20 (1378 m), water and sediment are negligible probably as a result of settling out. Density and viscosity are virtually constant down through sample

20, suggesting that total mixing of the different crude oils has taken place and that density stratification is not occurring. Distillation yields are somewhat variable, but also do not indicate any density stratification through sample 20, although this phenomenum is known to occur with time in surface tanks and even a-board tankers. The other data for samples one through twenty, however, exhibit a range in values, but the variation is random with no noticeable trends. This variation is typified by the sulfur and nitrogen content of the samples (Table 5), which suggest that segregation of the different crude oil streams stored, persists. The large range in copper values is attributed to contamination of samples by brass components within the collection chamber. No hydrogen sulfide was detected in any of the samples indicating the absence of microbial degradation. Microbial activity at the oil/brine interface would be inhibited by the hypersalinity of the brine.

TABLE 4 Summary of Analytical Data for Etzel Cavity K 117 Samples

Characteristic	Samples 1 - 20* (978 - 1378 m)	Sample 21 (1380 m)	Sample 22 (1382 m)
Water (Wt. %)	Negligible	3.9	4.6
Sediment (Wt. %)	Negligible	0.83	0.91
Density at 20°C (g/cm^3)	0.859 - 0.860	0.868	0.873
Mercaptans (mg/kg)	58 - 106	92	109
Sulfur (Wt. %)	1.81 - 2.12	1.83	1.80
Pour point ($^{\circ}$C)	-21 to -9	38	38
Salt (mg/kg)	85 - 186	14,586	24,983
Viscosity (cSt)			
at 25°C	10.62 - 11.57	ND**	ND
at 50°C	5.48 - 5.96	9.08	9.50
Neutralization number			
(mg KOH/g)	0.00 - 0.49	1.50	2.43
Vapor pressure (bar)	0.29 - 0.37	0.29	0.29
OD color	0.960 - 1.003	1.207	1.269
Nitrogen (Wt. %)	0.152 - 0.195	0.230	0.164
Trace metals (ppm)			
Copper	0.04 - 2.22	5.31	9.12
Iron	1.32 - 14.5	330	893
Vanadium	26.7 - 32.3	36.9	63.8
Nickel	16.0 - 19.8	18.2	25.0
Vanadium/Nickel	1.4 - 1.8	2.0	2.6
Distillation yields (Vol. %)			
Naphtha ($<191^{\circ}$C)	20.9 - 21.9	16.2	16.1
Distillate (191-343°C)	25.2 - 26.2	22.6	22.8
Gas oil (343-545°C)	23.0 - 27.3	23.1	23.2
Residuum ($>545^{\circ}$C)	23.8 - 28.3	33.8	33.1

*Range of results
**Not determinable

Immediately above the oil/brine interface a dense, viscous layer has formed. As can be seen from the data in Table 4, the two samples (21 and 22) collected from this layer have a moderately high sediment and water content, a high concentration of salt, are relatively enriched in the trace metals, have a large neutralization number, and a comparatively large OD color value. This latter data is suggestive of a higher concentration of asphaltic compounds (Thomassen, 1977), which is supported by the relatively higher residuum and lower naphtha distillation yields. This layer, which comprises only about 0.5 percent of the total volume of crude oil stored in the cavity, is thought to have formed in much the same way as sludge layers which accumulate in conventional surface tanks used for crude oil storage. The sediment and water inherently present in crude oil have settled out under the action of gravity. Concurrently, the trace metals, possibly as metalloporphyrins, and acidic colloids such as asphaltenes have flocculated or precipitated out of the crude oil. Although trace metals in crude oil are frequently associated with nitrogen (Hunt and O'Neal, Jr., 1965), no such relationship is evidenced by these data.

TABLE 5 Sulfur and Nitrogen Data for Etzel Cavity K 117 Samples

Depth (m)	Sulfur*	Nitrogen*
978	1.88	0.195
1064	1.88	0.166
1074	2.05	0.165
1084	1.83	0.156
1094	1.97	0.158
1104	1.87	0.153
1125	1.84	0.157
1145	1.81	0.159
1155	2.01	0.157
1180	1.91	0.153
1200	1.94	0.157
1240	2.12	0.154
1270	1.88	0.152
1315	1.90	0.155
1325	1.86	0.154
1340	1.86	0.155
1350	1.85	0.156
1370	1.85	0.154
1376	1.89	0.152
1378	1.90	0.158

*Values in weight percent

These analyses indicate that no deleterious changes in quality have occurred to the bulk of the crude oil stored in Etzel cavity K 117. Somewhat surprisingly no density stratification has occurred, but this may be because of thermal convection within the oil induced by the geothermal gradient within the salt mass. Although a sludge layer has formed, it comprises less than 0.5 percent of the total volume of crude oil stored in the cavity. This layer probably formed as a result of the impurities and certain colloidal compounds inherently present in crude oil settling or flocculating out over time.

Lesum Cavity L 103

Results of the analyses of the 10 crude oil samples collected from Lesum cavity L 103 are summarized in Table 6. Quality is quite uniform throughout the cavity as might be expected since only Arabian Light crude oil is stored. The deepest sample (No. 10) contains a minor amount of water and sediment and appreciable mercaptans and salt, which probably have settled out of the overlying crude oil, but in other respects is essentially of comparable quality to the other nine samples. No hydrogen sulfide was detected in any of the samples indicating the absence of microbial degradation. As in Etzel cavity K 117, microbial activity would be inhibited by the hypersalinity of the brine. No density stratification of the crude oil has occurred as evidenced by uniformity of the density data and distillation yields.

There is evidence supporting the presence of a dense, viscous sludge layer at the oil/brine interface, as discussed earlier, but samples were not obtained. This layer must comprise less than 0.5 percent of the total volume of crude oil in storage based on the incremental volume of the cavity between the lowest sample (751 m) and the level of the oil/brine interface (752 m).

TABLE 6 Summary of Analytical Data for Lesum Cavity L 103 Samples

Characteristic	Samples 1 – 8* (630 – 735 m)	Sample 9 (749 m)	Sample 10 (751 m)
Water (Wt. %)	Negligible	Negligible	0.2
Sediment (Wt. %)	Negligible	Negligible	0.03
Density at 20°C (g/cm^3)	0.850	0.850	0.851
Mercaptans (mg/kg)	49 – 72	84	195
Sulfur (Wt. %)	1.58 – 1.68	1.65	1.71
Pour point ($^{\circ}$C)	-27 to -22	-20	-17
Salt (mg/kg)	24 – 40	54	445
Viscosity (cSt)			
at 25°C	7.27 – 7.56	8.03	7.90
at 50°C	4.07 – 4.34	4.34	4.33
Neutralization number			
(mg KOH/g)	0.054 – 0.330	0.270	0.055
Vapor pressure (bar)	0.25 – 0.26	0.25	0.26
OD color	0.704 – 0.733	0.720	0.713
Nitrogen (Wt. %)	0.074 – 0.080	0.075	0.072
Trace metals (ppm)			
Copper	0.12 – 0.22	0.15	0.25
Iron	15.6 – 52.9	18.4	32.8
Vanadium	9.96 – 12.27	10.72	10.22
Nickel	4.25 – 4.53	4.33	4.35
Vanadium/Nickel	2.2 – 2.9	2.5	2.4

Distillation yields (Vol. %)				
Fraction:	<191°C	191–343°C	343–545°C	>545°C
Sample No.				
1	20.2	30.3	26.4	20.8
2	22.5	29.2	26.8	19.8
4	21.3	30.0	26.7	19.5
6	22.0	29.1	28.1	19.2
10	21.4	29.0	28.1	19.6

*Range of results

To reiterate, these analyses indicate that no deleterious changes in quality or density stratification have occurred to the crude oil, even though it has been in storage for over five years. Thermal convection within the oil, induced by the geothermal gradient of the salt mass, may be precluding density stratification and promoting homogeneity in quality.

Lesum Cavity L 104

Results of the analyses of the nine distillate fuel oil samples collected from

Lesum cavity L 104 are summarized in Table 7. There is little difference in quality among the samples, and the variations observed are probably attributable to
differences in quality originally present in the various batches of product stored.
All of the samples conform to the German standard specifications for the product.
Although not covered by these specifications, an apparent increase in salt content
has occurred in several of the samples. This increase does not seem to have had a
deleterious effect on the product's quality with respect to its corrosiveness, as
evidenced by the results of the copper strip test. Lambrich and Kuehne (1975), reporting on the experiences of Deutsche Shell AG in storing distillate fuel oil in
a solution cavity at Sottorf for 18 months, also noted a sporadic increase in salt
content. They postulated that this increase in salt content was due to thermal
convection cells in the oil induced by a geothermal gradient of up to 10°C in the
salt mass. Thermal convection cells should also cause an increase in water content,
but amounts of less than 0.1 weight percent would not be detected by the analytical
method used.

TABLE 7 Summary of Analytical Data for Lesum Cavity L 104 Samples

Characteristic	Samples 1 - 9* (730 - 850 m)
Water (Wt. %)	<0.1
Sediment (Wt. %)	<0.01
Flame point (^{o}C)	80 - 83
Pour point (^{o}C)	-15 to -7
Salt (mg/kg)	0 - 15
Viscosity (cSt)	
at 25°C	4.31 - 5.02
at 50°C	2.56 - 2.93
Density at 20°C (g/cm^{3})	0.839 - 0.840
OD color	0.077 - 0.140
Ash (Wt. %)	<0.005
Carbon residue (Wt. %)	0.02 - 0.08
Sulfur (Wt. %)	0.30 - 0.33
Copper strip corrosion	1b
Distillation (^{o}C)	
Vol. % off	
IBP**	198 -215
5	208 - 220
20	227 - 235
50	263 - 285
80	316 - 332
95	355 - 367

*Range of results
**Initial boiling point

Wilhelmine-Carlsglück Mine

As discussed earlier, the Wilhelmine-Carlsglück mine is different from Lesum and
Etzel in several important respects. First, it has a large horizontal extent, consisting of a series of galleries and chambers. Second, it is essentially dry and
has no known groundwater seepage. Third, storage is at virtually atmospheric pressure and a gaseous phase exists above the crude oil. Lastly, the crude oil is
pumped into the mine through a tubing string extending into the shaft sump. Because of the large horizontal extent of the mine, results derived from the analysis

of a single vertical series of samples cannot be construed as being necessarily representative of the entire volume of stored petroleum. Moreover, the relatively small vertical extent of the mine would have too small a geothermal gradient to induce thermal convection in the oil and density stratification could occur.

Gas samples. The gas phase in the mine shaft contains a relatively large proportion of light hydrocarbons in the range C_1 to C_8. Propane comprises 30 to 31 percent of the hydrocarbons present, with methane accounting for 9 to 10 percent, ethane 19 to 20 percent, the butanes 23 to 24 percent, and C_5 through C_8 hydrocarbons the remaining 15 to 18 percent. Although crude oil, when produced, is stabilized by removing the more volatile constituents such as methane and ethane, the process is not usually 100 percent efficient and fractional concentrations of these gases remain dissolved in the stream. During quiescent storage, methane and ethane become relatively more abundant in the gas phase than propane and the other hydrocarbons because of their significantly higher vapor pressure.

Crude oil samples. Results of the analyses of the four crude oil samples are presented in Table 8. Quality is quite uniform over the 20 m interval covered by the samples with only minor differences evident. The variation which does exist may

TABLE 8 Analytical Data for Wilhelmine-Carlsglück Mine Crude Oil Samples

Characteristic	Sample depth (m)			
	578.5	588.0	592.0	598.0
Water (Wt. %)	Neg.*	Neg.	Neg.	Neg.
Sediment (Wt. %)	Neg.	Neg.	Neg.	Neg.
Density at 20°C (g/cm^3)	0.908	0.908	0.908	0.908
Sulfur (Wt. %)	3.14	3.32	3.32	3.49
Pour point ($^{\circ}$C)	-36	-36	-36	-36
Salt (mg/kg)	23	26	23	23
Viscosity (cSt)				
at 25°C	51.1	44.8	45.4	44.0
at 50°C	19.0	17.3	17.9	17.0
Neutralization number				
(mg KOH/g)	0.22	0.22	0.17	0.17
Vapor pressure (bar)	0.28	0.32	0.28	0.28
OD color	1.233	1.224	1.221	1.220
Mercaptans (mg/kg)	74	67	49	46
Nitrogen (Wt. %)	0.234	0.212	0.202	0.202
Trace metals (ppm)				
Copper	0.08	0.09	0.07	0.09
Iron	1.7	1.7	2.3	2.7
Nickel	42.4	43.1	42.4	41.7
Vanadium	119	122	117	117
Distillation yields (Vol. %)				
Naphtha ($<191^{\circ}$C)	15.6	16.3	17.5	18.3
Distillate ($191-343^{\circ}$C)	21.3	21.8	21.5	21.7
Gas oil ($343-545^{\circ}$C)	24.6	25.2	25.3	25.8
Residuum ($>545^{\circ}$C)	36.3	34.8	33.8	32.8

*Negligible

be attributable to differences in quality originally present in the various batches of crude oil stored. No hydrogen sulfide was detected in any of the samples indicating the absence of microbial degradation, as would be expected since water does not naturally occur in the mine. Water settling out of the crude oil would become hypersaline through solution of the potash, and would inhibit microbial metabolism. Somewhat surprisingly there in no evidence of density stratification in the oil. Distillation yields are somewhat anomalous with naphtha increasing and residuum decreasing with depth, yet vapor pressure exhibits little variation. This may be due to differences in quality originally present when the oil was stored.

As discussed earlier, it was not possible to obtain a sample from the lowest level of the shaft and it is not known if a dense, viscous sludge layer analogous to the one encountered at Etzel exists.

The analyses indicate that, in the vicinity of the shaft, no deleterious changes in quality or density stratification have occurred, even though the crude oil has been in storage for five years. The relatively slight variations in quality between samples is probably attributable to differences originally present in the various batches of crude oil stored.

CONCLUSIONS

These studies have demonstrated conclusively that both crude oil and distillate fuel oil can be stored for prolonged periods in underground cavities without undergoing deleterious changes in quality. This fact is important to countries actively developing or contemplating strategic petroleum reserves, as well as to petroleum companies considering underground storage of crude oil and middle distillates as an alternative to more conventional means.

Crude oil stored for nearly six years in a salt dome solution-cavity, and for over five years in a converted potash mine gave no indications of having undergone any changes which would require special handling at a refinery. Quality of the oil in both of the reserves was essentially uniform throughout with no evidence of density stratification or microbial degradation, and no anomolously high salt content or acidity. A suite of crude oils stored in another salt dome solution-cavity for between two and three years, although uniform with respect to certain characteristics, exhibited some differences in quality. The variations observed are attributed to the inherently different properties of the crude oils as originally stored.

In both of the solution cavities containing crude oil; a dense, viscous sludge layer has formed at the oil/brine interface. This layer, which comprises only about 0.5 percent of the total volume of crude oil stored in the cavities, is thought to be analogous to the sludge layer which builds up, with time, in conventional crude oil tanks. There is no evidence that this layer results from incompatibility between the crude oil and the storage environment. Samples were not obtained from the lowest level of the potash mine and it is not known if a sludge layer has formed.

Distillate fuel oil stored for 19 months in a salt dome solution-cavity did not undergo any changes in quality which would effect its marketability or burning behavior. A slight increase in salt content was detected in some samples, but the corrosiveness of these samples was still within specification limits.

ACKNOWLEDGEMENT

The authors thank IVG, Mobil Oil AG, and Wintershall AG for permission to conduct sampling operations at their respective sites. In West Germany: Dr. Menz and his staff at the Preussag AG Erdöl und Erdgas Labor, Berkhöpen, performed the analyses; Prakla-Seismos GmbH, Hannover, developed the sampling device used at Lesum and Etzel; and F. Leutert GmbH, Adendorf, developed the sampling device used at Wilhelmine-Carlsglück. Dr. Richard E. Smith of the Strategic Petroleum Reserve Office, critically reviewed the manuscript and provided valuable comments. This work was

492

supported by the U. S. Department of Energy under contracts No. EL-78-C-01-7151 and DE-AC-01-80USO7151. The views and opinions of authors expressed herein do not necessarily state or reflect those of the United States Government or any agency thereof. Permission to publish has been granted by the Deputy Assistant Secretary. Strategic Petroleum Reserve.

REFERENCES

American Society for Testing and Materials (1977). <u>Annual Book of ASTM Standards</u>, Parts 23, 24, and 25. ASTM, Philadelphia.
Deutsches Institut für Normung (1976). <u>Mineralöl- und Brennstoffnormen, Taschenbucher 20, 32, und 57</u>. Beuth Verlag GmbH, Berlin.
Hunt, R. H., and M. J. O'Neal, Jr. (1965). In J. J. McKetta, Jr. (Ed.), <u>Advances in Petroleum Chemistry and Refining</u>, Vol. 10. Interscience, New York. Chap. 1, pp. 3-24.
Lambrich, K. H., and G. Kuehne (1975). Experiences gained in the creation and operation of caverns in salt domes with a large content of impurities. <u>Proceedings of the 9th World Petroleum Congress</u>, Tokyo.
Thomassen, A. R. (1977). Private communication.

FINNISH VIEWPOINTS AND RECENT DEVELOPMENTS IN OIL CAVERN TECHNOLOGY

A. Hakapää MSc(MinEng)[++] Y. Ignatius MSc(MechEng)[+]
+Neste Oy, Espoo, Finland
++Finncavern Ltd, Espoo, Finland

ABSTRACT

Current Finnish practice of storage of crude oil and oil products in unlined rock caverns is discussed with trends outlining future development. Commercial and strategic stockpile volume of hydro carbons underground already cover the consumption of about ten months. More storage caverns are being constructed.

KEYWORDS

Oil storage; unlined caverns; rock cavern technology; Finland.

FACTS ABOUT FINNISH ENERGY SUPPLY

Today, some 70% of Finland's energy needs are satisfied by imported fuels, some 45% by different oil products. In comparison with most European countries, the Finnish energy supply contains certain special features. With its area of 337 000 km^2, Finland is one of the middle-sized states in the world. More than one-third of the world's population living above 60° North live in Finland. The need for fuels is strongly concentrated on our long winter. About one-third of Finland's energy bill goes to heat buildings. Finnish industries, such as the wood-processing industry, chemical industry, and the metals industry, are in the main highly energy-intensive. Industry's share of Finland's annual energy bill is some 50%. As the country is sparsely populated (14 inhabitants/km^2), and as the transport distances are long, this of course increases the need for fuel.

STORAGE OF OIL PRODUCTS IN FINLAND

Commercial oil storage facilities

The commercial oil storage facilities are used partly for storing the crude oil and oil products of Neste, the national oil company, and partly for storing oil products of marketing companies. Owing to seasonal fluctuations in consumption, the amount of products stored must be kept rather high, at the level of about ten month's consumption.

Strategic Storage of Oil Products

In addition to the above storage facilities for over 20 years now, the Government of Finland has also carried on the so-called strategic storing of oil products. This storage is based on a special law, enacted in 1958.

In the years 1958 - 1972 the strategic storage of oil products was realized by resources included in the state budget. From the beginning of 1973, a special strategic storage levy has been collected from both private persons and companies, to finance the programme.

The strategic storage facilities are located around the country and new units are currently under construction.

The Finnish Government had chosen the underground storing method for storage facilities as early as in 1958. This method is also used by Neste, some marketing companies of oil sector, and some industrial enterprises and power plants.

TYPE OF STORAGE

The amounts of oil handled in all of the different phases are so large that the type of storage should be carefully studied both from economical and other viewpoints. The unlined rock caverns have been the most economical, safe, and practical storage method everywhere where suitable rock formations are available. This type of storage is surprisingly little used outside Scandinavia.

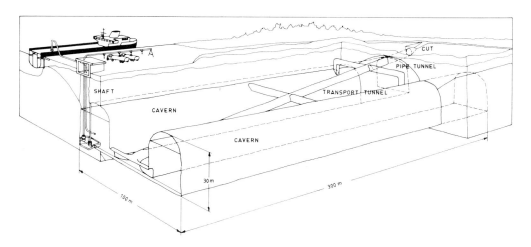

DESIGN AND CONSTRUCTION OF UNDERGROUND STORAGE FACILITIES

The engineering of every new governmental cavern is realized in two steps:

- feasibility study with cost estimate
- final and detailed technical plan and cost estimate

The feasibility study includes the following main activities:

- Each projected storage area is investigated by seismic soundings, and/or by diamond drilling in order to find the rock most suitable in terms of solidity, compactness, etc. These basic data are used in the layout as well as in the engineering of storage facilities. The investigations and feasibility study are performed either by some Finnish company specialized in geological surveys or by some engineering consultant expert in underground technology, as ordered by the Department of Energy of the Ministry of Trade and Industry.

- If the geological survey of the area results in more than one alternative as far as the rock characteristics are concerned, the total construction costs of different layout alternatives are compared in detail. This comparison is performed by the engineering consultant in question.

- The engineering consultant also draws up the preliminary plan for the excavation and the construction of oil caverns including preliminary engineering drawings for instrumentation and process equipment as well as the cost estimate for the approval of the Department of Energy.

Relative unit costs of storage facilities

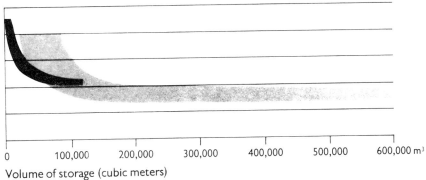

Volume of storage (cubic meters)

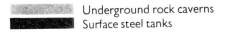 Underground rock caverns
Surface steel tanks

The capital investment for cavern storage is lower than for a tank farm if the volume is about 100.000 m³ and larger.

CHOICE OF CONTRACTORS

Contractors for state-owned rock storage facilities are selected by the Ministry of Trade and Industry, whose department for construction, The National Board of Public Construction has the owner's responsibility as contracting party and supervisor of a project. This organization issues the tender documents to qualified Finnish civil contractors.

In the Finnish Strategic Petroleum Reserve (SPR) programme there is already a very considerable volume of cavern space in use and there are more caverns to come.

THE USE OF UNDERGROUND STORAGE BY NESTE

To discuss the merits of the underground storage system we will describe the use of unlined underground caverns by the national oil company of Finland, Neste Oy. For better understanding the storage needs for different oils are discussed in detail.

Crude Oil

The company has no oil production of its own. All crude oil is purchased under long-term contracts. About two thirds is imported from the Soviet Union. This crude oil is piped to Ventspils harbour on the Baltic coast and transported by 25.000 dwt tankers all year around to our two refineries. The need for crude in the winter months is greater than in the summer. The rest of the crude is purchased mainly from the Middle East. For reasons of economy the tankers are about 250.000 dwt with about 100.000 tons of oil off-loaded before entering the Baltic Sea. These ships can navigate in the Baltic only during the eight ice-free months of the year. Therefore, relatively large quantities of crude oil have to be stored for winter use, when the harbours are closed to Middle East crude tankers. The need is 3-4 million tons. This crude oil is stored in unlined rock caverns below the refineries. This fairly large storage capacity also gives us possibilities to store separately different crudes and have long runs in the refinery using the same crude. The benefits of underground storage are: inexpensive storage, safety, no evaporation losses, best possible yield when processing differend crudes separately.

Fuel Oils

Heavy fuel oils are stored in underground caverns in temperatures of 70 - 90°C. The rock is a very good insulator and therefore heat losses are small. Underground caverns are used at the refinery and in distributing depots or by big consumers like power plants. The main benefit is lower storage cost. Light fuel oils are also stored underground. This product is the easiest of all to be stored underground. Main reasons for underground storage are again economy and safety.

Gasoline

Gasoline and diesel fuel are stored in underground caverns for safety reasons. If the quantity to be stored is more than 50.000 m^3 (300.000 bbl), the underground storage has proved cheaper than steel tanks.

LPG

In our country, there is a very large temperature range between the summer and winter months. The gasoline specification is different during the summer and winter seasons. More C-4 is to be added in the winter gasoline. It has been very advantageous to save butanes in the summer time for use in the winter. The economical underground cavern has made this seasonal storage possible. For safety reasons, we could hardly build above ground storage facilities for this service even in the event that this would have been economically feasible.

Capital investment for various types of LPG-storage facilities is shown below.

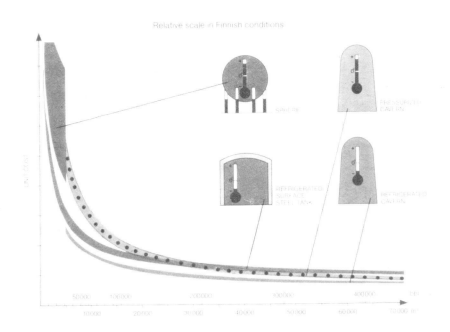

RECENT DEVELOPMENTS AND TRENDS

The economies inherent in cavern storage have become more and more apparent ever since the construction of the first caverns in the mid 1960´s. Each project has taught us more. The experience of several cavern units per year (on average) has encouraged a group of companies headed by the national oil concern Neste to establish a new company, Finncavern Ltd, for exportation of the cavern technology.

Finncavern Ltd undertakes turn-key cavern projects or parts thereof, including site investigations, feasibility studies, engineering, construction and start-up, even operation of the completed storage facility.

Below are discussed some developments and trends in the technology. All of them aim to reduce the cost of capital investment and/or operation and maintenance costs of the cavern storage.

Site Investigation

With proper implantation, the cavern project is viable. To achieve this, a combination of various methods of investigation is applied. Before undertaking extensive diamond drilling in the rock, it normally pays to do "non-destructive" testing of the bedrock. Seismic refraction sounding is one quite commonly used method. It is not difficult to carry out the actual field work, but skill in the proper interpretation of the results is of paramount importance. A good interpreter can find zones of discontinuity deep in the rock just as if he could x-ray the bedrock. In Finland the experience in seismic sounding is long and extensive.

Cavern Profiles

There has been a tendency towards larger cross sections. From small caverns, for instance 12 m wide and 20 m high, the cross sections have grown well beyond 500 m^2, with a width over 20 metres and height over 30 metres. This is due to increased experience in the assessment of rock quality. Such assessments are performed using detailed rock mechanical investigations and calculations, for instance utilizing the finite element method.

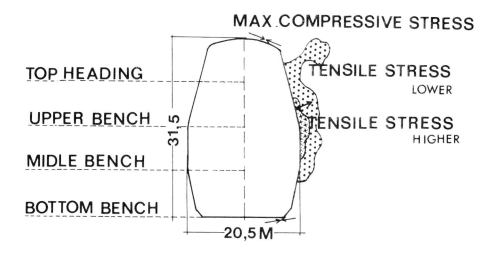

MAX COMPRESSIVE STRESS

TOP HEADING

TENSILE STRESS
LOWER

UPPER BENCH

TENSILE STRESS
HIGHER

MIDLE BENCH

BOTTOM BENCH

31,5

20,5 M

CROSS-SECTION 566 M^2

Above is an example of a major cavern cross section. Stress analysis is made by FEM-calculation.

In general, the size of cavern units and of project is on the increase, to minimize the burden of access tunnels. In cases of multicavern projects the tunneling volume is well below 10 per cent of the effective volume of the caverns.

Smoothwall Blasting

The known methods of controlled rock blasting have been specifically applied and modified to cavern excavation. Variations of pre-splitting and cushion blasting have been tested over the years in different types of rock. The result has been more stable and competent rock structure with reduced need for costly reinforcement. Also above ground structures have been protected by the use of controlled blasting methods.

Pump Room Design

For exploitation of the cavern storage, a common practice is to install submersible pumps suspended from their discharge pipes in the vertical pipe shaft.

Another alternative is to construct a dry pump room on the bottom level of the cavern and separated from it. A dry pump room would be built for instance for a commodity requiring heating, or when high pumping capacities are required, or when several caverns are to be exploited through the same pumps.

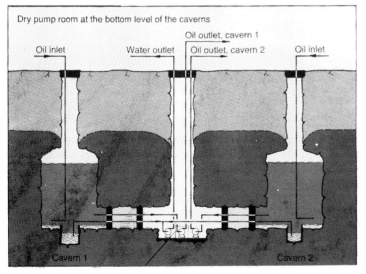

Shaft Design and Construction

Formerly, shafts used to be drilled and blasted, often applying the long hole drilling method. Since no longer holes than 40 - 50 metres could be drilled accurately, this method restricted the depth of a shaft.

In some recent cavern projects, the shafts have been bored by the raise boring method, where applicable. Of course, this method is not new in construction and mining industry. In this context, it has several advantages over the long hole method:
-it consumes less time and labor
-deeper shafts can be constructed accurately in one pass
-no damage or delay is caused by blasting
-consequently, no extensive reinforcing is required
-the total cost is competetive.

CONCLUSION

The benefits of underground oil storage are numerous and the system can be favourably utilized in addition to strategic storage, in many cases where buffer capacity is needed. A well planned and executed site investigation is essential to reduce the capital investment cost of any storage project.

We in Finland believe this technology offers wide scope for use in an abundance of applications in hard rock all over the world.

MOTIVATION AND PARAMETERS FOR THE USE OF UNDERGROUND SPACE

J. Vered-Weiss
Deputy Managing Director Engineering
Petroleum Services Ltd., P.O. Box 24229 Tel-Aviv Israel

ABSTRACT

The paper describes the essential parameters of underground space utilization for fuel storage and specifically development of a new technique for the storage of refined petroleum products ready for use.

KEYWORDS

Parameters and criteria, fuel distillate storage system, impermeabilization, quality control.

I.   1. Safety and security
        2. Economy in relation to ground utilization
        3. Environmental considerations in regard of pollution and aesthetics
        4. Geological conditions

            Priorities are interrelated with specific conditions and require-
            ments of the area or country.

II.   Political and economic factors have increased the need for the storage
        of all types of hydrocarbons in those countries lacking sufficient re-
        sources of fuel and gas.

        Within the wide range of the possible use of subsurface space the
        underground storage of hydrocarbons has been greatly accelerated in
        many countries.

III.   Widely differing techniques have been developed, depending on the geo-
        logical, geotechnical and logistical conditions in the relevant areas
        on one hand, and the type of hydrocarbons to be stored on the other
        hand.

IV.   For Israel, depending on the import of hydrocarbons for much of her re-
        quirements, storage of fuel reserve is essential.

Above ground storage in conventional steel tanks has been the practice in the past.

For some years the possibilities of underground storage have been studied and explored through drilling and core analysis and is now being translated into practice in several projects.

The initiation of the program started with the determination of desirable locations based on the following criteria:

(a) Logistical
    Demand, type and quantity, flow and existing piping network

(b) Environmental
    Safety   -   Center of population
    Security   -   Strategic location
    Pollution   -   Protection of utilizable aquifers
    Geology and geotechnique - Suitable stable formations
    Types of fuel - Crude oil, LPG and natural gas and refined
                           products with reservation of quality for long
                           time storage

Starting from these criteria and based on a country-wide geological evaluation of available data specific areas were selected with indications of correspondence with the required conditions and more detailed study conducted in order to obtain analytical data in respect of rock strength and stratification as well as mineralogical analysis and hydrological data. In parallel models of suitable storage systems were established for the different types of fuel to be stored and pilot plants designed for the sensitive refined products.

Preliminary programs were worked out for submission to the licensing authorities.

V.   A project of this type nearing completion is an LPG storage cavern horseshoe shape dimension 7m width and 10m height with a length of 300m in silicified chalk at a depth of 70m below rock surface.

This cavern has been lined with cast concrete and a polymer lining and will operate at a pressure of 6 bar at constant temperature of 20°C. This type of storage will be in operation in late summer 1981 and will also serve as a pilot project for a big fuel distillate underground storage. The main problem is the impermeabilization of the rock in a cavern, free of water, in order to preserve fuel distillate standard quality. Research of different types of polymers and their compatibility with and suitability for the fuel distillate storage are being conducted and bacteriological research with the aim of preventing bacteria and fungal growth is also being executed on a laboratory scale.

A further operational pilot plant is in the early erection stage, to be lined with a different polymer and the before-mentioned big scale storage is in the design stage.

Steel lining has been considered but the problems of corrosion and cost have relegated this solution to a low priority.

This type of lining has been used in the past but with qualified success. I have been elaborating on this type of storage for the reason that there are no agreed solutions as far as we know so far.

The aspect of storing crude oil has been studied and will be implemented on the basis of known storage system.

VI. The parameters and basic techniques studied for the purpose of hydrocarbon storage are opening up a very wide field of application of underground storage for a great variety of targets and will be utilized in the future. All this in a country which has an interesting history of underground works in early periods between 1000 BC and 200 AD.